HIGH GAIN, HIGH POWER FREE ELECTRON LASER: PHYSICS AND APPLICATION TO TEV PARTICLE ACCELERATION

North-Holland
Delta Series

NORTH-HOLLAND
AMSTERDAM • OXFORD • NEW YORK • TOKYO

High Gain, High Power Free Electron Laser: Physics and Application to TeV Particle Acceleration

Proceedings of the I.N.F.N. International School on Electromagnetic Radiation and Particle Beams Acceleration
Varenna, Italy, June 20-25, 1988

Edited by

R. Bonifacio
University of Milano, Italy

L. De Salvo Souza
University of Milano, Italy

C. Pellegrini
Brookhaven National Laboratory, U.S.A.

1989

NORTH-HOLLAND
AMSTERDAM • OXFORD • NEW YORK • TOKYO

© Elsevier Science Publishers B.V., 1989

All rights reserved. No part of this publication may be reproduced, stored in a retrieval system, or transmitted, in any form or by any means, electronic, mechanical, photocopying, recording or otherwise, without the prior written permission of the publisher, North-Holland Physics (a section of Elsevier Science Publishers B.V.).

Special regulations for readers in the U.S.A.: This publication has been registered with the Copyright Clearance Center Inc. (CCC), Salem, Massachusetts. Information can be obtained from the CCC about conditions under which photocopies of parts of this publication may be made in the U.S.A. All other copyright questions, including photocopying outside of the U.S.A., should be referred to the publisher, unless otherwise specified.

No responsibility is assumed by the publisher for any injury and/or damage to persons or property as a matter of products liability, negligence or otherwise, or from any use or operation of any methods, products, instructions or ideas contained in the material herein.

ISBN: 0 444 87395 3

Published by:

North-Holland Physics
(Elsevier Science Publishers B.V.)
P.O. Box 103
1000 AC Amsterdam
The Netherlands

Sole distributors for the U.S.A. and Canada:

Elsevier Science Publishing Company, Inc.
655 Avenue of the Americas
New York, N.Y. 10010
U.S.A.

Library of Congress Cataloging-in-Publication Data

```
I.N.F.N. International School of Electromagnetic Radiation and
   Particle Beams Acceleration (1988 : Varenna, Italy)
    High gain, high power free electron laser : physics and
 application to TEV particle acceleration : proceedings of the
 I.N.F.N. International School on Electromagnetic Radiation and
 Particle Beams Acceleration, Varenna, Italy, June 20-25, 1988 /
 edited by R. Bonifacio, L. De Salvo Souza, C. Pellegrini.
      p.   cm. -- (North-Holland delta series)
    Includes index.
    ISBN 0-444-87395-3 (U.S.)
    1. Particle beams--Congresses.  2. Particle acclerators-
 -Congresses.   I. Bonifacio, R.  II. De Salvo Souza, L. (Lucia)
 III. Pellegrini, C.  IV. Istituto nazionale de fisica nucleare.
 V. Title.  VI. Series.
   QC793.3.B4I5  1988
   539.7'2--dc20                                          89-3317
                                                             CIP
```

Printed in The Netherlands

INTRODUCTION

During the past few years the physics and technology of charged particle beams, on which electron-positron linear colliders in the TeV region, storage rings from synchrotron radiation sources and Free Electron Lasers are based, has seen a remarkable development. Also, radiation and particle beams are essential instruments for the most recent developments in atomic physics, as for instance trapping and selective ionization of atoms. These systems are also finding increasing applications in industry, for instance one can mention the use of x-ray lithography for the production of chips at the submicron level.

In Italy programs in these fields are either in an advanced state of construction or are being started at the Frascati National Laboratories, the Legnaro National Laboratory, the University of Milan, ENEA and Trieste. Industrial companies, like Ansaldo, are actively involved in international programs of construction of large accelerators and related components, like superconducting magnets.

Similar situations exist in all industrialized countries and also developing countries, like Brazil, China, India, Korea and Taiwan. They are now constructing, or preparing to construct, accelerators for synchrotron radiation production, high energy physics and Free Electron Lasers.

The INFN "International School on Electromagnetic Radiation and Particle Beams; Physics and Applications", wants to address the physics and technology issues of this field, train young people and, at the same time, provide a forum for discussions on recent advances for the scientists active in this field. Correspondingly the School was divided into two sections: one tutorial, the other a workshop.

The first Course of the School was dedicated to a special subject: "High Gain, High Power Free Electron Lasers; Physics and Applications to TeV Particle Acceleration". The Course was held in Villa Cipressi, Varenna, Italy, from June 20 - 25, 1988. This subject reflects the recent interest in TeV electron positron colliders, the possibility offered by Free Electron Lasers to power them and the developments in the production of high brightness electron beams. Another subject closely related is synchrotron radiation from high brightness electron beams.

The School was made possible by the support of the INFN (Istituto Nazionale di Fisica Nucleare) and the Physics Department of the University of Milan, and we want to thank them for it. We also wish to acknowledge the excellent work of the School Secretary, G. Posadinu, and to thank Prof. E. Sindoni, Manuela Boscolo and the young people of the ELFA group for their invaluable help.

Rodolfo Bonifacio, Lucia De Salvo Souza and Claudio Pellegrini

TABLE OF CONTENTS

Introduction v

Organization ix

List of participants xi

SCHOOL

A high brightness electron accelerator and its particle beam
physics experimental program
    C. Pellegrini 1

Emittance growth in laser-driven rf electron guns
    K.-J. Kim 25

One-dimensional theory of a free-electron laser amplifier:
Steady-state and superradiance
    R. Bonifacio, F. Casagrande, G. Cerchioni, L. De Salvo Souza
    and P. Pierini 35

Selected topics in FELs
    E.T. Scharlemann 95

High gradient accelerators for linear light sources
    W.A. Barletta 127

WORKSHOP

Linear collider regimes
    U. Amaldi 171

The CLC project and the design for an e^+e^- collider
    S. van der Meer 185

Plasma assisted inverse free electron laser
    J.L. Bobin 197

Radiation from fine, intense, self-focussed beams at
high energy
    W.A. Barletta and A.M. Sessler   211

The ELFA project: Guidelines for a high-gain FEL with short
electron bunches
    R. Bonifacio, I. Boscolo, F. Casagrande, G. Cerchioni,
    R. Corsini, L. De Salvo Souza, D. Fadini, M. Ferrario,
    C. Maroli, P. Pierini and N. Piovella   221

POSTERS

Tapering and self-tapering in a free electron laser amplifier
    R. Bonifacio, F. Casagrande, M. Ferrario, P. Pierini and
    N. Piovella   227

Bistability in free electron lasers
    R. Bonifacio, F. Castelli and L. De Salvo Souza   243

Slippage and superradiance in a high-gain FEL: Linear theory
    R. Bonifacio, C. Maroli and N. Piovella   259

Three-dimensional effects by beat wave excitation in
magnetoactive plasmas
    F. Esposito, R. Fedele, G. Miano and V.G. Vaccaro   275

Propagation of a short rf pulse train in an iris-loaded waveguide
    L. Ferrucci, C. Pagani and L. Serafini   283

Numerical integration of transient particle and field equations in
axi-symmetrical cavities
    L. Serafini, C. Pagani, L. Ferrucci, L. Muda and A. Peretti   293

Author index   307

Subject index   309

Directors of the School

Rodolfo BONIFACIO (University of Milan)
Claudio PELLEGRINI (Brookhaven National Laboratory)

Scientific Advisory Committee

Nicola CABIBBO (President of INFN) - Chairman

William BARLETTA (Lawrence Livermore National Laboratory)
Gianpaolo BELLINI (University of Milan)
Piero DAL PIAZ (Legnaro National Laboratory)
Antonino PULLIA (University of Milan)
Andrew SESSLER (Lawrence Berkeley National Laboratory)
Sergio TAZZARI (Frascati National Laboratory)

Local Organizing Committee

Rodolfo BONIFACIO (University of Milan) - Chairman

Ilario BOSCOLO (University of Milan)
Federico CASAGRANDE (University of Milan)
Enio SINDONI (University of Milan)

Lucia DE SALVO SOUZA (University of Milan) - Scientific Secretary

LIST OF PARTICIPANTS

BRASIL

DE SALVO SOUZA Lucia
Università di Milano
Dipartimento di Fisica
Via Celoria 16
20133 Milano
Italy
tel (02)-2392-268

FRANCE

ROBIN J.
Laboratoire de Physique et Optique
  Corpuscolaire
Université Pierre et Marie Curie
Tour 12, Et.5, 4 Place Jussieu
725252 Paris Cedex 05
tel (1)-4325-2885

CARLOS Pierre J.
CEN Saclay
DPHN/HE
91191 Gif Sur Yvette Cedex
tel (33)-69087479

FAUGERAS Paul
CERN
SPS Division
1211 Geneve 23
Switzerland
tel (022)-834636

ITALY

AMALDI Ugo
CERN
1211 Geneve 23
Switzerland
tel (22)-833027

BELLINI Gianpaolo
Università di Milano
Dipartimento di Fisica
Via Celoria 16
20133 Milano
tel (02)-2392-370

BELLOMO Giovanni
Università di Milano
Dipartimento di Fisica
Via Celoria 16
20133 Milano
tel (02)-2392-574

BONIFACIO Rodolfo
Università di Milano
Dipartimento di Fisica
Via Celoria 16
20133 Milano
tel (02)-2392-268

BOSCOLO Ilario
Università di Milano
Dipartimento di Fisica
Via Celoria 16
20133 Milano
tel (02)-2392-264

CASAGRANDE Federico
Università di Milano
Dipartimento di Fisica
Via Celoria 16
20133 Milano
tel (02)-2392-264

CASTELLI Fabrizio
Università di Milano
Dipartimento di Fisica
Via Celoria 16
20133 Milano
tel (02)-2392-230

CERCHIONI Giovanna
Università di Milano
Dipartimento di Fisica
Via Celoria 16
20133 Milano
tel (02)-2392-236

CORSINI Roberto
Università di Milano
Dipartimento di Fisica
Via Celoria 16
20133 Milano
tel (02)-2392-236

DIVIACCO Bruno
Sincrotrone Trieste
Padriciano 99
34012 Trieste
tel (040)-2260531

ESPOSITO Filippo
INFN - Sezione di Napoli
Università di Napoli
Dip. di Scienze Fisiche
Padiglione 20
Mostra d'Oltremare
80125 Napoli
tel (081)-7253409

Participants

FADINI Daniele
Università di Milano
Dipartimento di Fisica
Via Celoria 16
20133 Milano
tel (02)-2392-236

FEDELE Renato
INFN - Sezione di Napoli
Università di Napoli
Dip. di Scienze Fisiche
Padiglione 20
Mostra d'Oltremare
80125 Napoli
tel (081)-7253409

FERRARIO Massimo
Università di Milano
Dipartimento di Fisica
Via Celoria 16
20133 Milano
tel (02)-2392-232

FERRUCCI Luca
INFN - Sezione di Milano
Via Celoria 16
20133 Milano
tel (02)-2392-232

GIOVE Dario
INFN-LASA
Via F.lli Cervi 201
20090 Segrate (MI)
tel (02)-2392-559

MICHELATO Paolo
INFN-LASA
Via F.lli Cervi 201
20090 Segrate (MI)
tel (02)-2392-559

NAPPI Ciro
CNR
Istituto di Cibernetica CNR
Via Toiano 6
80072 Arcofelice (NA)

PAGANI Carlo
INFN-LASA
Via F.lli Cervi 201
20090 Segrate (MI)
tel (02)-2392-226

PATTERI Piero
INFN
Laboratori Naz. di Frascati
Via Enrico Fermi
00044 Frascati (Roma)
tel (06)-9403435

PELLEGRINI Claudio
National Synchrotron Light Source
Brookhaven National Laboratory
Upton Long Island, N.Y., USA
tel (516)-2824635

PIERINI Paolo
Università di Milano
Dipartimento di Fisica
Via Celoria 16
20133 Milano
tel (02)-2392-265

PIOVELLA Nicola
Università di Milano
Dipartimento di Fisica
Via Celoria 16
20133 Milano
tel (02)-2392-230

PRATI Paolo
ANSALDO RICERCHE
ARI-CPI
Corso Perone 25
16152 Genova
tel (010)-6558448

ROSSI Lucio
Dipartimento di Fisica di Milano - LASA
Via F.lli Cervi 201
20090 Segrate (MI)
tel (02)-2392-569

SERAFINI Luca
INFN-LASA
Via F.lli Cervi 201
20090 Segrate (MI)
tel (02)-2392-228

SINDONI Elio
Università di Milano
Dipartimento di Fisica
Via Celoria 16
20133 Milano
tel (02)-2392-267

STAGNO Vincenzo
Università di Bari
Dipartimento di Fisica
Via Amendola 173
70126 Bari
tel (080)-243183

TAZZARI Sergio
INFN
Laboratori Naz. di Frascati
Via Enrico Fermi
00044 Frascati (Roma)
tel (06)-9423567

Participants

VARIALE Vincenzo
INFN Università di Bari
Dipartimento di Fisica
Via Amendola 173
70126 Bari
tel (080)-243190

THE NETHERLANDS

HASELHOFF E.H.
University of Twente
Department of Quantum Electronics
P.O. Box 217
7500 AE Enschede

VAN DER MEER Simon
CERN
EP Division
1211 Geneve 23
Switzerland
tel (22)-832915

UNITED KINGDOM

McNEIL Brian
University of Twente
P.O. Box 217
7500 AE Enschede
The Netherlands
tel (0331)-53899111

USA

BARLETTA William
Lawrence Livermore National Laboratory
P.O. Box 808, L-626
Livermore, CA
tel (415)-6705

HALBACH Klaus
University of California
Lawrence Berkeley Laboratory
1 Cyclotron Road
MS80-101
Berkeley, CA 94720
tel (415)-4865868

YU Li Hua
National Synchrotron Light Source
Brookhaven National Laboratory
Upton Long Island, N.Y.
tel (516)-2825012

KIM Kwang-Je
University of California
Lawrence Berkeley Laboratory
MS80-101
Berkeley, CA 94720
tel (415)-4867224

SCHARLEMANN Ernst T.
Lawrence Livermore National Laboratory
P.O. Box 808, L-626
Livermore, CA
tel (415)-4225795

SESSLER Andrew
University of California
Lawrence Berkeley Laboratory
1 Cyclotron Road
Berkeley, CA 94720
tel (415)-4865024

WEST GERMANY

CIRKEL Hans Jurgen
Siemens AG UB KWU
Hammerbacher Strasse 12
D-8520 Erlangen
tel (09131)-18-4846

HEINRICHS Horst
University of Wuppertal
Gauss-str. 20
5600 Wuppertal 1
tel (202)-4392753

Participants

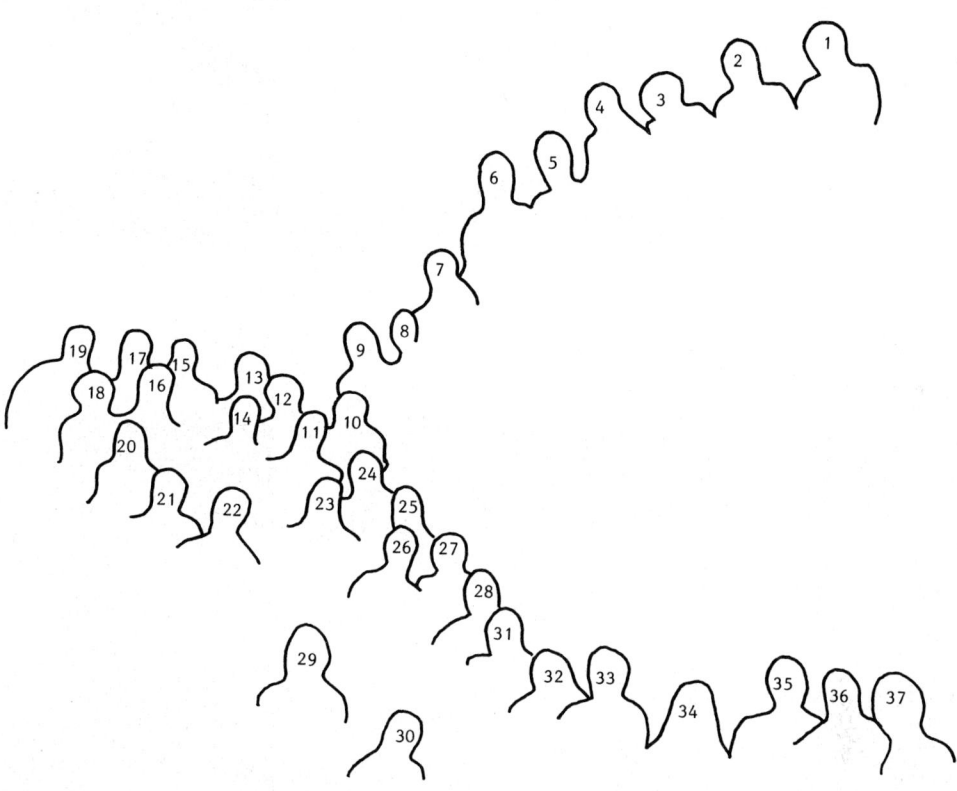

LIST FOR RECOGNIZING THE PARTICIPANTS

1 - Daniele FADINI	14 - Paolo PRATI	26 - Luca FERRUCCI
2 - Federico CASAGRANDE	15 - Fabrizio CASTELLI	27 - Roberto CORSINI
3 - Pierre CARLOS	16 - Lucio ROSSI	28 - Piero PATTERI
4 - Paul FAUGERAS	17 - Bruno DIVIACCO	29 - Lucia de SALVO SOUZA
5 - Ilario BOSCOLO	18 - Luca SERAFINI	30 - Giovanni POSADINU
6 - Eltjo HASELHOFF	19 - Dario GIOVE	31 - Rodolfo BONIFACIO
7 - Vincenzo VARIALE	20 - Giovanna CERCHIONI	32 - Brian Mc NEIL
8 - Kwang Je KIM	21 - Vincenzo STAGNO	33 - Claudio PELLEGRINI
9 - Horst HEINRICHS	22 - Andrew SESSLER	34 - Manuela BOSCOLO
10 - William BARLETTA	23 - Renato FEDELE	35 - Giovanni BELLOMO
11 - Nicola PIOVELLA	24 - Paolo MICHELATO	36 - Massimo FERRARIO
12 - Hans Jurgen CIRKEL	25 - Ciro NAPPI	37 - Paolo PIERINI
13 - Ernst T. SCHARLEMANN		

A HIGH BRIGHTNESS ELECTRON ACCELERATOR AND ITS PARTICLE BEAM PHYSICS

EXPERIMENTAL PROGRAM

C. Pellegrini

Brookhaven National Laboratory
Center for Accelerator Physics

1. INTRODUCTION

In recent years there has been a growing interest in the production of high brightness electron beams, for use as Free Electron Laser (FEL) drivers and also in high luminosity linear colliders. In addition there has been an increasing effort to develop new methods of particle acceleration, and this has led to the need of studying the interaction between high power electromagnetic radiation, in particular in the IR or millimeter region, and relativistic electron beams.

Since existing accelerators do not produce beams with the required brightness, some new programs have been initiated in a few laboratories to build accelerators on which one can carry out research and develop new high brightness electron sources, and study particle beam physics and the interaction of these beams with electromagnetic radiation. One such facility, the Accelerator Test Facility (ATF), is under construction at the Brookhaven National Laboratory, and is scheduled to start an experimental program in the beginning of 1990.

In this paper we will review the main characteristics of the ATF, and we will also give a short discussion of the experimental program to be carried out with it. This program will utilize the unique possibility offered by the ATF, of providing a high brightness electron bunch, synchronized with high power laser pulses. Part of the program will be based on the study of the interaction between the laser radiation and the electrons, in particular laser acceleration of particle beams, and the non linear dynamics of electrons in a strong electromagnetic field. Another part will utilize the high brightness of the electron beam to study the physics of Free Electron Lasers in the high gain regime, with the aim of establishing a basis for the the development of FELs in the Soft X-ray region.

The ATF design and construction is being done by many scientists and engineers, and in this paper I am reporting the results of their work. They are: K. Batchelor, T.S. Chou, R.C. Fernow, J. Fischer,

J. Gallardo, H.G. Kirk, R.B. Palmer, J. Sheehan, T. Srinivasan-Rao, S. Ulc, A. Van Steenbergen, and M. Woodle of Brookhaven National Laboratory; I. Bigio, and N. Kurnit of Los Alamos National Laboratory; K. T. Mc Donald of Princeton University. The original idea of an ATF and the initial effort to make it a reality are mainly due to R. B. Palmer.

2. GENERAL DESCRIPTION OF THE ATF

The ATF is a linac-laser complex for research in laser acceleration and in the generation of coherent radiation from electron beams. It consists of:

1) a 5 MeV, high brightness, RF electron gun;

2) a 50 to 100 MeV, S band, linac;

3) a NdYag laser, producing 6 ps long pulses, with peak power of 200 MW;

4) a Carbon Dioxide laser, with peak power of 100 GW, in a 6 ps pulse.

The gun can be operated with a photocathode driven by the NdYag laser; the same Yag is also used to switch the 6 ps CO_2 pulse, thus providing synchronized picosecond long pulses of electrons and laser light. A schematic drawing of the system is given in Fig. 1. A layout of the linac and experimental area is given in Fig. 2 and 3. The main beam parameters are given in Table 1. There are two main mode of operations: a low current, very small emittance mode, to be used mainly for laser acceleration studies and the study of non linear electromagnetic effects; a high current mode, to be used mainly for FEL studies.

To characterize a particle beam we use quantities like the beam energy and peak current. However they do not give a full description of the beam and we must add other quantities which can give a measure of the beam density in the six dimensional position-velocity phase space. We will indicate with x, y and z the horizontal, vertical and longitudinal coordinates, with x', y' their derivatives respect to z, (giving the angles of the particle trajectory respect to the z axis), and with γ the beam energy in rest mass units, mc^2. To characterize the beam six dimensional phase space density we use the emittance, the brightness and the longitudinal brightness.

The normalized beam emittance for each one of the three degrees of freedom, horizontal, x, vertical, y, and longitudinal, z, of a beam, is defined as

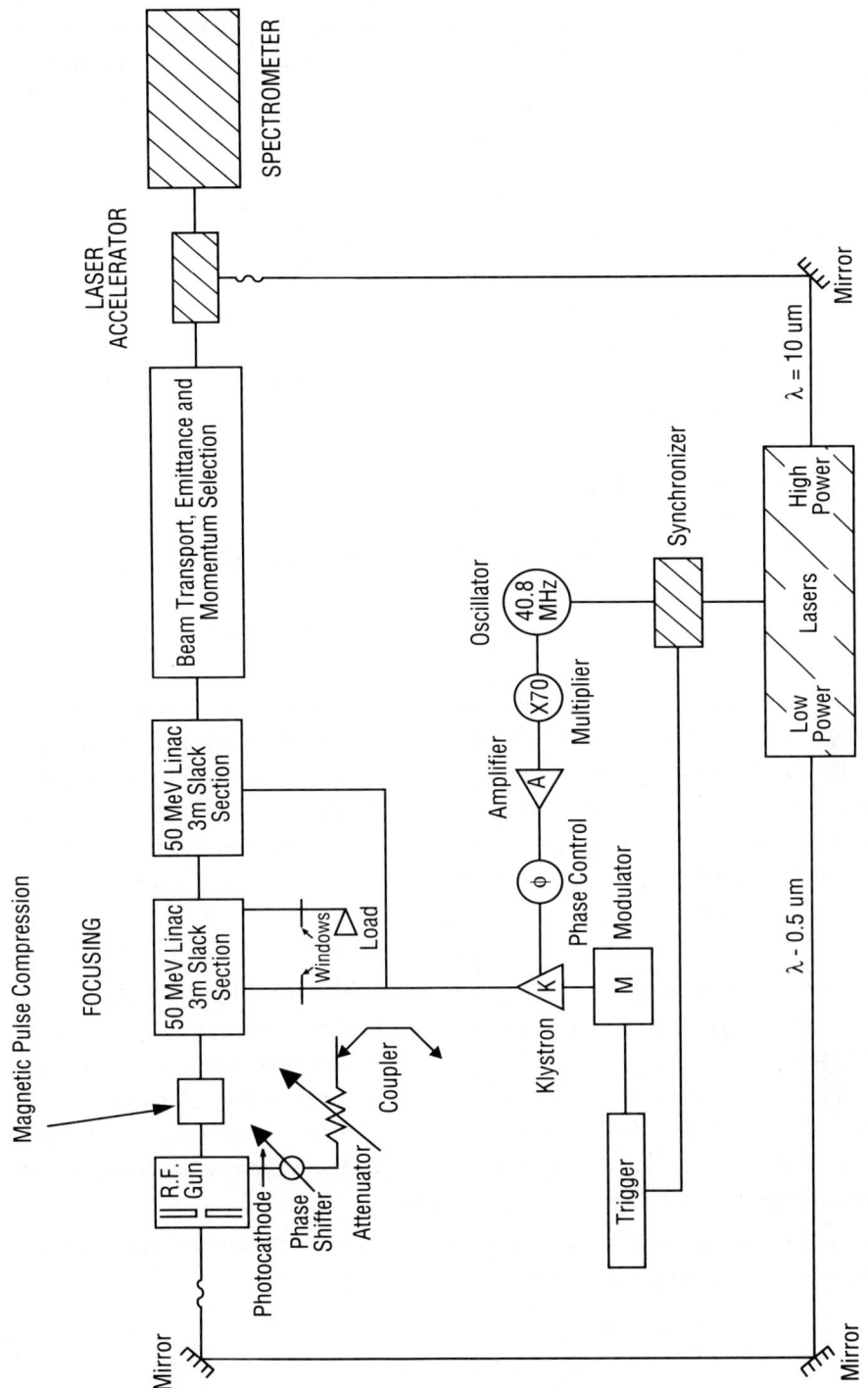

Figure 1. Schematic diagram of the Accelerator Test Facility

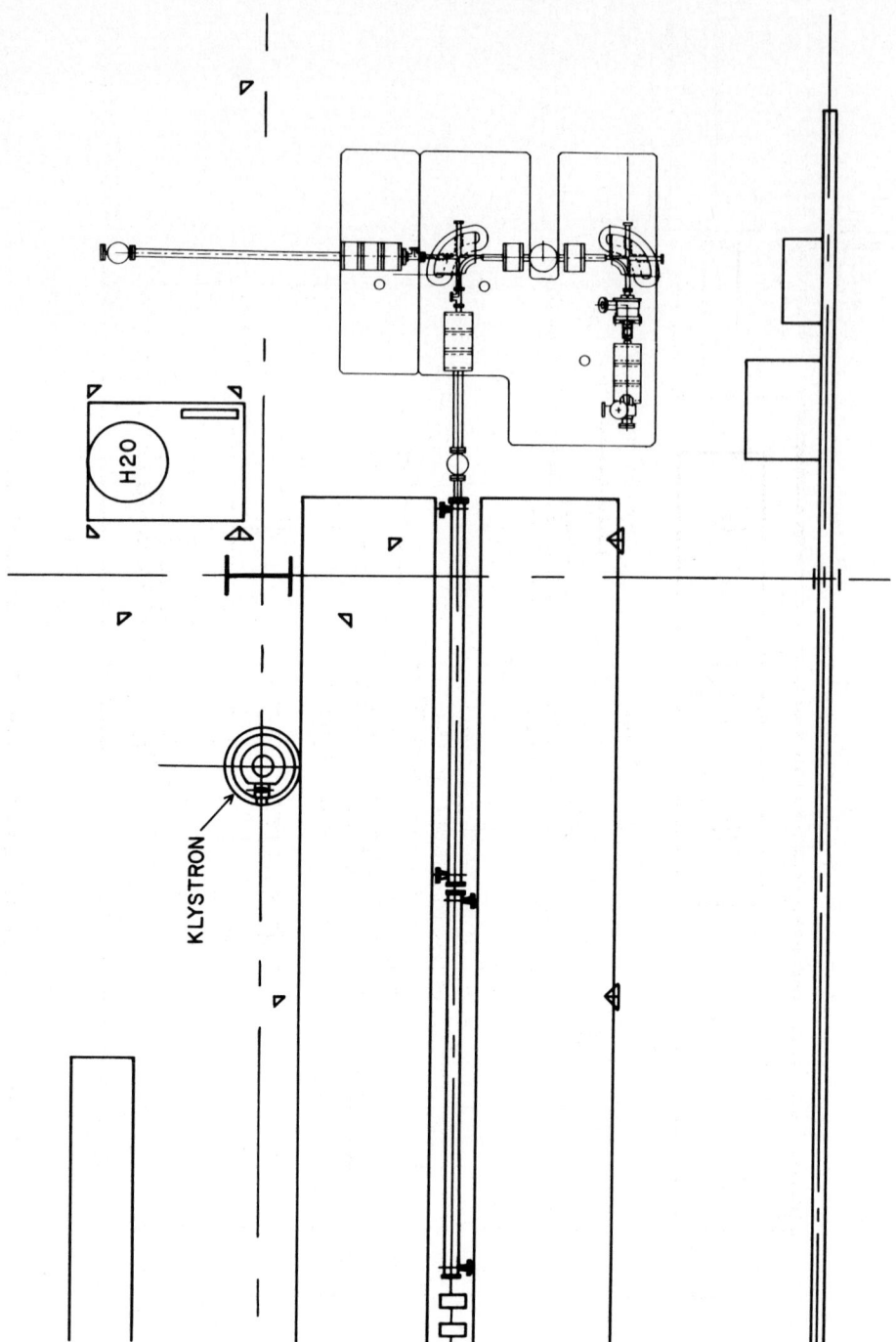

Figure 2. Layout of the lineac, RF gun and the transport line from the gun to the lineac

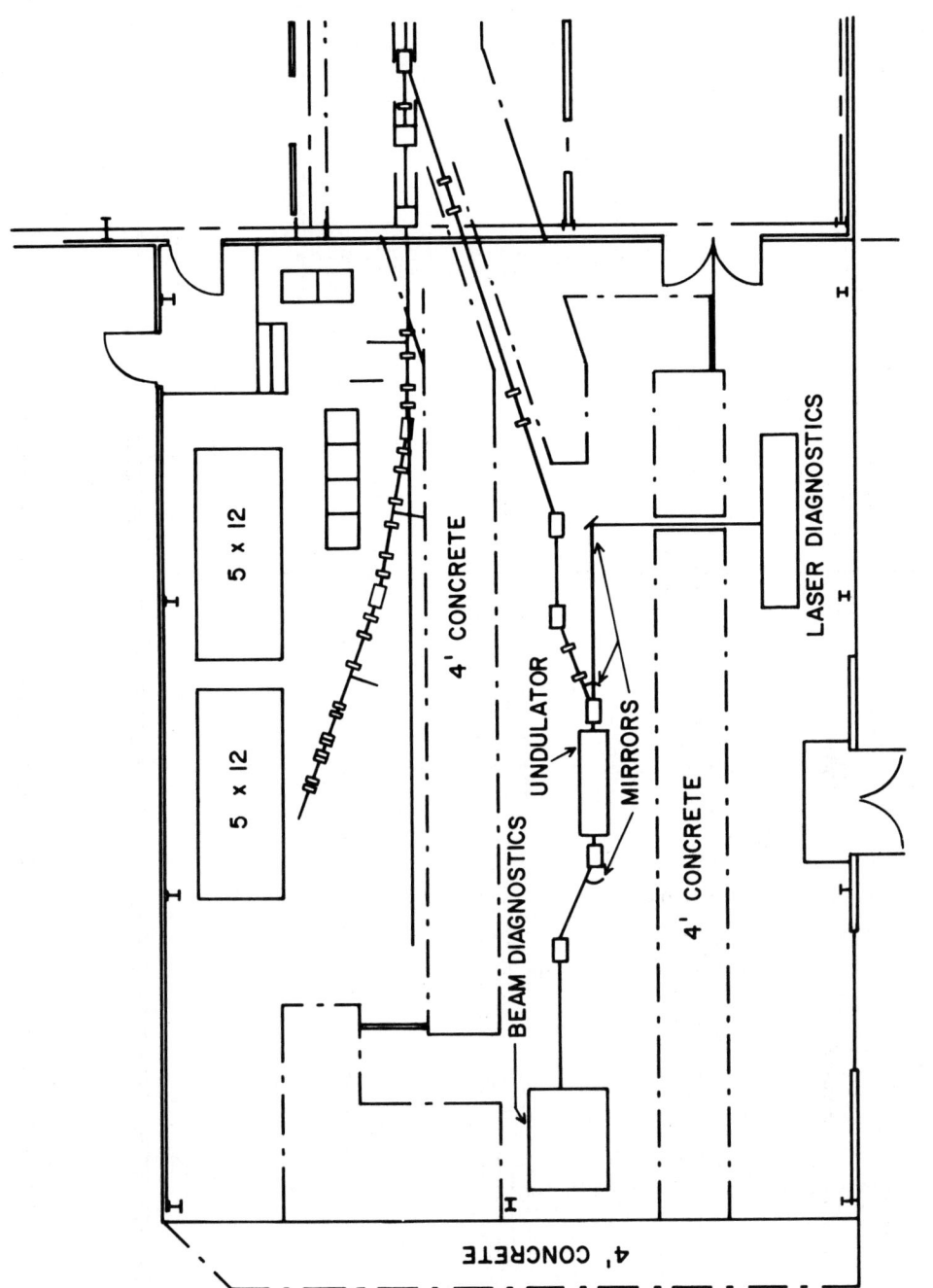

Figure 3. Layout of the ATF experimental area

TABLE 1

Linac Design Parameters

	High Current	Low current
Energy, MeV	50	50
Repetition rate, Hz	3	3
Peak Current, A	100	.01
Energy spread, rms, %	0.3	0.1
Emittance ($\gamma\sigma_x\sigma_{x'}$), m rad	5×10^{-6}	10^{-8}
Beam Brightness, A/m^2	4×10^{12}	1×10^{14}
Electron pulse length, rms, ps	2.5	2.5
Longitudinal emittance ($\sigma_E\sigma_L\gamma$), m	2.25×10^{-6}	7.5×10^{-7}
Longitudinal Brightness, A	4×10^4	12
Electron bunch separation, ns	12.5	12.5
Klystron pulse length, μs	5	5
Maximum number of bunches/klystron pulse	100	100
Average current in klystron pulse, mA	50	0.005

$$\varepsilon_{N\,x_i} = \gamma(<x^2><x'^2> - <xx'>^2)^{1/2} \tag{2.1}$$

These quantities are conserved for a beam subject to linear, time independent forces. Under the same conditions the emittance also represents the phase space area in the plane x, $\gamma x'$.

In addition to being a conserved quantity, at least when subject to the simple force described above, the emittance is also a measure of how well the beam can be utilized in applications like FELs or colliders. A small emittance is needed to produce large luminosity in a collider [1], or to operate an FEL at short wavelength [2].

When the beam is propagating under the action of non linear or time dependent forces the emittance defined in (2.1) can grow, as we will discuss for instance in section 3.1.

The beam Brightness given in Table 1 is defined as the ratio of the beam peak current to the product of the horizontal and vertical normalized rms emittances:

$$B = \frac{I_p}{\varepsilon_{N\,x}\,\varepsilon_{N\,y}} \tag{2.2}$$

The Longitudinal Brightness is the ratio of the bunch charge to the normalized longitudinal emittance

$$B_L = \frac{eNc}{\sqrt{(2\pi)}\varepsilon_{N,L}} \quad (2.3)$$

When the normalized emittances are invariant quantities characterizing the beam, the brightness and the longitudinal brightness are also invariants describing both the beam six dimensional phase-space and its charge. For this reason they are commonly used to characterize the beam "quality". Their largest value up to now has been obtained at the SLC damping ring at SLAC [3], and on an RF driven photocathode at Los Alamos [4], and is on the order of 2×10^{11} A/m^2 for the Brightness, and 70 A for the Longitudinal Brightness, in the SLC case, and respectively 10^{12} A/m^2 and 1200 A for the Los Alamos case.

The level of performance indicated in Table 1 is beyond what achieved up to now; although our calculations indicate that it should be possible to obtain these beam parameters, it is reasonable to assume that this will require a long period of commissioning and studies. In effect this R&D on the ATF beam is an integral part of our program.

To simplify the initial commissioning of the ATF we have also the capability of operating the system initially with a thermoionic cathode, although with reduced performances. We have estimated this initial system performance, and it is given in Table 2.

TABLE 2

Linac Characteristics for Initial Operation

Energy, MeV	50
Peak Current, A	10
Energy spread	.005
Emittance ($\gamma\sigma_x\sigma_{x'}$), m rad	3×10^{-5}
Electron pulse length, ps	10
Electron bunch separation, ns	.35
Klystron pulse length, μs	5.

The beam energy will be 50 MeV for initial operation; the addition of a second Klystron will allow to increase it to 100 MeV.

3. RF GUN

The development of the high brightness electron gun is an important part of the ATF program. The research to be done with the ATF, and more in general the future development of linear colliders, short wavelength FEL, and new methods of acceleration, depends strongly on the capability of producing small emittance, high peak current electron beams.

TABLE 3

RF GUN RESULTS

	Stanf.	LANL
Cathode	Thermal	Photo (Cs_3Sb)
Frequency, GHz	3	1.3
Energy, MeV	.8	1
Peak Current, A	10	100
Energy spread	<1%	6%
Emittance ($\gamma\sigma_x\sigma_{x'}$), mm mrad	20	5
Electron pulse length, ps	10	53

For this reason we are now dedicating a large effort to the gun design and to development of good instrumentation to measure the beam characteristics.

The RF gun design is based on the results of Los Alamos [4] and Stanford [5], and is shown in Fig. 4. This technique has produced the smallest emittance from a gun and the highest brightness. Some of the results obtained at Los Alamos and Stanford are given in Table 3.

This gun design has these advantages:

a) Using an RF field one can obtain a large electric field near the cathode, of the order of 30 to 100 MeV/m; this helps to control space charge effects, and brings the electrons to relativistic velocity in a small distance.

b) With a laser driven photocathode one can also control the time structure of the electron beam, spacing the bunches by several RF cycles, and obtaining larger peak currents for the same average current.

c) By changing the laser spotsize one can also control the charge per bunch and the emittance.

The main characteristics of the BNL gun are given in Table 4, and the design is shown in Fig. 4.

The gun design is based on a structure resonating in a π mode, with one and a half cell [6]. This structure has the advantage of minimizing the emittance growth due the RF field, as discussed in section 3.1. The electric field in this structure, calculated by H. Kirk using the electromagnetic code MAFIA, is shown in Fig. 5. The cathode is at the half cell wall, the position for maximum accelerating field. The field in this structure is of the standing wave type, to maximize the accelerating field for given input power. In estimating the electron beam dynamics and the beam emittance given in Table 1, we have assumed an accelerating field on the cathode of 100 MV/m. The maximum field obtainable experimentally might be smaller because of breakdown, and the beam characteristics would be correspondingly decreased. However experiments done at SLAC [7], on an S-band, disc loaded, standing wave structure, have shown that it is possible to reach surface fields of 300 MV/m before breakdown, for short RF pulses, approximately 2.5μs long. In our case, for a pulse length of 5μs, we have a maximum surface field of about 120 MeV/m.

TABLE 4

RF Gun Design Parameters

Structure type	Resonant side coupled
Structure inner diameter, mm	83.08
Structure length, mm	78.75
Number of cells	1+1/2
Operating frequency, GHz	2.856
Beam Energy, MeV	4.85
Beam aperture, mm	20
Shunt Impedance, $M\Omega/m$	57
Cavity Q	11,800
Max. Surface Electric Field, MV/m	118
Average accelerating gradient, MV/m	66.6
Electric field on cathode, MV/m	100
Cavity Stored Energy, J	3.5
Cavity Peak Power, MW	5.3

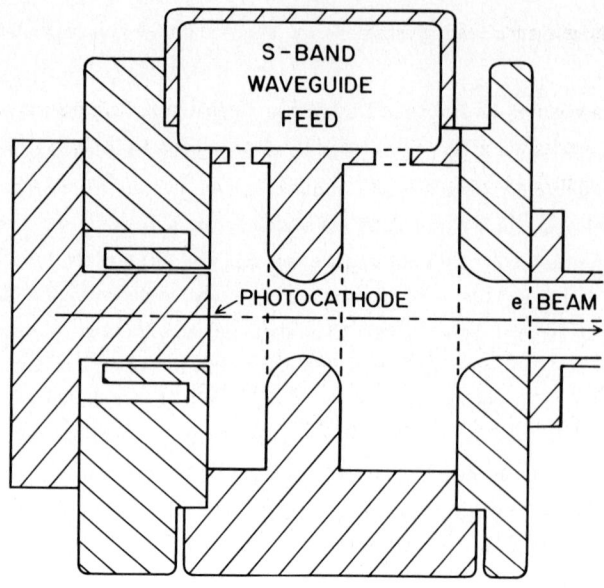

Figure 4. Schematic drawing of the RF gun

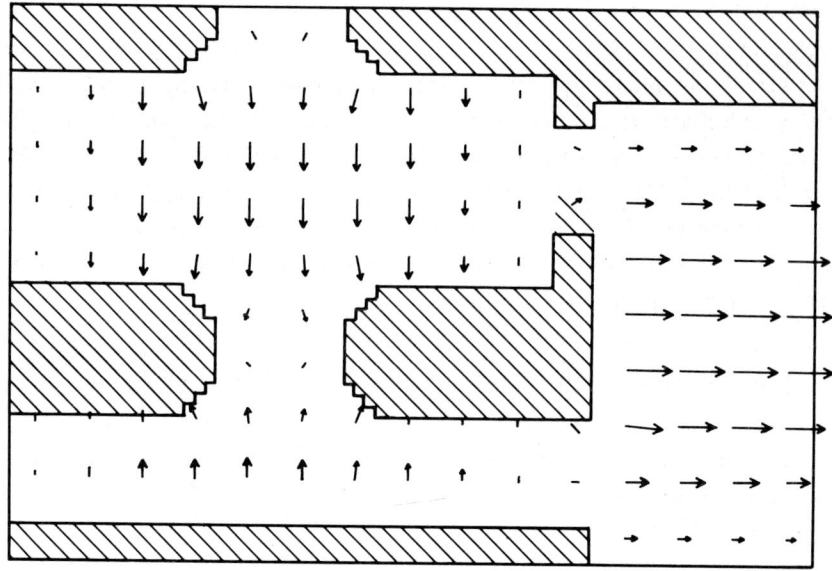

Figure 5. Electric field in the RF gun

3.1. RF field in the gun and beam emittance

As shown in Fig. 4 and 5, the gun consists of one and a half cell, with the cathode at the half cell position to obtain a maximum accelerating electric field on it. The RF field configuration is a " π " mode [6], with the longitudinal electric field given by

$$E_z = E_0 \cos(kz)\sin(\omega t + \phi_0) \tag{3.1}$$

where $k = 2\pi/\lambda$, $\omega = kc$, and λ is the RF wavelength; ϕ_0 is the RF phase at which the electron leaves the cathode and starts to be accelerated; the cathode position is z=0. The length of each cell is $\lambda/2$. We assume the longitudinal field to be independent of the transverse coordinates, in which case the radial electric field and the azimuthal magnetic field are given by

$$E_r = -\frac{r}{2}\frac{\partial E_z}{\partial z} \tag{3.2}$$

$$H_\theta = \frac{r}{2c}\frac{\partial E_z}{\partial z} \tag{3.3}$$

r and θ being cylindrical coordinates. In this approximation the transverse fields are linear. The main advantage of this mode is that the radial force is maximum at the gap, (see Fig. 5), but the particles cross this gap when $\sin(\omega t + \phi_0) \approx 0$.

The effect of this fields and of the space charge field on the beam emittance has been analyzed with the codes Parmela and Mask. An analytical calculation has also been done by K.J. Kim [8]. The results of this calculation give a good approximation to the numerical results and allow us to use some simple formulae to describe the beam emittance. The fields (3.2), (3.3) produce a linear, time dependent, transverse force. The increase in emittance due to the RF comes through the time variation of the transverse force seen by different electrons; it has a minimum when the electron exit phase corresponds to $\omega t + \phi_0 \approx \pi/2$. The beam emittance depends on the accelerating electric field; it is convenient to introduce the quantity α

$$\alpha = \frac{eE_0}{2mc^2 k} \tag{3.4}$$

For an electron beam with a gaussian cylindrical density distribution and rms radius and length σ_x, σ_z, the minimum emittance is given by

$$\varepsilon_x^{RF} = \alpha k^3 \sigma_x^2 \sigma_z^2 \frac{1}{\sqrt{2}} \tag{3.5}$$

The longitudinal emittance is given by

$$\varepsilon_z^{RF} = \sqrt{3}[<\gamma_f>-1]\, k^2\sigma_z^3 \qquad (3.6)$$

The space charge effects can be described in terms of the beam peak current, I, and the charge density distribution. For a cylindrical Gaussian beam the important quantity is

$$A = \frac{\sigma_x}{\sigma_z} \qquad (3.7)$$

The transverse electric and magnetic forces tends to cancel as $1/\gamma^2$, for $\gamma \gg A$.

The transverse and longitudinal space charge induced emittances can be written as

$$\varepsilon_i^C = \frac{\pi\, I\, \mu_i(A)}{4\alpha k \sin\Phi_0 I_A}, \qquad i=x \text{ or } z \qquad (3.8)$$

where I_A is the Alfven current (17,000 A), and the the two form factors are approximately given by

$$\mu_x = \frac{1}{5+3A} \quad , \quad \mu_z = \frac{1}{1+4.5A+2.9A^2} \qquad (3.9)$$

The two emittance terms are not indipendent; an upper limit on the total emittance can be obtained by summing linearly the RF and space charge terms.

For our case, if we assume the reference value for the field at cathode given in Table 4, 100MV/m, and we assume also an rms bunch length of 2.5 ps, an rms radius of 3mm, and a charge of 1nC, we obtain from (3.5),(3.6) and (3.8) a peak current of 160 A, $\varepsilon_x^{RF}=1.3\times10^{-6}m\ rad$, $\varepsilon_z^{RF}=2.2\times10^{-5}m$, $\varepsilon_x^C=4.5\times10^{-6}m\ rad$, $\varepsilon_z^C=1.1\times10^{-6}m$.

These results are consistent with the numerical calculations using Mask or Parmela. At 100 MeV/m of accelerating field we expect a transverse emittance of approximately $5\times10^{-6}\ m\ rad$. Achieving such a high field without breakdown in the RF gun is an open question. A more conservative value is 50 MeV/m. At this level of accelerating field the emittance is approximately doubled, and we can expect $\varepsilon_x \approx 10^{-5}\ m\ rad$.

3.2. Gun to linac transport

The beam transport system from the gun to the linac is being designed to provide a good matching to the linac, the possibility of measuring the longitudinal and transverse beam emittance, and the capability of longitudinal bunch compression to maximize the beam peak current. A schematic view is shown in Fig.6.

A high brightness electron accelerator

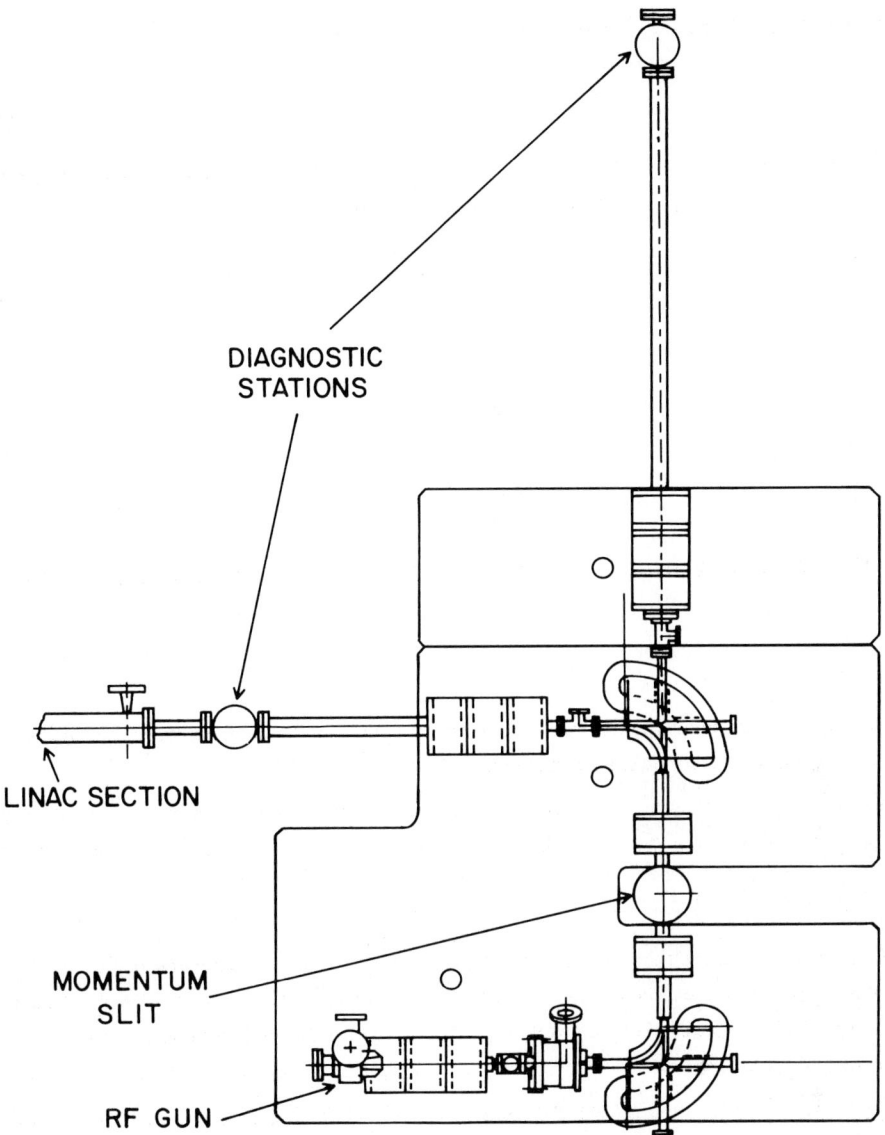

Figure 6. Detailed view of the gun to lineac beam transport line

At the gun exit the beam has a rather large angular divergence, caused by the defocusing RF force present in this region. To control the beam we need a focusing element as near as possible to the gun exit. We have chosen to use a quadrupole triplet, for the flexibility it provides. This triplet is followed by two 90 degrees bending and by another triplet which focuses the beam through the linac sections. The system is designed to have a large energy dispersion between the two dipoles, and zero dispersion outside. The two quadrupoles between the dipoles are used to make the dispersion zero between the second dipole and the linac. In the high energy dispersion region, at the middle point between the two dipoles, there will be a slit to select the beam energy and energy spread. The first triplet is set to provide a focus for the horizontal and vertical betatron oscillations at the slit position. For the calculated transverse beam emittance the energy resolution provided by the slit is ±0.1%.

The transverse emittance will be measured in a diagnostic box placed between the second quadrupole triplet and the linac section. The measurement can be done by measuring the beam spot size for the different strength of the triplet.

To measure the energy spread and the bunch length one can switch off the second dipole and propagate the beam through another quadrupole to a final detector. In the region between the first triplet and dipole there will be an RF cavity phase locked to the linac, and which can give a vertical momentum to the electrons. The field in the cavity will be phased so that the particle corresponding to the beam central energy will cross at zero phase, and not be deflected vertically. The particle crossing at an earlier or later time will be deflected vertically in opposite directions. By observing the beam spot size at the end of this beam line one can determine the electron pulse duration from the vertical size and the energy spread from the horizontal size. In effect on this detector one can display the beam longitudinal phase space. The expected time resolution of this system is of the order of one picosecond, and the power needed in the deflecting cavity is 100 KW.

In the design of this transport line one has to estimate the effects of space charge, and non linear fields in the magnets, both of which can blow up the beam transverse emittance. We have used these calculations to set up an upper limit on the amount of non linear terms acceptable in the magnets, which can produce geometric and chromatic aberrations. With our present design we estimate an emittance blow up not larger than a factor of two.

3.3. Photocathode

The choice of the photocathode material is another critical element. To obtain the beam brightness of Table 1 we need, for the high current case, a laser spot size of 3 millemeters, a current density of about 500 A/cm^2, and an effective cathode temperature of 0.2 eV. For the low current case the spot size is 20μm, and the corresponding current density is about 3 KA/cm^2.

Another important quantity characterizing the photocathode is its quantum efficiency. For the high current case we want to produce about 10^{10} electrons, and, for a quantum efficiency of η, we will need

$10^{10}/\eta$ photons. If we operate in the UV region, at $\lambda = 260 nm$, we need a pulse energy, at the quadrupled frequency, of $\approx 10^{-8}/\eta$. For a quantum efficiency of 10^{-4}, this is 0.1 mJ.

Several types of photocathodes are being studied, at BNL and other laboratories. The Cs_3Sb cathode used at LANL has good current density, about 600 A/cm^2, good quantum efficiency, about .1, but requires very good vacuum, and can have a short lifetime [4]. Alternatives being studied at BNL include metal cathodes. These offer some attractive characteristics: high current density, and resistance to damage. Their disadvantage is the low, 10^{-4} or smaller, quantum efficiency, requiring a large energy in the laser pulse.

For this reason the RF gun has the capability of changing the cathode, so that we will be able to experiment with different materials, and surface characteristics.

4. THE LINAC

The linac is based on the SLAC design and uses two accelerating cavities, each three meters long, to accelerate the beam up to an energy of 50 MeV using a single 20 MW klystron, and 100 MeV with two klystrons. The main characteristics of the linac sections are given in Table 5.

For a travelling-wave constant gradient type structure the accelerating field is given by [9]

$$E_0 = (2\alpha_0 r P_0)^{1/2} \tag{4.1}$$

where α_0 is the structure attenuation constant at the guide input, r is the shunt impedance and P_0 the input power. The shunt impedance per unit length, r, is defined as

$$r = \frac{E^2}{-dP/dz} \tag{4.2}$$

and the attenuation as

$$\alpha = \frac{-dP/dz}{2P} \tag{4.3}$$

The other quantity of interest is

$$Q = \frac{\omega P}{v_G(-dP/dz)} \tag{4.4}$$

where v_G is the group velocity.

For a constant gradient linac the attenuation changes along an accelerating section of length l as

$$\alpha = \frac{1-e^{-2\tau}}{2l[1-\frac{z}{l}(1-e^{-2\tau})]} \quad (4.5)$$

where τ is the total attenuation The same is true for the group velocity, v_G, and the filling time, $\tau_f = l/v_G$. The power at the accelerating section end is related to the input power by

$$P(l) = P_0 e^{-2\tau} \quad (4.6)$$

With this notations we can also write the energy gain per section as

$$\Delta E = e(1-e^{-2\tau})^{\frac{1}{2}}(P_0 rl)^{\frac{1}{2}} \quad (4.7)$$

TABLE 5

Accelerating Section Design Parameters.

Structure Type, Travelling Wave

Input Power, MW	20
Energy Gain (unloaded), MeV	48
Energy Gain (loaded, 50 mA), MeV	46
Operating Frequency, GHz	2.856
Shunt Impedance, $M\Omega/m$	80
Attenuation, nepers	.57
Filling time, average, μs	0.84
Q	15,000
Operating Mode	$2\pi/3$
Structure Diameter, mm	82.5
Beam Aperture, mm	23
Disk Spacing, mm	35
Length, mm	3050
Input β	.995

For our linac sections we have: $\tau=0.57$, $r=80M\Omega/m$, and $l=3m$, giving

$$\Delta E = 12.8 \, P_{0(MW)}^{1/2} \quad MeV \tag{4.8}$$

for zero accelerated current.

For 20 MW input power per section we can obtain up to 57 MeV/section at zero current.

4.1. Beam loading and bunch to bunch energy energy spread

The beam energy spread has two contributions, one due to the bunch length and one due to beam loading. If the electron bunch phase angle relative to the RF field is $\theta=0$, the energy spread due to the bunch length is [9]

$$\frac{\sigma_E}{E} = 1-\cos(\frac{2\pi\sigma_z}{\lambda}) \tag{4.9}$$

where σ_z is the rms bunch length. For $\sigma_z = 2.5ps$, this gives $\sigma_E/E \approx 0.1\%$, and $\sigma_E \approx 50 KeV$ at $E = 50 MeV$.

If we run the system in a steady state, and I is the beam average current, the beam energy is reduced to

$$\Delta E = \Delta E_0 - \frac{eIrl}{2}(1-2\tau e^{-2\tau}/(1-e^{-2\tau})) \tag{4.10}$$

With the numerical values of Table 2 this becomes

$$\Delta E = 12.8 \, (P_{0,MW})^{1/2} - 56 \, I_{Amp} \, , \quad MeV \tag{4.11}$$

The steady state is reached after two filling times, or, in our case $1.7\mu s$. Before reaching this steady state situation there will be a difference in energy from bunch to bunch.

Additional energy changes from bunch to bunch can be produced by changes in the amplitude or phase of the RF field from the Klystron. A change in phase by $\Delta\Phi$ will produce an energy change

$$\frac{\Delta E}{E} = \frac{\cos(\Phi+\Delta\Phi)-\cos\Phi}{\cos\Phi} \tag{4.12}$$

For $\Phi=0$ and small changes this can be rewritten as $\Delta E/E \approx (\Delta\Phi^2)/2$. The requirement $\Delta E/E \approx <10^{-3}$ leads to $\Delta\Phi<.05$, or 2.5 degrees.

The accelerating field amplitude must be stable to better than 10^{-3} during the micropulse and from pulse to pulse.

4.2. Single bunch energy spread and emittance

Assuming a rms bunch length of about 2.5 ps the energy spread produced by the difference in phase between particles is, for a central phase of zero degrees, is about 10^{-3}. For an output energy of 50 MeV this corresponds to a rms spread of 50 KeV.

In addition we must consider the effect of the longitudinal wakefield [9]. For a SLAC structure and our bunch length, the peak value of the beam induced voltage is about $0.15 MV/m/10^{10}e^-$. For a nC charge in a bunch this would give a peak beam induced field of 90 KeV/m. For a six meter long accelerating section this produces a spread on the order of 540 KeV, or about 1% of a final 50 MeV energy. This effect is clearly much larger than the spread due to the finite bunch length, and is too large for most FEL experiments or other applications.

This effect can be partially reduced by shifting the phase at which the bunch crosses the RF, and using the RF time dependence to compensate for the wake field [9]. For a 1 nC charge one needs a RF phase shift of about 3.6 degrees, to obtain an energy spread of 3×10^{-3}.

To reduce the energy spread to the 10^{-3} level we will need to use the full linac acceleration to 100 MeV in two sections, and reduce the charge below the 1 nC value.

We have also considered the effect of wake fields on the transverse emittance in crossing the linac [9]. For our linac length of 6m, 1nC charge, 50 MeV final energy, and an initial emittance of 5×10^{-6} *m rad*, it is sufficient to inject the beam in the linac with a transverse displacement error smaller than 0.15 mm to have an increase in emittance smaller than 20%.

5. LASER SYSTEM

The laser system has to satisfy several requirements:
a. produce a 6 ps pulse to extract electrons from the photocathode; depending on the type of material used for the cathode one needs either frequency doubled or quadrupled radiation; the energy per pulse must be enough to extract 1 nC of electrons, also in the case of a quantum efficiency which can be as low as 10^{-4} for a metal cathode; this would require about 10^{14} photons with a 5 eV energy or about 0.1 mJ in 6 ps, i.e. 13 MW of peak power;
b. "cut" a 6 ps slice from a CO_2 laser pulse, illuminating two germanium plates, as shown in Fig. 7;
c. produce a 6 ps, 100 GW pulse of CO_2 radiation, synchronized to the electron bunch accelerated in the linac.

A schematic view of the laser system, designed and built at Los Alamos National Laboratory by I. Bigio and N. Kurnit, is given in Fig. 7. A more detailed view of the Nd-Yag laser is given in Fig. 8.

PICOSECOND CO_2 LASER SOURCE FOR THE LASER ACCELERATOR TEST FACILITY

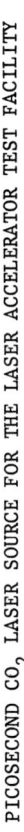

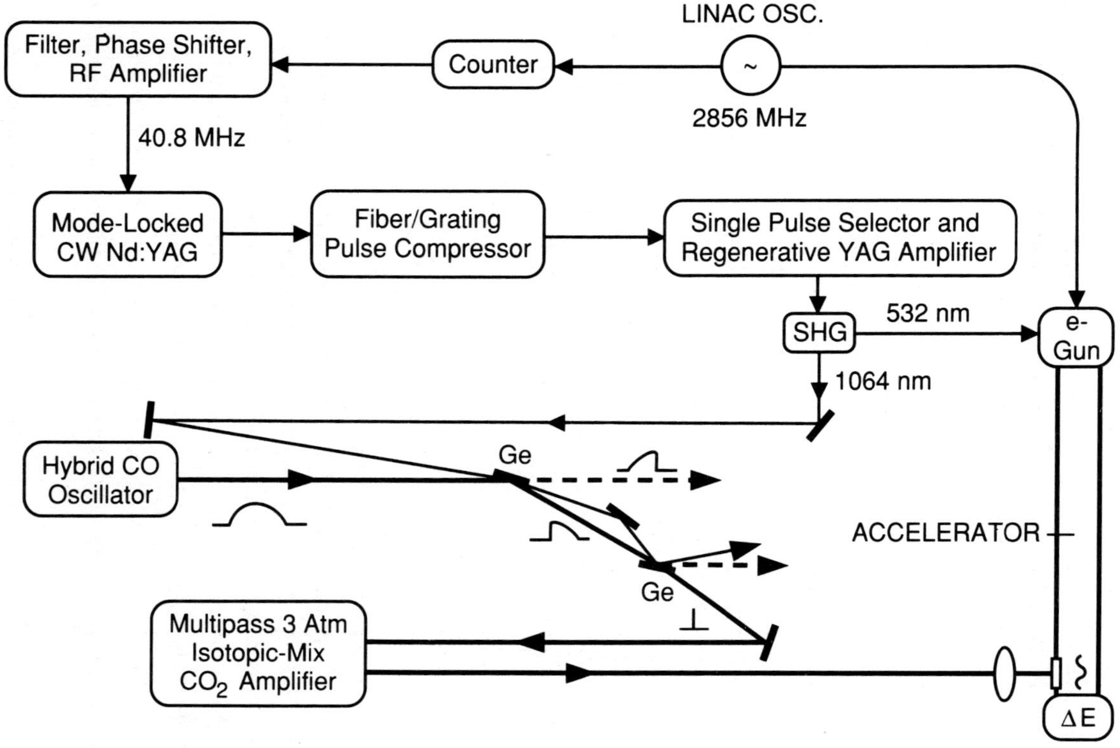

Figure 7. Scheme of the ATF laser system

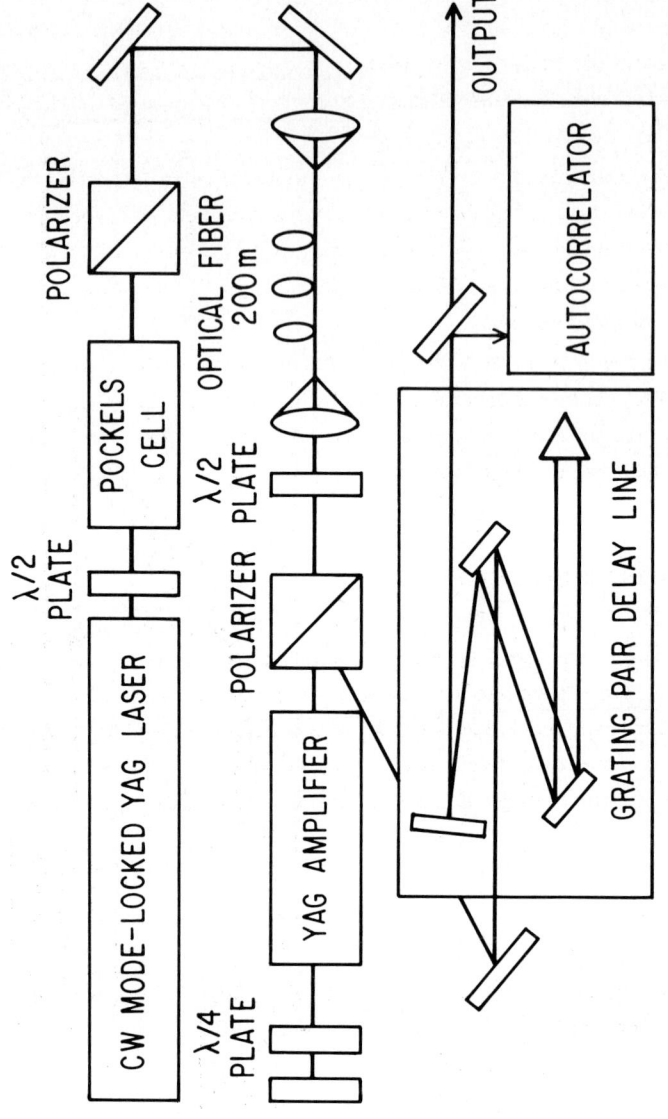

Figure 8. Scheme of the Yag laser with the pulse compression system

To extract the electrons at the right phase for acceleration and minimum beam emittance, one drives the CW Yag oscillator with a 40.8 MHz frequency generator. The same signal is also multiplied by 70 to produce the signal driving the klystron and the linac. To avoid fluctuations in the beam energy we must keep the jitter in the laser pulse relative to the RF in the linac to a value not larger than one picosecond. This jitter can be produced in the electronic system driving the RF, or in the system triggering the mode locked laser. Keeping the jitter to such a low value presents many problem that will have to be solved.

The output of the Yag mode-locked oscillator is a train of pulses about 80 ps long and separated by 12.5 ns. After compression in the optical fiber-pulse compressor system a train of these pulses can be amplified in the regenerative amplifier. The maximum number of the pulses that it is possible to amplify is 300, for a total time duration of $3.75\mu s$. We will also have the possibility of isolating and amplifying any single laser pulse and operate the linac with a single electron bunch.

The germanium plates can be switched from transparent to reflective only once during the $1.25\mu s$ Yag amplifier pulse. Thus the experiments using the interaction of the CO_2 laser pulse and the electrons can only use one Yag pulse and one electron bunch. For the experiment, like FELs, not using the CO_2 light, one can utilize up to 300 electron bunches.

6. THE ATF EXPERIMENTAL PROGRAM

The ATF will offer some unique possibilities for studies on the interaction of electromagnetic radiation and particle beams. The experimental program now being prepared takes advantage of the the beam characteristics. The program is mainly oriented in two directions: i) Laser Acceleration of particle beams using the very high electric field provided by high power lasers; ii) production of coherent radiation from relatistic electrons using the FEL mechanism, and exploiting the large ATF beam brightness.

6.1. Laser acceleration

Laser acceleration studies at the ATF utilize the unique combination of high brightness electron beams and high power lasers, available at this facility.

Three types of experiments are being prepared:

1) laser acceleration on a microlinac structure [10]; uses on open, periodic, linac-like material structure on which one shines the CO_2 light to produce the accelerating field; the characteristic dimension of the structure is equal to the laser wavelength, $10\mu m$; one expects accelerating fields as high as 1 GeV/m.

2) Inverse FEL [11]; uses the interaction between the laser light and a beam moving through an undulator magnet to transfer energy to the electrons; with our system, using the CO_2 laser and a 60 cm long undulator, one can double the beam energy to 100 MeV. Larger accelerating fields, of the order of 600 MeV/m, can be obtained by adding a low density plasma in the undulator, to produce a "FEL-Plasma wave accelerator" [12].

3) Inverse Cherenkov acceleration [13]; uses the interaction of particles and radiation in a gas to transfer energy from the laser to the beam; also in this case one can expect accelerating fields as high as 1 GeV/m.

6.2. FEL

Free Electron Lasers (FELs) have been successfully operated from the millimeter region to the IR, visible and near UV. The FEL has very favourable scaling laws for extension to the Soft X-Ray region, if some theoretical expectations, like optical guiding, Self Amplified Spontaneous Emission, and non linear saturation behaviour, are confirmed experimentally, and if we can produce electron beams with sufficiently high brightness and small energy spread [2].

The ATF will allow us to study both the beam production and the FEL physics. The beam characteristics expected from the ATF, and given in Table 1, are sufficient to produce coherent radiation in the 500 Angstrom region. However the initial beam energy only allows to produce radiation in the IR or visible region. We can still study the relevant FEL physics in the IR region, at a few micrometer, and, if the results confirm our expectations, increase the beam energy at a later time.

With the 50 MeV beam and an undulator with a period of about 3 centimeters, one can build a single pass, Self Amplified Spontaneous Emision, IR FEL with a gain large enough to study the FEL physics necessary for a UV to soft X-ray FEL.

In addition to the FEL experiments one can also utilize the non linear Compton backscattering of CO_2 photons on the electrons to provide a picosecond sources of X-Rays [14], with about 10^8 photons per pulse.

Work performed under the auspices of the US Department of Energy.

REFERENCES

1) See for instance C. Pellegrini, Electron-Positron Colliders in the TeV Region, in Proceedings of the 1985 International Symposium on Lepton and Photon Interactions at High Energies, M. Konuma and K. Takahashi eds., Kyoto 1985.
2) C. Pellegrini, Nucl. Instr. and Meth., A272, 1988, p. 364.
3) A. M. Hutton et al., IEEE Trans. Nucl. Sci., NS-32, 1985, p. 1659.
4) J. S. Fraser et al., Photocathodes in Accelerator Applications, in Proc. 1987 IEEE Particle Accelerator Conference, E. Lindstrom and L. S. Taylor eds., p. 1705, Washington 1987.
5) S.V. Benson et al., Nucl. Instr. and Meth., A250 (1986), p. 39
6) K. T. McDonald, Design of the Laser-Driven RF Electron Gun for the BNL Accelerator Test Facility, Princeton Preprint DOE/ER/3072-43, 1988.
7) J. E. Wang and G.A. Loew, Progress Report on New RF Breakdown Studies in an S-Band Structure at SLAC, in Proc. 1987 IEEE Particle Accelerator Conference, E. Lindstrom and L. S. Taylor eds., p. 1705, Washington 1987.
8) K.-J. Kim, RF and Space-Charge Effects in Laser-Driven RF Electron Guns, Lawrence Berkeley Laboratory Report LBL-25807, 1988.
9) See for instance P. B. Wilson et al, High Energy Electron Linacs; Applications to Storage Ring RF Systems and Linear Colliders, in Physics of High Energy Particle Accelerators, R.A. Carrigan, F.R. Huson and M. Month eds., p. 450, AIP Conf. Proc. Series, Vol. 87, 1982.
10) R. B. Palmer, Particle Accelerators, 11, 1980, p. 81.
11) E. D. Courant, C. Pellegrini and W. Zakowicz, Phys. Rev. A, 32, 1985, p. 2813.
12) J. L. Bobin, Laser Wiggler Beat Wave, in Laser Acceleration of Particles, C. Joshi and T. Katsouleas eds., AIP Conf. Proc., Vol.130, p. 345, 1985.
13) J. R. Fontana, Inverse Cherenkov Acceleration, in Laser Acceleration of Particles, C. Joshi and T. Katsouleas eds., AIP Conf. Proc., Vol.130, p. 357, 1985.
14) R. C. Fernow et al, Proposal for an Experimental Study of non linear Thomson Scattering, Princeton University Preprint DOE/ER/3072-39, 1986.

EMITTANCE GROWTH IN LASER-DRIVEN RF ELECTRON GUNS

Kwang-Je Kim

Accelerator and Fusion Research Division, Lawrence Berkeley Laboratory, 1 Cyclotron Road, Berkeley, CA 94720

We present a simple analysis for the evolution of the electron-beam phase space distribution in laser-driven rf guns. In particular, formulas are derived for the transverse and longitudinal emittances at the exit of the gun. The results are compared and found to agree well with those from simulation.

1. INTRODUCTION

Laser-driven rf electron guns[1] are potential sources of high-current, low-emittance, short bunch-length electron beams, which are required for many advanced accelerator applications, such as free-electron lasers and injectors for high-energy machines. In such guns the design of which was pioneered at Los Alamos National Laboratory[1] and is currently being developed at several other laboratories[2,3,4], a high-power laser beam illuminates a photo-cathode surface placed on an end wall of an rf cavity. The emitted electrons are accelerated immediately to a relativistic energy by the strong rf field in the cavity. The main advantages of this type of gun are that the time structure of the electron beam is controlled by the laser, eliminating the need for bunchers, and that the electric field in rf cavities can be made very strong, so that the degrading effects due to space-charge repulsion can be minimized.

A calculation of emittances in rf guns must include effects due to the time variation of the rf field over the duration of the acceleration period and over the duration of the electron pulse, and those due to the space-charge repulsion. Because a rigorous analysis of these effects is complicated, explicit calculations of the emittances have usually been based on simulation codes. Here we present an approximate but simple analytic calculation[5] that retains the main physical effects.

2. ANALYSIS OF RF EFFECTS[5]

The structure of a rf gun cavity is schematically illustrated in Fig. 1. The cavity consists of ($\frac{1}{2}$ + n)-cells, the first cell being a half-cell so that electrons see the maximum accelerating field as they are emitted from the cathode. The accelerating field in the cavity will be assumed to be of the following simple form:

$$E_z = E(z) \cos kz \sin (\omega t + \phi_0); \quad E(z) = E_0 \theta(z_f - z) \ . \tag{1}$$

Here λ is the rf wavelength, $k = 2\pi/\lambda$, c is the velocity of light, $\omega = ck$, ϕ_0 is the rf phase as the electron leaves the cathode (located at $z = 0$) at $t = 0$, E_0 is the peak accelerating field, θ is the step function, and $z_f = (n + \frac{1}{2})\lambda/2$ is the location of the cavity exit. When E_z is independent of transverse coordinates, the transverse fields are linear in transverse coordinates and are given by

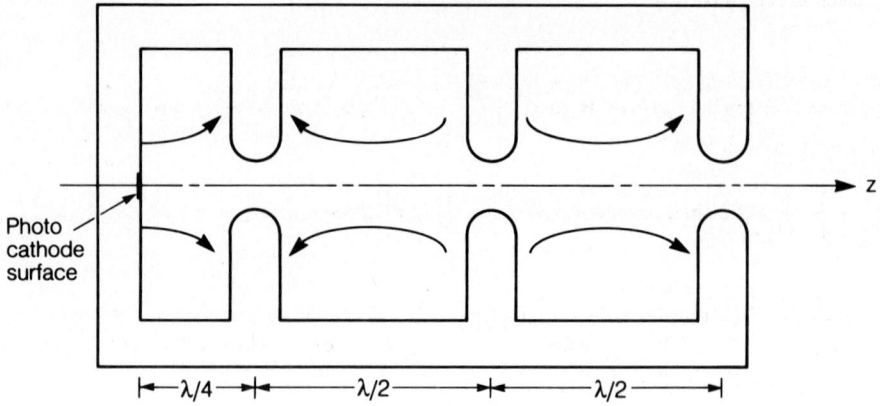

FIGURE 1
Schematics of the rf laser gun.

$$E_r = -\frac{r}{2}\frac{\partial}{\partial z} E_z , \quad cB_\theta = \frac{r}{2c}\frac{\partial}{\partial t} E_z . \qquad (2)$$

Here r and θ are the cylindrical coordinates. The fields, Eq. (1) and Eq. (2), can be considered as a π-mode excitation of an ideal cavity.[2]

The equations for rf acceleration for $z < z_f$ are

$$\frac{d\gamma}{dz} = \frac{eE_0}{2mc^2}[\sin(\phi) + \sin(\phi + 2kz)] , \qquad (3)$$

$$\frac{d\phi}{dz} = k\left[\frac{\gamma}{\sqrt{\gamma^2-1}} - 1\right] . \qquad (4)$$

Here γ is the electron energy in units of its rest energy and $\phi = \omega t - kz + \phi_0$. Near the cathode where $kz \ll \phi$, it follows from Eq. (3) that

$$dz \approx \frac{1}{2\alpha k \sin\phi_0} d\gamma , \qquad (5)$$

where

$$\alpha = eE_0/2mc^2 k \qquad (6)$$

is a dimensionless parameter representing the acceleration gradient.

An approximate solution of Eqs. (3) and (4) can be obtained based on the fact that the RHS of Eq. (4) is significantly different from zero in the region where the electrons are still nonrelativistic, i.e., near the cathode surface. The result is that, at the exit of the (½ + n)-cell cavity, the phase φ approaches the value

$$\phi_f = \frac{1}{2\alpha \sin\phi_0} + \phi_0 , \qquad (7)$$

and the electron energy becomes

$$\gamma_f = 1 + \alpha\,[(n + \tfrac{1}{2})\,\pi\,\sin\phi_f + \cos\phi_f]\ .\qquad(8)$$

The force in the radial direction is given by $F_r = e\,(E_r - \beta c B_\theta)$, which, using Eqs. (1) and (2), can be written in the following form:

$$F_r = er\left\{ -\frac{1}{2c}\frac{d}{dt}\left[E(z)\sin kz\cos(\omega t + \phi_0)\right] - \frac{1}{2}\left[\frac{d}{dz}E(z)\right]\sin\phi \right.$$

$$\left. + \frac{(\beta-1)}{2}\left[\frac{d}{dz}E(z)\right]\sin kz \cos(\omega t + \phi_0)\right\}\ .\qquad(9)$$

The transverse momentum is obtained by integrating Eq. (9) with respect to time t. We assume that the transverse deflection is small so that the radius r can be regarded as constant. The first term in the above is a total derivative of an expression that vanishes at the cathode surface and outside the cavity, thus its contribution to the transverse momentum vanishes. The contribution from the third term is small because of the $(\beta - 1)$-factor. Thus the main contribution comes from the second term, which is important only in the region where $dE(z)/dz$ is non-vanishing, i.e., near the cavity exit. With $E(z)$ given by the step function (see Eq. (1)), the dimensionless transverse momentum at the cavity exit in the Cartesian coordinate, $p_x = \beta\gamma x'$, where x' is the angle in the x-direction, becomes

$$p_x = (\alpha\,k\,\sin\phi_f)x\ .\qquad(10)$$

Here we have assumed that the transverse momentum vanishes at the cathode surface. Equations (8) and (10) are the basis for our discussion of the rf-induced emittances. The phase ϕ_f varies over the length of the electron bunch, $\phi_f = \langle\phi_f\rangle + \Delta\phi$; $\Delta\phi = -k\Delta z$, where $\langle\phi_f\rangle$ is the average phase. Thus the transverse phase-space distribution consists of a collection of lines with different slopes corresponding to different $\Delta\phi$, as illustrated in Fig. 2.

The normalized transverse emittance is defined as[6]

$$\varepsilon_x = \sqrt{\langle p_x^2\rangle\langle x^2\rangle - \langle p_x x\rangle^2}\ ,\qquad(11)$$

where the angular brackets refer to the average values. It can be shown that the rms emittance defined by Eq. (11) is invariant[7] when the force is linear. The emittance increase in rf guns is due to the ϕ-dependence of the focusing force [Eq. (10)] and the nonlinearity of the space-charge force.

Assuming that the electrons' density distribution is Gaussian, with rms transverse and longitudinal lengths given respectively by σ_x and σ_z, we find from Eqs. (9) and (11) that the emittance is at a minimum for $\langle\phi_f\rangle = \pi/2$ and

$$\varepsilon_x^{rf} = \alpha\,k^2\,\sigma_x^2\,\sigma_z\,|\cos\langle\phi_f\rangle|\ ;\ \langle\phi_f\rangle \neq \pi/2\ ,\qquad(12)$$

$$\varepsilon_x^{rf} = \alpha\,k^3\,\sigma_x^2\,\sigma_z^2/\sqrt{2}\ ;\ \langle\phi_f\rangle = \pi/2\ .\qquad(13)$$

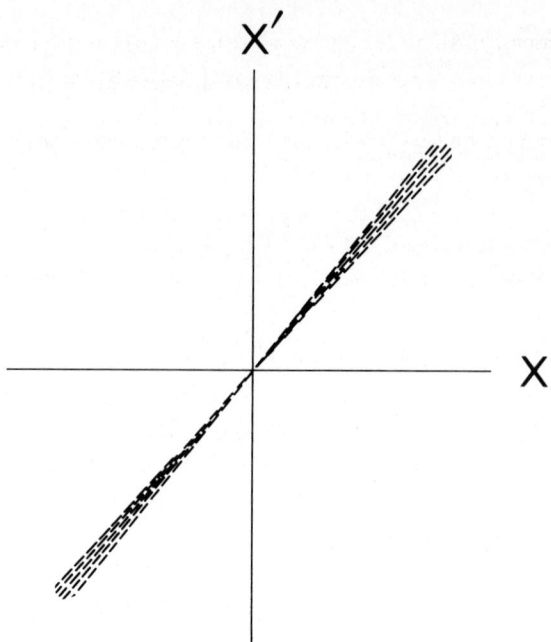

FIGURE 2
Electron distribution in transverse phase space due to time-dependent focusing of the rf field.

In the above, the superscript rf refers to the fact that we are considering the rf-induced effects.

The longitudinal emittance is defined as

$$\varepsilon_z = \sqrt{\langle(\Delta p_z)^2\rangle \langle(\Delta z)^2\rangle - \langle\Delta p_z \Delta z\rangle^2} \; , \tag{14}$$

where $\Delta p_z = \Delta(\beta\gamma) \approx \Delta\gamma$ for the relativistic case $\beta \approx 1$. We write $\gamma_f = \langle\gamma_f\rangle + \Delta\gamma$, and expand Eq. (8) in $\Delta\phi = -k\Delta z = \phi_f - \langle\phi_f\rangle$. Setting $\langle\phi_f\rangle = \pi/2$ to minimize the transverse emittance, we obtain

$$\Delta\gamma = \alpha k \Delta z - \frac{1}{2}\left[\langle\gamma_f\rangle - 1\right] k^2 (\Delta z)^2 - \frac{\alpha k^3}{3!}(\Delta z)^3 + \cdots \; . \tag{15}$$

The nonlinear terms in the above contribute to the longitudinal emittance, and one obtains to the lowest order

$$\varepsilon_z^{\rm rf} = \sqrt{3}\left[\langle\gamma_f\rangle - 1\right] k^2 \sigma_z^3 \; . \tag{16}$$

3. ANALYSIS OF THE SPACE-CHARGE EFFECTS[5]

A repulsive force attributable to space charge causes the emittance to increase. To study this effect, we assume that all electrons are moving with the same velocity, v, in the z-direction. In the reference frame moving with the electrons, the electromagnetic interaction is completely described by a purely electrostatic field $\mathbf{E}'$. The field components in the laboratory frame which give rise to the x- and z-components of the force are given by the Lorentz transformation $E_x = \gamma E_x'$, $B_y = \gamma(\beta/c) E_x'$, $E_z = E_z'$. Here, B_y is the magnetic field. The components of the force are

$$F_x = e(E_x - vB_y) = \frac{e}{\gamma} E_x' \, , \, F_z = eE_z' \, . \tag{17}$$

In the following, we assume the charge distribution to be cylindrically symmetric, so that we do not need to consider F_y separately.

The field $\mathbf{E}'$ is a function of γ since the bunch dimension in the moving frame, σ_x' in the transverse direction and σ_z' in the longitudinal direction, are related to the corresponding bunch dimension in the laboratory frame σ_x and σ_z by $\sigma_x' = \sigma_x$ and $\sigma_z' = \gamma\sigma_z$; that is, the bunch in the moving frame appears to be elongated by a factor γ. It can be shown that the force given by Eq. (17) vanishes as γ^{-2} for $\gamma \gg A$, where $A = \sigma_x/\sigma_z$ is the aspect ratio. Thus we can write $\mathbf{E} = \mathbf{f}(\gamma)/\gamma^2$, where $\mathbf{f}(\gamma)$ is a function slowly varying in γ for $\gamma \gg A$. The contribution to the dimensionless momentum due to the space-charge force is given by

$$(p_x, p_y, \Delta p_z) \equiv \mathbf{p} = \frac{1}{mc^2} \int \frac{1}{\gamma^2 \beta} \mathbf{f}(\gamma) \, dz \, . \tag{18}$$

We evaluate Eq. (18) approximately by noting that the factor $1/\gamma^2\beta$ in the integrand decreases rapidly as γ becomes large. Thus we replace $\mathbf{f}(\gamma)$ by $\mathbf{f}(1)$ and dz by Eq. (5). The result is $\mathbf{p} = (\pi/2)(1/E_0 \sin \phi_0) \mathbf{E}^{sc}$, where $\mathbf{E}^{sc}$ is the electrostatic field due to the charge distribution at rest in the laboratory frame. From this and from Eqs. (11) and (14), we obtain the space-charge induced emittances as follows:

$$\varepsilon_i^{sc} = \frac{\pi}{4} \frac{1}{\alpha k} \frac{1}{\sin \phi_0} \frac{I}{I_A} \mu_i(A) \, ; \, i = x \text{ or } z \, , \tag{19}$$

where I is the peak current, $I_A = 17{,}000$ Amp known as the Alfvén current, and the functions $\mu_i(A)$ are defined in terms of the normalized field $\mathscr{E}_i = (4\pi\varepsilon_0/n_0) E_i^{sc}$ ($n_0 =$ the line density) by

$$\mu_x(A) = \sqrt{\langle \mathscr{E}_x^2 \rangle \langle x^2 \rangle - \langle \mathscr{E}_x \cdot x \rangle^2} \, , \, \mu_z(A) = \sqrt{\langle \mathscr{E}_z^2 \rangle \langle \Delta z^2 \rangle - \langle \mathscr{E}_z \cdot \Delta z \rangle^2} \, . \tag{20}$$

Note that $\mu_i(A)$, being dimensionless, are functions of A only. For the Gaussian charge distribution, these functions are approximately given by

$$\mu_x(A) \sim \frac{1}{3A + 5} \, , \, \mu_z(A) = \frac{1.1}{1 + 4.5A + 2.9A^2} \, . \tag{21}$$

The functions $\mu_x(A)$ and $\mu_z(A)$ are shown in Fig. 3 and Fig. 4, respectively.

How do the rf-induced and the space-charge induced emittance add? It turns out that there are correlations between those two effects so that the total emittance ε_i is given by

$$\sqrt{(\varepsilon_i^{rf})^2 + (\varepsilon_i^{sc})^2} < \varepsilon_i < \varepsilon_i^{rf} + \varepsilon_i^{sc} \, . \tag{22}$$

A more detailed derivation of the results in Sections 2 and 3, as well as the application of these results to the case where the electron distribution is uniform in a cylinder, can be found in Ref. 5.

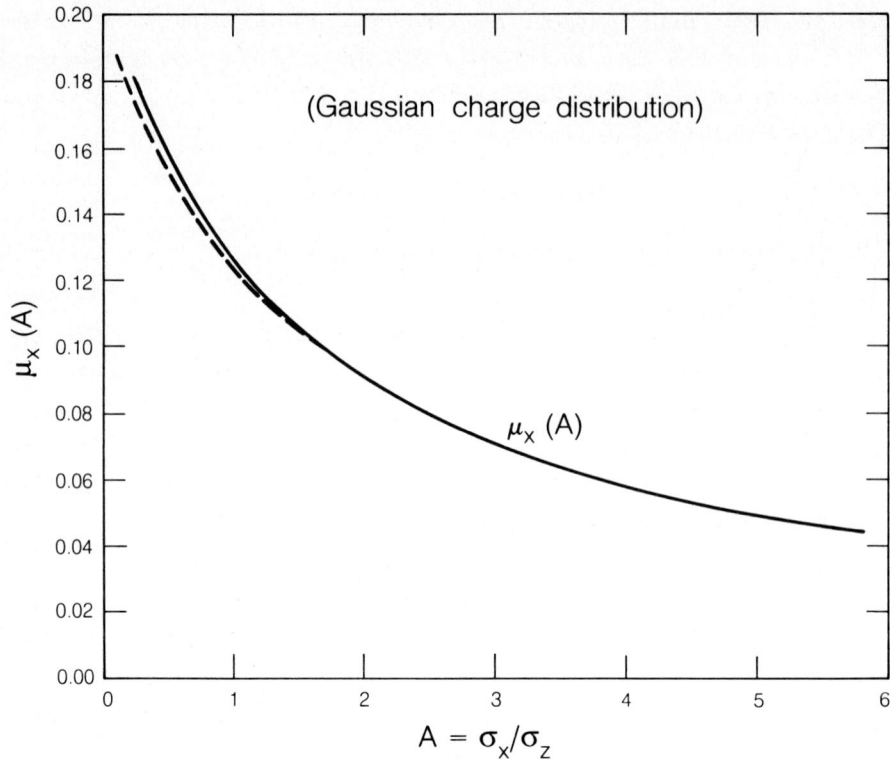

FIGURE 3
The transverse space-charge factor $\mu_x(A)$. The dotted line is the approximate function given by Eq. (21).

4. COMPARISON WITH SIMULATION

We consider the (½ + 1)-cell gun under construction at Brookhaven National Laboratory and compare the analysis of this paper with the numerical simulation by McDonald[2], hereafter referred to as KM. The gun parameters are

$$eE_0 = 100 \text{ MV/m}, \quad \lambda = 10.5 \text{ cm} \ .$$

The dimensionless rf strength α [Eq. (6)] corresponding to this case is $\alpha = 1.64$. The optimum initial phase calculated from Eq. (7) by demanding that the final phase ϕ_f be 90° is $\phi_0 = 71°$ as compared to $\phi_0 = 68°$ in KM. The final γ calculated from Eq. (8) by setting n = 1, $\phi = 90°$, and the above value of α is $\gamma_f = 8.7$ as compared to $\gamma_f = 9.2$ in KM.

The rms bunch length is $\sigma_z = 0.6$ mm or $\sigma_\phi = k\sigma_z = 3.6 \times 10^{-2}$. From Eq. (18), the corresponding rms energy spread is $\sigma_E = mc^2\sigma_{\Delta\gamma} = 30$ keV as compared to $\sigma_E = 17$ keV in KM.

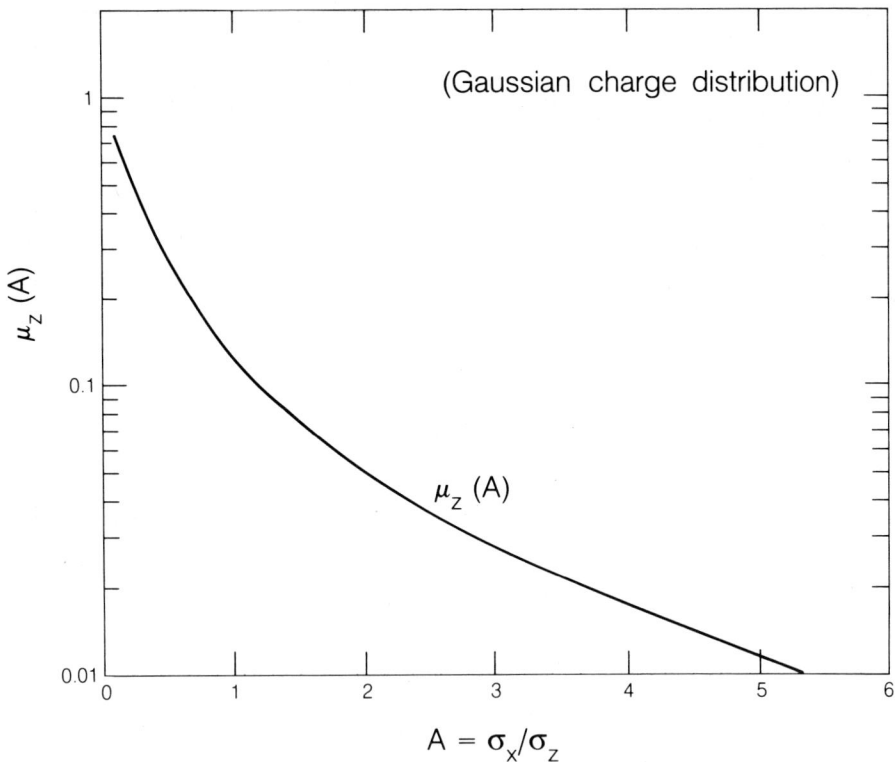

FIGURE 4
The longitudinal space-charge factor $\mu_z(A)$.

The rms beam transverse size is $\sigma_x = 3.5$ mm. From Eq. (10), we obtain the transverse angular divergence at the exit of the gun to be $p_x = \alpha k \sigma_x/\gamma \approx 40$ mrad, which is in rough agreement with Fig. 4 in KM.

The rf contribution to the transverse emittance from Eq. (13) is 1.1 mm-mrad, as compared to 1.4 mm-mrad in KM (Table 1). The transverse space-charge factor corresponding to the aspect ratio $A = 3.5/0.6 \approx 6$ is, from Fig. 3, $\mu_x \sim 4 \times 10^{-2}$. From Eq. (19) and using $I = c \times 1 \text{ nC}/\sqrt{2\pi}\,\sigma_z \approx 200$ A, we obtain $\varepsilon_x^{sc} = 4$ mm-mrad, as compared to $\varepsilon_x^{sc} = 6.2$ mm-mrad obtained by KM.

The longitudinal space-charge factor for $A = 6$ is about 0.01 from Fig. 4. The quantity $\varepsilon_z^{sc}/\gamma_f$ determined from Eq. (19) is about 1×10^{-5} cm. The longitudinal emittance due to the rf effect is about three times larger than the space-charge emittance. The total phase-space area is in rough agreement with the phase-space area indicated by Fig. 5 of KM.

The agreement of our simple theory with the simulation calculation is encouraging. Further comparison is reported in Ref. 8.

5. CONCLUSIONS

In this paper, we have reviewed an approximate but simple theory of electron beam dynamics in laser-driven rf guns and derived formulas for physically interesting quantities such as emittances. These formulas should be useful in selecting initial parameters for the design of the gun.

There are several effects which are not taken into account in this simple treatment, such as non-linearities of the rf field, higher space harmonics, image-charge effects, etc. The fact that the results of the simple theory agree reasonably well with those of detailed simulation suggests that those effects are small. There are also contributions to the emittances from the photo-emission process at the cathode surface. These contributions, which are easy to incorporate into our expressions, are usually much smaller than those considered in this paper.

The rms emittances defined by Eq. (11) and Eq. (14) are known to be conserved when the focusing forces are linear and when the space-charge force is negligible[7]. It may be nevertheless possible to reduce the rms emittances by the combined effect of the nonlinear space charge force and a suitable linear focusing arrangement[9].

ACKNOWLEDGEMENTS

The author thanks R. Miller, A. Sessler, Y.-J. Chen, K.M. McDonald and S. Chattopadhay for useful discussions. This work was supported by the Director, Office of Energy Research, Office of Basic Energy Sciences, Materials Sciences Division, U.S. Department of Energy under Contract No. DE-AC03-76SF00098.

REFERENCES

1) J.S. Fraser et al., "Photocathodes in Accelerator Applications," Proc. 1987 IEEE Particle Accelerator Conference, IEEE Cat. No. 87CH.2387-9, 1705 (March, 1987).

2) K.T. McDonald, "Design of the Laser-Driven RF Electron Gun for the BNL Accelerator Test Facility," Princeton Preprint DOE/ER/3072-43 (March, 1988); submitted to IEEE trans. on Electron Devices.

3) S. Chattopadhay et al., "Conceptual Design of a Bright Electron Injector based on a Laser-Driven Photo-Cathode RF Gun," paper submitted to 1988 Linear Accelerator Conference.

4) H. Chaloupka et al., "A Proposed Super Conducting Photoemission Source of High Brightness," paper submitted to the European Particle Accelerator Conference, Rome, Italy (June 7–11, 1988).

5) K.-J. Kim, "RF and Space-Charge Effects in Laser-Driven RF Electron Guns," LBL-25807 (August, 1988); Nucl. Inst. Methods (in press).

6) P.M. Lapostolle, "Possible Emittance Increase Through Filamentation Due to Space Charge in Continuous Beam," IEEE Trans. Nucl. Sci. **18**, 1101 (1971).

7) F.J. Sacherer, IEEE Trans. Nucl. Sci. 18 (1971) 1105.

8) Y.-J. Chen, "Simulations of High-Brightness RF Photocathode Guns for LLNL-SLAC-LBL 1 GeV Test Experiment," LLNL pub. UCRL-99675 (August, 1988).

9) B.E. Carlsten and R.L. Sheffield, Photoelectric injection design considerations, paper submitted to 1988 Linac Conference.

One-Dimensional Theory of a Free-Electron Laser Amplifier: Steady-State and Superradiance

R.Bonifacio, F.Casagrande, G.Cerchioni, L.De Salvo Souza, P.Pierini

*Dipartimento di Fisica dell'Università
and Istituto Nazionale di Fisica Nucleare
Via Celoria 16, 20133 Milano, Italy*

Abstract

We discuss the one-dimensional theory of a single-pass, constant- wiggler FEL amplifier. First we introduce the basic physics of the system and derive the fundamental relations. Next we include more advanced topics, such as slippage and propagation in a waveguide. In particular, we illustrate the two novel high-gain dynamical regimes of weak superradiance and strong superradiance, in which the electrons emit cooperative synchrotron radiation with peak power proportional to the square of the beam current.

1. INTRODUCTION

The Free-Electron Laser (FEL), invented[1] and operated[2] in the seventies, after being considered for some years as an almost exotic device, is now renewing in the scientific community the enthusiasms arisen by the laser since its advent in the sixties. In fact, the FEL is a source of powerful and tunable coherent radiation, potentially able to cover those regions of the electromagnetic spectrum which are not accessible to conventional lasers, such as the FIR and the XUV. Many experiments have been carried out, and many more are being developed or at least planned all over the world. Time is already mature for both basic and applied research with FEL facilities in many different areas such as the spectroscopy of atomic, molecular or condensed matter, biology and medicine, controlled thermonuclear fusion, particle acceleration, and so on. While the flow of theoretical studies is more and more increasing, the basic physical principles are well understood and begin to be included in university courses in quantum electronics, optics, particle accelerators or plasma physics. Really, the FEL has grown up from the overlapping of such different branches of physics and engineering.

The results of this burst of theoretical and experimental activity is widely spread out in the literature. The first aim of this paper is to provide an introduction to the basic physics of the FEL; hence its tone is pedagogical, at least as far as some frontier topics are discussed. Furthermore, we mostly focus on a class of FEL devices, the single-pass amplifiers, and in particular on the high-gain regime of operation. This choice i) reflects the theoretical activity developed by our group, ii) can be connected to a series of relevant experiments performed or in progress, and to an experimental activity hopefully close to start in Milano[3], iii) is dictated by the possible relevant role of the FEL in the next generation of particle accelerators, where the FEL could help in obtaining very high gradients of accelerating field for reasonably long linear colliders capable to operate in the TeV range with extremely high luminosities[4]. On the other hand, we recommend a number of Conference Proceedings[5], Special Issues[6] and even books[7], besides many review papers which cannot even be listed here[8], for emphasis on different FEL schemes or regimes (e.g., FEL oscillators), as well as historical introductions, or global overviews of the state of the art.

The structure of this paper is as follows. Section 2 is intended as an FEL primer; thus it is as pedagogical as possible and nearly without equations. Secs. 3–5 are also rather pedagogical, even if obviously more technical. In Sec. 3 we introduce the one-dimensional model of an FEL amplifier which is basic to our treatment, together with a parametrization which proves very useful in the discussion of the FEL physics. Sec. 4 is devoted to a general discussion of the steady-state physics, i.e. neglecting propagation effects, including the Compton and the Raman regimes, the linear stability analysis, the collective instability and the exponential gain. In Sec. 5 we present a number of results for steady-state in the high-gain Compton regime, and derive some basic results of the small-gain regime in a suitable limit of our treatment.

In Secs. 6, 7 some more advanced or even novel topics are discussed; hopefully, also these sections should be quite accessible to the broadest audience. In Sec. 6 slippage and propagation effects are included in the model, which can give rise to a physics quite different from that of the steady-state. Under proper conditions the FEL can operate in a superradiant regime in which the cooperative emission from bunched electrons can generate radiation pulses with peak power proportional to the squared electron density. Strictly connected with superradiance, also a spiking regime due to a trailing edge instability is discussed. In Sec. 7 the main FEL equations are generalized to the relevant case of amplifiers with a waveguide. Results are presented concerning tuning, slippage and gain. Other relevant FEL topics in the 1D theory, like e.g. tapering, are discussed by us elsewhere in these Proceedings[9].

2. INTRODUCTION TO FEL PHYSICS

The purpose of this Section is to provide an introduction to the fundamentals of FEL physics. It is addressed to readers with no background in this field; therefore we have tried to stress the basic physical concepts with a minimum display of equations.

2a.1. FEL "SPONTANEOUS EMISSION"

It is well-known that a charge which is really "free" cannot radiate because of the energy-momentum conservation; in FEL physics, "free" electrons means unbound electrons. Equally well-known is the fact that an accelerated charge does radiate. This radiation is particularly intense and confined around the instantaneous velocity vector if the motion of the charge is relativistic[10]. Furthermore, for a given applied force, the radiation emitted due to a transverse force is greater than that due to a parallel force by a factor γ^2 where

$$\gamma = \left(1 - \beta^2\right)^{-1/2} = \left(1 - \frac{v^2}{c^2}\right)^{-1/2} \tag{2.1}$$

is the electron energy, γmc^2, in rest mass units. This radiation is called synchrotron radiation[10], and is also discussed in detail elsewhere in these Proceedings[11,12]. Here we stress only that the brightness and monochromaticity of synchrotron radiation are greatly enhanced if, like in the FEL, a beam of relativistic electrons is injected along the axis of an undulator or wiggler, i.e., an insertion device which provides a magnetostatic field $\vec{B}_w$ which is transverse and spatially periodic along the wiggler axis.

Wigglers are arranged in linear or circular configurations, whose simplest realizations are two arrays of permanent magnets with alternating polarities (Fig. 2.1a), or two helical coils with current circulating in opposite directions (Fig. 2.1b), respectively. In both cases, the particles are periodically deflected by the Lorentz force (Gaussian units)

$$\vec{F} = \frac{e}{c}\vec{v} \times \vec{B}_w \tag{2.2}$$

and execute a wiggle or a helical motion, respectively, thus emitting synchrotron radiation which is linearly or circularly polarized.

This is the FEL "spontaneous emission", because no radiation field is excited before the electrons are injected into the magnetic structure. Its main features are the following: i) the intensity is proportional to the electron current, that is, the radiation is incoherent; ii) it is confined in a narrow cone around the z-axis (within an angle $\simeq 1/\gamma\sqrt{N_w}$, where N_w is the number of wiggler periods[11]; iii) it is narrow-band radiation, with an on-axis spectral distribution

$$\frac{d^2 I}{d\Omega d\omega} \propto \text{sinc}^2\left(\pi N_w \frac{\omega - \omega_s}{\omega_s}\right) \tag{2.3a}$$

where $\text{sinc}(x) \equiv \sin x / x$. Hence, the spectrum is peaked around a "spontaneous" frequency $\omega_s = 2\pi c/\lambda_s$ where, for on-axis radiation,

$$\lambda_s = \frac{1 - \beta_\parallel}{\beta_\parallel}\lambda_w \tag{2.3b}$$

with $\beta_\parallel \equiv v_\parallel/c$ and λ_w the undulator period, and its full line-width at half height is

$$\frac{\Delta\omega}{\omega} \simeq \frac{1}{N_w} \tag{2.3c}$$

The results (2.3a–c) are most easily explained in the (average) longitudinal electron rest frame, that is the most physical reference frame to describe the FEL process. Here, each

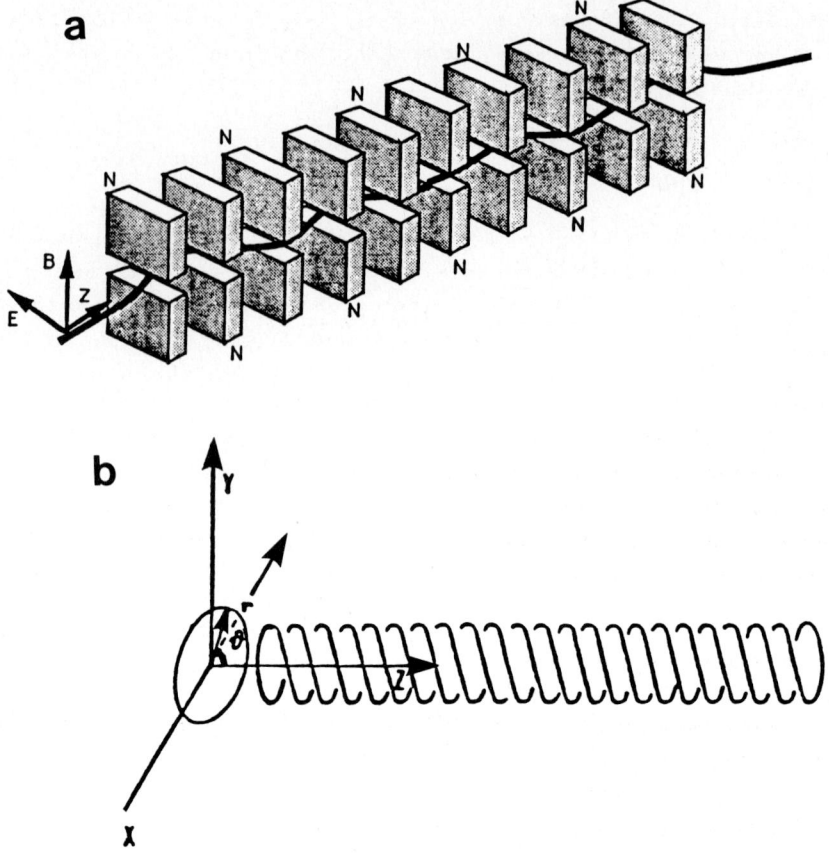

Fig. 2.1 Planar (a) and Helical (b) Wiggler

electron "sees" the N_w- period wiggler as an N_w-period counterpropagating radiation field, with Lorentz-contracted wavelength $\lambda'_w = \lambda_w/\gamma_\parallel$ where $\gamma_\parallel^2 = 1/(1-\beta_\parallel^2)$. Hence, it oscillates N_w times, emitting a wave-packet of length $N_w \lambda'_w$, peaked at $\lambda'_s \simeq \lambda'_w$. In other terms, it acts as a "relativistic mirror" (Fig. 2.2) where the radiation is reflected by Compton back-scattering.

In this process we neglect the Compton shift, in agreement with the assumption of a purely classical (relativistic) description in which $\hbar\omega \ll \gamma mc^2$ or $\lambda \gg \lambda_c$, where λ_c is the Compton wavelength. From this picture the spectral distribution (2.3a-c) is easily explained,

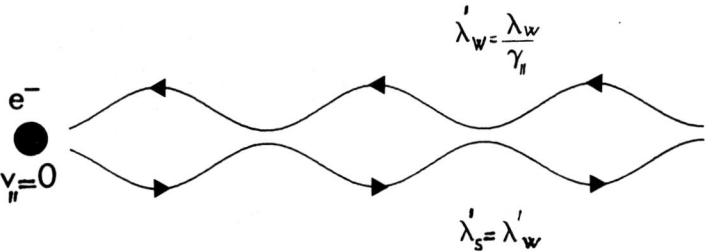

Fig. 2.2 The electron acts as a relativistic mirror in its rest frame of reference.

because: i) the Fourier transform of a plane-wave truncated after N_w oscillations is a sinc^2-function, ii) its linewidth is N_w^{-1}, iii) it is peaked at the wavelength λ_s (eq.(2.3b)). Only this last point requires some elaboration. Really, when back-scattered radiation of wavelength λ'_s in the moving frame is observed in the laboratory frame (as synchrotron radiation), its wavelength is reduced by a factor $\gamma_\parallel$ due to the Doppler effect. In this way one obtains the relation of up-frequency conversion for the peak wavelength λ_s in the laboratory frame

$$\lambda_s \simeq \frac{\lambda_w}{2\gamma_\parallel^2} \qquad (2.3b')$$

which is a good approximation to the exact relation (2.3b) in the limit $(1 - \beta_\parallel) \ll 1$. Really, the result (2.3b) can be derived in the laboratory frame looking for a radiation wavelength such that either the synchrotron radiation emitted by an electron at each wiggler period is in phase with that emitted in any other period, so that there is positive interference, or the radiation slips over the electron by exactly one wavelength while the electron advances by one wiggler period, so that the relative phase remains constant.

We shall obtain the result (2.3b) when deriving the FEL dynamic equations. Here we remark that eq.(2.3b') describes the fundamental FEL tunability. To better appreciate this point, we anticipate the expression of the magnitude of the electron transverse velocity

$$|\vec{\beta}_\perp| \simeq \frac{a_w}{\gamma}, \qquad (2.4)$$

where

$$a_w = e\lambda_w B_w / 2\pi m c^2 \simeq 0.93 B_w(\text{T})\lambda_w(\text{cm}) \qquad (2.5)$$

is the so-called undulator parameter and B_w is the r.m.s. wiggler field. Thus from

$$\frac{1}{\gamma^2} = 1 - \beta_\parallel^2 - \beta_\perp^2 \simeq \frac{1}{\gamma_\parallel^2} - \frac{a_w^2}{\gamma^2}$$

it follows that (2.3b') can be written as

$$\lambda_s \simeq \lambda_w \frac{1+a_w^2}{2\gamma^2} \qquad (2.3b'')$$

which shows that the peak wavelength λ_s can be changed by varying either the electron energy (γ) or the undulator parameters (λ_w, B_w). Just to quote an example, with $\lambda_w = 2.5$ cm, $a_w = 1$ and $W = 25$ MeV one obtains $\lambda_s \simeq 10$ μm, i.e., infrared radiation.

Note that the insertion device is generally called undulator if $a_w \leq 1$, wiggler if $a_w > 1$. From eq.(2.4), $a_w \simeq |\vec{\beta}_\perp|\gamma$ so that an observer on-axis sees the radiation emitted by the electrons along all their trajectories only in the undulator case.

2b. FEL "STIMULATED EMISSION"

Let us consider now the case in which a radiation field, of wavelength $\lambda \simeq \lambda_s$, copropagates with the electron beam along a constant-parameter wiggler. From eq.(2.3b") one can define a resonant electron energy, γ_r:

$$\gamma_r = \sqrt{\frac{\lambda_w(1+a_w^2)}{2\lambda}} \qquad (2.6)$$

We shall see that for electrons injected into the wiggler with the resonant energy (2.6), the relative phase between the transverse oscillations of electrons and radiation remains constant. Depending on the value of this phase, each electron can: i) give energy to the field and decelerate, that is, "stimulated emission" which provides "gain"; ii) take energy from the field and accelerate, that is, "absorption". Thus if in a beam of electrons the former process prevails over the latter, the device can amplificate an external field, like in the first FEL amplifier experiment[2a] (Fig. 2.3a), or the spontaneous emission, either in the presence of an optical cavity, as in the first FEL oscillator experiment[2b] (Fig. 2.3b), or even in a single-pass configuration[13]. This oversimplified picture of the FEL gain process is entirely at the single-particle level. However, electrons can communicate each other via the common radiation field, or even directly for high enough current density. An FEL can operate in the high-gain regime just due to this possibility.

2c. HIGH-GAIN AMPLIFIERS

The problem with many electrons is that at the wiggler entrance the electron beam provided by any particle accelerator exhibits a random longitudinal distribution, besides some energy spread and a finite emittance. For sake of simplicity let us consider an electron pulse long enough that the slippage of radiation over the electrons can be neglected. Then we realize that the electron phases with respect to the radiation phase are randomly distributed over each radiation wavelength. Hence, if we think of a nearly monoenergetic and resonant electron beam, on average half electrons will decelerate and half electrons will accelerate (Fig. 2.4a), with the result of no net gain. As we shall see, for "short" wigglers and "low" current

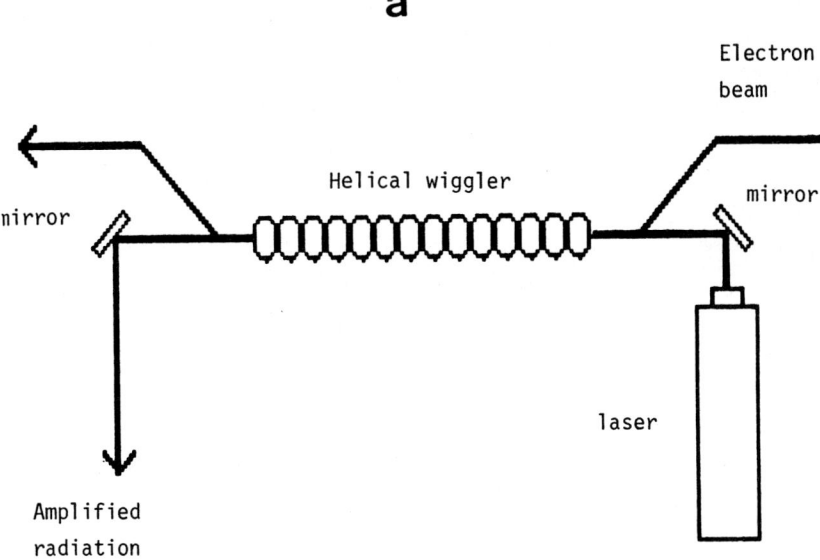

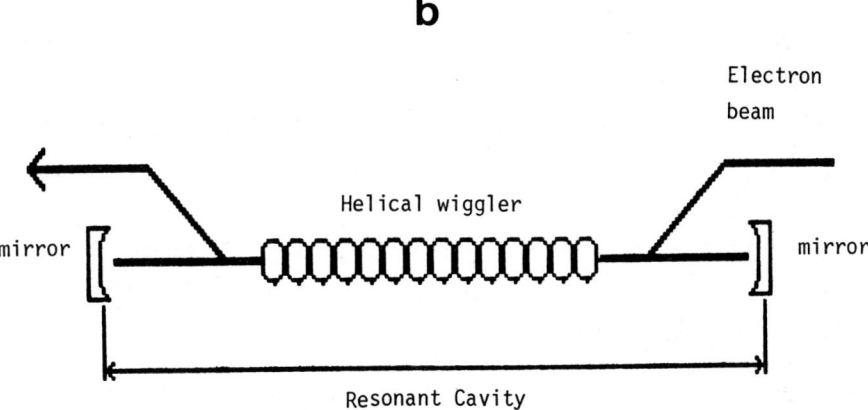

Fig. 2.3 Scheme of the first FEL experiments: a) FEL amplifier (ref. 2a), b) FEL oscillator (ref. 2b)

such that the single-particle picture applies, we can obtain amplification by the injection of particles with average energy slightly above resonance, $\langle\gamma\rangle_0 > \gamma_r$, such that gain (slightly) prevails over absorption, as previously discussed.

However, if the wiggler is "long" enough and the electron current is "high" enough, energy modulation becomes space modulation, i.e., phase modulation: the electrons self-bunch on the scale of a radiation wavelength (Fig. 2.4b).

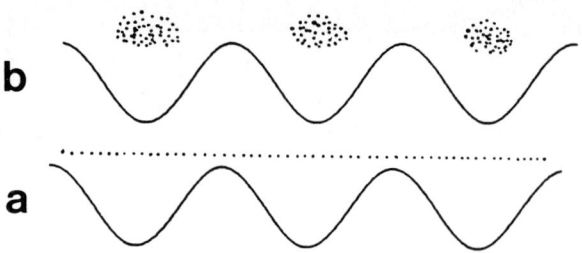

Fig. 2.4 a) Unbunched beam in the ponderomotive (radiation + wiggler) field. b) Bunched beam as a result of FEL dynamics.

Furthermore, this bunching turns out to occur around a phase corresponding to gain. Since most of the electrons have nearly the same phase, they emit coherent collective synchrotron radiation (or Compton back-scattering). It is a very nice example of a cooperative effect in radiation-matter interaction.

We shall see the existence of a collective instability for the system[14], which leads to electron self-bunching and to exponential growth of radiation until nonlinear saturation effects set a limit on the conversion of electron kinetic energy into radiation energy, i.e., on the efficiency of the FEL process. As we shall see, in the saturated high-gain regime the radiated power from N electrons scales as $N^{4/3}$, instead of N as in the previous case of nearly uncoupled particles. With a variable-parameter or tapered wiggler, the scaling goes as $N^{5/3}$ [9]. On the other hand, when propagation effects and slippage are relevant, we shall show that under proper conditions a novel superradiant regime may occur, in which the radiation scales as N^2. This regime has not yet been demonstrated experimentally. On the contrary, in conditions of negligible slippage, high-gain FEL amplifiers and oscillators have been operated in several laboratories[13,15,16]. In particular, a Berkeley-Livermore collaboration has first obtained microwave FEL radiation with peak power in the 100 MW range[13], which has been upgraded to the GW range in a second experiment with a tapered wiggler[15], where an impressive 40% efficiency has been obtained.

In the following we shall concentrate only on the theory of FEL amplifiers. From the discussion in this section it turns out that a proper theory to describe a high-gain FEL amplifier has to satisfy two basic conditions: i) it must be a many-particle theory, which includes the single-particle picture only as a limit case. ii) the electron dynamics cannot be described with a fixed radiation (plus wiggler) field; on the contrary, it is essential a self-consistent coupling of the particle dynamics, given e.g. by the relativistic Newton-Lorentz equations, with the field dynamics, given e.g. by the Maxwell equations with a source term due to the electron transverse current.

One can set up a self-consistent scheme for the FEL dynamics (Fig. 2.5), like the so called Maxwell-Bloch scheme in laser physics[36], but in a completely classical framework. This is certainly the most pedagogical scheme, but is not at all unique; e.g., one could use a fully Hamiltonian approach (see (17) and later on), or a Maxwell-Vlasov distribution function treatment[7].

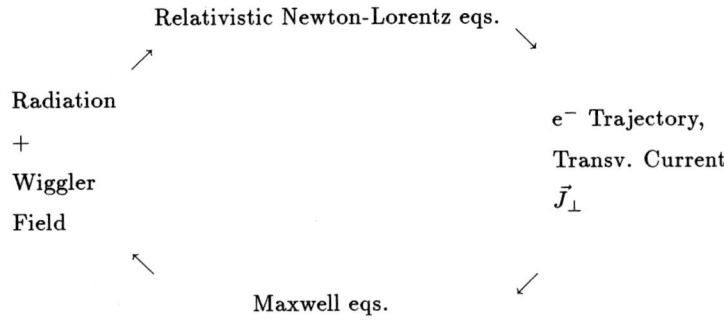

Fig. 2.5 Classical FEL theory: the self-consistent scheme

3. ONE-DIMENSIONAL FEL EQUATIONS AND THE "UNIVERSAL" SCALING

In this section we derive the dynamic equations for a single-pass FEL amplifier with a constant parameter wiggler. We discuss a model which, though idealized, contains the basic physics of the system. The main approximations are: i) we consider a one-dimensional dynamics, ii) we use the slowly-varying envelope approximation (SVEA) in the field equation. Three-dimensional effects are considered e.g. in Ref. 12. Variable parameter wigglers are treated e.g. in Refs. 9,12.

3a. ELECTRON EQUATIONS

We start from the dynamic equations for N relativistic electrons subjected to two given fields: a magnetostatic (wiggler) field, $\vec{B}_w = \nabla \times \vec{A}_w$, and a co-propagating radiation field, $\vec{E} = -(1/c)\partial \vec{A}/\partial t$ and $\vec{B} = \nabla \times \vec{A}$. The evolution of the momentum and the energy of each electron is ruled by the Newton-Lorentz equations

$$\frac{d(\gamma m \vec{v})}{dt} = e\left[\vec{E} + \frac{\vec{v}}{c} \times \left(\vec{B}_w + \vec{B}\right)\right] \qquad (3.1)$$

$$\frac{\mathrm{d}(\gamma mc^2)}{\mathrm{d}t} = e\vec{E}\cdot\vec{v}_\perp \tag{3.2}$$

Eq. (3.1) determines the electron trajectory as mostly due to $\vec{B}_w$; in particular, electrons injected parallel to the wiggler axis (z-axis) get a transverse component, so that they can couple to the radiation field and exchange energy, as described by eq. (3.2).

We consider a helical wiggler field which is transverse and spatially periodic with a period λ_w on axis, described by the vector potential

$$\vec{\mathcal{A}}_w(z) = \frac{\tilde{a}_w}{\sqrt{2}}\left(\hat{e}e^{-ik_w z} + \text{c.c.}\right) \tag{3.3}$$

where $\tilde{a}_w$ is real and constant, $k_w = 2\pi/\lambda_w$ and $\hat{e} = (\hat{x}+i\hat{y})/\sqrt{2}$, $\hat{e}\cdot\hat{e}=0$, $\hat{e}\cdot\hat{e}^*=1$. The case of a linear wiggler is outlined in Sec. 7.

Also, we consider a plane-wave radiation field of wavelength λ, co-propagating with the electrons along the z-axis, described by the vector potential

$$\vec{\mathcal{A}}(z,t) = -\frac{i}{\sqrt{2}}\left[\hat{e}\tilde{a}(z,t)e^{i(kz-\omega t)} - \text{c.c.}\right] \tag{3.4}$$

where $\tilde{a}(z,t)$ is a complex, slowly-varying amplitude, whose dependence on transverse coordinates, x and y, is neglected in the plane-wave and paraxial approximation, and $k = 2\pi/\lambda = \omega/c$.

The total potential experienced by the electron is

$$\vec{\mathcal{A}}_{tot}(z,t) = \vec{\mathcal{A}}_w(z) + \vec{\mathcal{A}}(z,t) \tag{3.5}$$

First of all, we can solve for the transverse electron motion due to the Lorentz force. From eq.(3.1) we find for the transverse momentum

$$\frac{\mathrm{d}}{\mathrm{d}t}(\gamma m\vec{v}_\perp) = -\frac{e}{c}\left[\frac{\partial\vec{\mathcal{A}}_{tot}}{\partial t} - \left(\vec{v}\times\nabla\times\vec{\mathcal{A}}_{tot}\right)_\perp\right] = -\frac{e}{c}\frac{\mathrm{d}\vec{\mathcal{A}}_{tot}}{\mathrm{d}t} \tag{3.6}$$

Let us now introduce the dimensionless wiggler vector potential

$$\vec{a}_w(z) \equiv \frac{a_w}{\sqrt{2}}\left(\hat{e}e^{-ik_w z} + \text{c.c.}\right) \tag{3.7a}$$

where a_w is the undulator parameter (2.5),

$$a_w \equiv \frac{e\tilde{a}_w}{mc^2} = \frac{e\lambda_w B_w}{2\pi mc^2} \tag{3.7b}$$

and the dimensionless radiation vector potential

$$\vec{a}(z,t) = -\frac{i}{\sqrt{2}}\left[\hat{e}a(z,t)e^{i(kz-\omega t)} - \text{c.c.}\right] \tag{3.8a}$$

where
$$a \equiv \frac{e\tilde{a}}{mc^2} = \frac{e\lambda E}{2\pi mc^2} \tag{3.8b}$$

Then eq. (3.6) can be written as

$$\frac{d}{dt}\left(\gamma\vec{\beta}_\perp\right) = -\frac{d}{dt}\left(\vec{a}_w + \vec{a}\right) = -\frac{d}{dt}\vec{a}_{tot} \tag{3.9}$$

which gives for perfect injection on-axis ($\vec{\beta}_\perp(0) = 0$) and negligible initial field,

$$\vec{\beta}_\perp = -\frac{\vec{a}_{tot}}{\gamma} \tag{3.10}$$

which generalizes the approximated relation (2.4).

By (3.10) we can now derive the electron energy equation. From eq. (3.2)

$$\frac{d\gamma}{dt} = \frac{e}{mc^2}\vec{E}\cdot\vec{v}_\perp = -\frac{e}{mc^2}\frac{\partial\vec{A}_{tot}}{\partial t}\cdot\vec{\beta}_\perp = \frac{1}{\gamma}\frac{\partial\vec{a}_{tot}}{\partial t}\cdot\vec{a}_{tot} = \frac{1}{2\gamma}\frac{\partial|\vec{a}_{tot}|^2}{\partial t} \tag{3.11}$$

Now, from eqs. (3.7), (3.8),

$$|\vec{a}_{tot}(z,t)|^2 = a_w^2 - ia_w\left[a(z,t)e^{i\theta(z,t)} - \text{c.c.}\right] + |a(z,t)|^2 \tag{3.12}$$

where
$$\theta = (k + k_w)z - \omega t \tag{3.13}$$

is the electron phase in the combined "ponderomotive" (radiation + wiggler) field, which propagates in the z-direction at the phase velocity

$$v_p = \frac{\omega}{k + k_w} \tag{3.14}$$

By substitution of (3.12) into (3.11) and with the paraxial approximation $d/dt \simeq v_\parallel d/dz \simeq c d/dz$, the electron energy equation can be written as

$$\frac{d\gamma}{dz} = -\frac{k}{2}\frac{a_w}{\gamma}\left(ae^{i\theta} + \text{c.c.}\right) \tag{3.15}$$

The r.h.s. of eq.(3.15) is a "ponderomotive" force whose strength is proportional to the electron-radiation coupling ($|a|a_w/\gamma \propto |\vec{E}\cdot\vec{v}_\perp|$). This force is responsible for electron acceleration ($d\gamma/dz > 0$) or deceleration ($d\gamma/dz < 0$) and eventually bunching. Note that if a were constant both in modulus and in phase, (3.15) would reduce to $d\gamma/dz \propto \cos\theta$, i.e., the pendulum equation (we shall see that $d\gamma/dz \propto d^2\theta/dz^2$)[18].

In order to derive an equation for the phase θ, we start simply differentiating eq. (3.13):

$$\frac{d\theta}{dz} = k_w - k\left(\frac{1}{\beta_\parallel} - 1\right) \tag{3.16}$$

Note that the electron phase with respect to the ponderomotive field remains constant, i.e., $d\theta/dz = 0$, if $k = k_w \beta_\| / (1 - \beta_\|)$ or

$$\lambda = \lambda_w \frac{1 - \beta_\|}{\beta_\|} \tag{3.17}$$

that is the exact (on-axis) resonance condition (eq.(2.3b)). This relation is obtained also by imposing that $v_p = v_\|$. Hence we get one more interpretation of the resonance condition; namely, an electron has the resonant energy γ_r (eq.(2.6)) if it is injected with longitudinal velocity equal to the phase velocity of the ponderomotive field.

The approximated resonance relation

$$\lambda \simeq \lambda_w \frac{1 + a_w^2}{2\gamma^2} \tag{3.18}$$

follows from (3.17) in the limit $1 - \beta_\| \ll 1$ and in the approximation $\vec{a}_{tot} \simeq \vec{a}_w$, i.e., neglecting the radiation contribution to the electron transverse velocity like in (2.4).

In this approximation, from

$$\beta_\|^2 = 1 - \beta_\perp^2 - \frac{1}{\gamma^2} \simeq 1 - \frac{1 + a_w^2}{\gamma^2}$$

one can derive, in the limit $\gamma^2 \gg 1 + a_w^2$,

$$\frac{1}{\beta_\|} \simeq 1 + \frac{1 + a_w^2}{2\gamma^2} \tag{3.19}$$

which inserted in eq. (3.16) gives

$$\frac{d\theta}{dz} = k_w - \frac{1 + a_w^2}{2\gamma^2} k \tag{3.20}$$

By introducing the electron resonant energy γ_r (eq.(2.6)), eq. (3.20) becomes

$$\frac{d\theta}{dz} = k_w \left(1 - \frac{\gamma_r^2}{\gamma^2}\right) \tag{3.20'}$$

However, if the approximation $\vec{a}_{tot} \simeq \vec{a}_w$ is dropped and the exact result (3.12) is used, there are additional contributions to the electron trajectory due to the radiation field[17] and thus corrections to the phase equation (3.20'). The final form is

$$\frac{d\theta}{dz} = k_w \left(1 - \frac{\gamma_r^2}{\gamma^2}\right) + \frac{k}{2\gamma^2} \left[i a_w \left(a e^{i\theta} - \text{c.c.}\right) - |a|^2\right] \tag{3.21}$$

3b. FIELD EQUATION

In the second part of the self-consistent scheme for FEL dynamics we look for the evolution equation of the radiation field, $\vec{A}$, as determined by the electron transverse current $\vec{J}_\perp$. The wave equation in the one-dimensional approximation reads

$$\left(\frac{\partial^2}{\partial z^2} - \frac{1}{c^2}\frac{\partial^2}{\partial t^2}\right)\vec{A}(z,t) = -\frac{4\pi}{c}\vec{J}_\perp(z,t) \tag{3.22}$$

where $\vec{J}_\perp$ is the average over the electron beam cross section of the transverse density current

$$\vec{J}_\perp(\vec{x},t) = e\sum_{j=1}^{N}\vec{v}_\perp \delta(\vec{x}-\vec{x}_j(t)) \tag{3.23}$$

By substituting the result (3.10) for the transverse velocity $\vec{v}_\perp$, projecting eq.(3.22) on the $\hat{e}$-direction, and using the dimensionless vector potentials (3.7), (3.8), we obtain

$$\left(\frac{\partial^2}{\partial z^2} - \frac{1}{c^2}\frac{\partial^2}{\partial t^2}\right)a(z,t)e^{i(kz-\omega t)} = \\ i\frac{4\pi e^2}{mc^2}n_\perp\sum_{j=1}^{N}\frac{a_w e^{-ik_w z}-iae^{i(kz-\omega t)}}{\gamma}\delta(z-z_j(t)) \tag{3.24}$$

where $n_\perp$ is the transverse electron number density, i.e. the total number of electrons divided by the transverse beam section. Note that the transverse current (the r.h.s. of eq.(3.24)) has a main contribution due to the wiggler, plus a minor contribution due to the radiation field.

Now, in the slowly-varying envelope approximation (SVEA)

$$\left|\frac{\partial a}{\partial z}\right| \ll |ka| \qquad \left|\frac{\partial a}{\partial t}\right| \ll |\omega a| \tag{3.24a}$$

second-order derivatives can be neglected and

$$\left(\frac{\partial^2}{\partial z^2} - \frac{1}{c^2}\frac{\partial^2}{\partial t^2}\right)ae^{i(kz-\omega t)} \simeq 2ik\left(\frac{\partial}{\partial z}+\frac{1}{c}\frac{\partial}{\partial t}\right)ae^{i(kz-\omega t)} \tag{3.24b}$$

Furthermore, as the complex variable a is slowly-varying on the scale of a radiation wavelength λ, it can be driven only by a transverse current averaged over a volume with longitudinal dimension $\ell_\parallel$ several λ's long. Hence, we average the transverse current also with respect to z, thus obtaining with (3.24a) and (3.13)

$$\left(\frac{\partial}{\partial z}+\frac{1}{c}\frac{\partial}{\partial t}\right)a = \frac{1}{2k}\frac{4\pi e^2}{mc^2}\frac{n_\perp}{\ell_\parallel}\left\{a_w\sum_j\frac{e^{-i\theta_j}}{\gamma_j} - ia\sum_j\frac{1}{\gamma_j}\right\} \tag{3.25}$$

Now, introducing the total electron number density n, and defining the average over a sample of N electrons of any electron dynamical variable f

$$\langle f(\theta,\gamma)\rangle \equiv \frac{1}{N}\sum_{j=1}^{N} f(\theta_j,\gamma_j) \tag{3.26}$$

the wave equation can be written as

$$\left(\frac{\partial}{\partial z}+\frac{1}{c}\frac{\partial}{\partial t}\right)a = \frac{k}{2}\left(\frac{\omega_p}{\omega}\right)^2\left(a_w\langle\frac{e^{-i\theta}}{\gamma}\rangle - ia\langle\frac{1}{\gamma}\rangle\right) \tag{3.27}$$

where

$$\omega_p = \sqrt{\frac{4\pi e^2 n}{m}} \tag{3.28}$$

is the plasma frequency. Note that we assume space matching of the radiation and the electron beams (plane-wave and charged sheets in one dimension), whereas for different cross sections a filling factor should be introduced.

3c. SPACE-CHARGE

We have derived the dynamic eq.(3.27) for the complex field amplitude, and eqs.(3.15), (3.21) for the electron energy and phase. Now, when we write these equations for the j-th electron of the beam, we must add the space-charge force acting on that electron due to the longitudinal self-field created by electron density fluctuations. According to refs.17,19, the space-charge contribution to the energy eq.(3.15) is

$$\left(\frac{d\gamma_j}{dz}\right)_{sc} = -ik\left(\frac{\omega_p}{\omega}\right)^2\left(\langle e^{-i\theta}\rangle e^{i\theta_j} - \text{c.c.}\right) \tag{3.15'}$$

Clearly, the space-charge force is appreciable if the electron current is "higher", so that the plasma frequency ω_p is "big"; this point will be discussed in the next section. Furthermore, the space-charge effect would vanish not only (trivially) if one considers a single electron, but also if the electron phases were homogeneously distributed in such a way that the electron bunching is zero, where the bunching is described just by the quantity in the r.h.s. of (3.15')

$$b \equiv \langle e^{-i\theta}\rangle \qquad (0 \le |b| \le 1) \tag{3.29}$$

Note that if we sum the space-charge force over the electrons, the result is zero; really, it is an internal force with respect to the electron system.

Now the electron dynamic equations (3.15), (3.21) can be rewritten for the j-th electron, with inclusion of space-charge (eq.(3.15')), and in the paraxial approximation ($d/dz \simeq \partial/\partial z + (1/\bar{v}_\parallel)\partial/\partial t$, where $\bar{v}_\parallel \equiv \langle v_\parallel\rangle_\circ$ is the "bulk" velocity of the macroscopic electron distribution):

$$\left(\frac{\partial}{\partial z}+\frac{1}{\bar{v}_\parallel}\frac{\partial}{\partial t}\right)\theta_j = k_w\left(1-\frac{\gamma_r^2}{\gamma_j^2}\right)+\frac{k}{2\gamma_j^2}\left[ia_w\left(ae^{i\theta_j}-\text{c.c.}\right)-|a|^2\right] \tag{3.30}$$

$$\left(\frac{\partial}{\partial z}+\frac{1}{\bar{v}_\parallel}\frac{\partial}{\partial t}\right)\gamma_j = -\frac{ka_w}{2\gamma_j}\left(ae^{i\theta_j}+\text{c.c.}\right)-ik\left(\frac{\omega_p}{\omega}\right)^2\left(\langle e^{-i\theta}\rangle e^{i\theta_j}-\text{c.c.}\right) \tag{3.31}$$

3d. UNIVERSAL SCALING

The system of coupled evolution equations (3.27), (3.30), (3.31) can be set in a dimensionless form by introducing the following variables and parameters[14,17]:

$$\Gamma_j \equiv \frac{1}{\rho} \frac{\gamma_j}{\gamma_r}$$

$$A = \frac{\omega}{\omega_p \sqrt{\rho \gamma_r}} a \qquad \left(|A|^2 = \frac{1}{\rho} \frac{|E|^2/4\pi}{n \gamma_r m c^2} \right)$$

$$\bar{z} = 2k_w \rho z$$

$$\bar{t} = 2k_w \rho t \qquad (3.32)$$

$$\rho = \frac{1}{\gamma_r} \left(\frac{a_w}{4} \frac{\omega_p}{c k_w} \right)^{2/3}$$

$$\sigma = 4\rho \frac{1 + a_w^2}{a_w^2}$$

In eq.(3.31), the complex field amplitude A is such that $\rho |A|^2$ gives the ratio between the energy densities of radiation and of the resonant electron beam; $\bar{z}$, $\bar{t}$ are scaled coordinates, in particular, $\bar{z}$ is a dimensionless length defined in the range $0 \leq \bar{z} \leq 4\pi \rho N_w$ for $0 \leq z \leq L_w$; ρ is the fundamental FEL parameter ($\rho \simeq 0.136 \gamma_r^{-1} J^{1/3} B_w^{2/3} \lambda_w^{4/3}$ [SI units], with J the electron current density), and σ the space-charge parameter (see eq.(3.33b)) below). By the scaling (3.32), eqs.(3.27), (3.30), (3.31) become:

$$\left(\frac{\partial}{\partial \bar{z}} + \frac{1}{\bar{v}_\parallel} \frac{\partial}{\partial \bar{t}} \right) \theta_j = \frac{1}{2\rho} \left(1 - \frac{1}{\rho^2 \Gamma_j^2} \right) + \left[i \frac{1}{\rho} \left(A \frac{e^{i\theta_j}}{\Gamma_j^2} - \text{c.c.} \right) - \frac{\sigma}{2} \frac{|A|^2}{\Gamma_j^2} \right] \qquad (3.33a)$$

$$\left(\frac{\partial}{\partial \bar{z}} + \frac{1}{\bar{v}_\parallel} \frac{\partial}{\partial \bar{t}} \right) \Gamma_j = -\frac{1}{\rho} \left(A \frac{e^{i\theta_j}}{\Gamma_j} + \text{c.c.} \right) - i\sigma \left(\langle e^{-i\theta} \rangle e^{i\theta_j} - \text{c.c.} \right) \qquad (3.33b)$$

$$\left(\frac{\partial}{\partial \bar{z}} + \frac{1}{c} \frac{\partial}{\partial \bar{t}} \right) A = \frac{1}{\rho} \langle \frac{e^{-i\theta}}{\Gamma} \rangle - i \frac{\sigma}{2} A \langle \frac{1}{\Gamma} \rangle \qquad (3.33c)$$

Note that, for fixed wiggler parameters and given initial conditions, the whole system (3.33) depends only on one electron-beam parameter, the FEL parameter ρ ("universal scaling").

4. STEADY-STATE: COMPTON AND RAMAN REGIMES, LINEAR STABILITY ANALYSIS, COLLECTIVE INSTABILITY AND EXPONENTIAL GAIN

4a. STEADY-STATE EQUATIONS AND HAMILTONIAN MODEL

Under a standard transformations of coordinates

$$\bar{z}' = \bar{z} \qquad \bar{t}' = \bar{t} - (\bar{z}/\bar{v}_\parallel) \tag{4.1}$$

the differential operators in the l.h.s. of the FEL equations (3.33) change as follows:

$$\left(\frac{\partial}{\partial \bar{z}} + \frac{1}{\bar{v}_\parallel}\frac{\partial}{\partial \bar{t}}\right) \Rightarrow \frac{\partial}{\partial \bar{z}'} \tag{4.2a}$$

$$\left(\frac{\partial}{\partial \bar{z}} + \frac{1}{c}\frac{\partial}{\partial \bar{t}}\right) \Rightarrow \frac{\partial}{\partial \bar{z}'} - \frac{1}{c\bar{\beta}_\parallel}(1 - \bar{\beta}_\parallel)\frac{\partial}{\partial \bar{t}'} \tag{4.2b}$$

Clearly, in free space, propagation effects can be neglected if the velocity difference between radiation and electrons (slippage) is not appreciable during the interaction in the wiggler. This is the case in a suitably defined "long-bunch" and "short-wiggler" limit, to be discussed in Sec. 6. In that limit, the derivative with respect to time in (4.2b) can be neglected, so that only the space dependence is left in all equations; one can follow the "steady-state" evolution of the system as it moves along the wiggler z-axis. The equations of motion differ only by the scaling from those e.g. in refs.20,21. As we shall see in section 7 the slippage can be eliminated in a waveguide designed so that the group velocity of light v_g equals the bulk electron velocity $\bar{v}_\parallel$. Now we introduce a slightly different scaling which is suitable for the discussion of the collective FEL instability and the high-gain regime. Namely, we redefine the electron phase θ and the complex field amplitude A in terms of the detuning parameter δ as follows:

$$\tilde{\theta}_j = \theta_j - \delta \bar{z} \qquad \tilde{A} = A e^{i\delta \bar{z}}$$

$$\delta = \frac{1}{2\rho}\frac{\langle\gamma\rangle_\circ^2 - \gamma_r^2}{\gamma_r^2} \tag{4.3a}$$

then the scaling is like in (3.32), provided that one performs the simple substitutions

$$\gamma_r \to \langle\gamma\rangle_\circ, \qquad k_w \to \left(\frac{\gamma_r}{\langle\gamma\rangle_\circ}\right)^2 k_w \tag{4.3b}$$

By (4.3a), (4.3b) the steady-state FEL equations can be written as:

$$\frac{d\tilde{\theta}_j}{d\bar{z}} = \frac{1}{2\rho}\left(1 - \frac{1}{\rho^2\Gamma_j^2}\right) + i\frac{1}{\rho}\left(\tilde{A}\frac{e^{i\tilde{\theta}_j}}{\Gamma_j^2} - c.c.\right) - \frac{\sigma}{2}\frac{|\tilde{A}|^2}{\Gamma_j^2} \qquad (4.4a)$$

$$\frac{d\Gamma_j}{d\bar{z}} = -\frac{1}{\rho}\left(\tilde{A}\frac{e^{i\tilde{\theta}_j}}{\Gamma_j} - c.c.\right) - i\sigma\left(\langle e^{-i\tilde{\theta}}\rangle e^{i\tilde{\theta}_j} - c.c.\right) \qquad (4.4b)$$

$$\frac{d\tilde{A}}{d\bar{z}} = \frac{1}{\rho}\langle\frac{e^{-i\tilde{\theta}}}{\Gamma}\rangle + i\left(\delta - \frac{\sigma}{2}\langle\frac{1}{\Gamma}\rangle\right)\tilde{A} \qquad (4.4c)$$

Note that the detuning parameter δ appears in the field eq. (4.4c).
Two constants of motion can be derived from eqs. (4). One of them is

$$\langle\Gamma\rangle + |\tilde{A}|^2 = \text{const.} \qquad (4.5)$$

which in the original variables reads: $n\langle\gamma\rangle mc^2 + |E|^2/4\pi = \text{const}$, that is, energy conservation. However, there is another and phase-dependent constant of motion, namely

$$H = \frac{1}{2\rho}\sum_{j=1}^{N}\left(\Gamma_j + \frac{1}{\rho^2\Gamma_j^2}\right) + \frac{i}{\rho}\left(\tilde{A}^*\sum_{j=1}^{N}\frac{e^{-i\tilde{\theta}_j}}{\Gamma_j} - c.c.\right) + \\ -\left(\delta - \frac{\sigma}{2}\langle\frac{1}{\Gamma}\rangle\right)N|\tilde{A}|^2 + \sigma N|\langle e^{-i\tilde{\theta}}\rangle|^2 = \text{const.} \qquad (4.6)$$

Actually, a fully canonical treatment for both electrons and radiation leads to a Hamiltonian model of the FEL, including high-density effects, which can describe high-gain amplifiers operating both in the Compton or in the Raman regimes[17] as we shall show in this section. Notice that if we introduce real and imaginary parts of the field variable, setting

$$\tilde{A} \equiv (\tilde{\theta}_\circ + i\Gamma_\circ)/\sqrt{2N} \qquad (4.7)$$

we can write a dimensionless Hamiltonian for $(2N+2)$ canonically conjugate electron and field variables[17]

$$\tilde{H}(\tilde{\theta}_\circ, \tilde{\theta}_1, \ldots, \tilde{\theta}_N; \Gamma_\circ, \Gamma_1, \ldots, \Gamma_N; \bar{z}) = \\ \frac{1}{2\rho}\sum_{j=1}^{N}\left(\Gamma_j + \frac{1}{\rho^2\Gamma_j^2}\right) + \frac{1}{\rho}\sqrt{\frac{2}{N}}\left(\Gamma_\circ\sum_{j=1}^{N}\frac{\cos\tilde{\theta}_j}{\Gamma_j} + \tilde{\theta}_\circ\sum_{j=1}^{N}\frac{\sin\tilde{\theta}_j}{\Gamma_j}\right) + \\ -\frac{1}{2}\left(\delta - \frac{\sigma}{2}\langle\frac{1}{\Gamma}\rangle\right)\left(\tilde{\theta}_\circ^2 + \Gamma_\circ^2\right) + \sigma N|\langle e^{-i\tilde{\theta}}\rangle|^2 \qquad (4.6')$$

The steady-state FEL eqs. (4.4) coincide with the Hamilton eqs.

$$\frac{d\tilde{\theta}_j}{d\bar{z}} = \frac{\partial\tilde{H}}{\partial\Gamma_j} \quad ; \quad \frac{d\Gamma_j}{d\bar{z}} = -\frac{\partial\tilde{H}}{\partial\tilde{\theta}_j} \qquad (j = 0, 1, \ldots, N) \qquad (4.4')$$

where $j = 0$ ($j \neq 0$) refers to field (electron) variables.

4b. LINEAR STABILITY ANALYSIS AND COLLECTIVE INSTABILITY

The initial condition for eqs. (4.4) with no field excitation and unbunched, monoenergetic electron beam

$$\tilde{A}_\circ = 0 \qquad \langle e^{-i\tilde{\theta}}\rangle_\circ = 0 \qquad (\Gamma_j)_\circ = \frac{1}{\rho} \qquad (4.8)$$

is an equilibrium condition for the system. Despite the apparent complexity of the equations, the stability of the state (4.8) can be investigated via a simple linear stability analysis[17] in terms of three collective variables (for the linear regime), i.e., the field amplitude $\tilde{A}$ (to be treated as a small quantity) and the electron variables

$$X \equiv \langle e^{-i\tilde{\theta}_\circ}\delta\tilde{\theta}\rangle \qquad Y \equiv \langle e^{-i\tilde{\theta}_\circ}\delta\Gamma\rangle \qquad (4.9)$$

The linearized system reads (dot=d/d$\bar{z}$):

$$\begin{aligned}\dot{X} &= Y + i\rho\tilde{A} \\ \dot{Y} &= -\tilde{A} - \sigma X \\ \dot{\tilde{A}} &= -iX - \rho Y + i\delta_1\tilde{A}\end{aligned} \qquad (4.10)$$

where δ_1 is an "effective" detuning

$$\delta_1 = \delta - \left(\frac{\sigma}{2}\right)\rho \qquad (4.11)$$

By a two-fold differentiation we obtain the following closed third-order differential equation for the linear dynamics of the field:

$$\dddot{\tilde{A}} - i\delta_1\ddot{\tilde{A}} + (\sigma - 2\rho)\dot{\tilde{A}} - i[1 + \sigma(\rho^2 + \delta_1)]\tilde{A} = 0 \qquad (4.12)$$

Searching roots of the form: $\tilde{A}(\bar{z}) \propto \exp(i\lambda\bar{z})$, we derive the cubic dispersion relation, including high-density effects and radiative corrections,

$$(\lambda^2 - \sigma)(\lambda - \delta_1) + 2\rho\lambda + 1 + \rho^2\sigma = 0 \qquad (4.13)$$

The system is stable if eq. (4.13) admits three real roots, unstable if admits one real root ($\lambda = \lambda_1$) and two complex-conjugate roots ($\lambda = \lambda_2 \pm i\lambda_3$). In the latter case, the field amplitude grows exponentially along the wiggler, as far as nonlinear saturation effects become dominant, as

$$\tilde{A}(\bar{z}) \propto \exp(\lambda_3\bar{z}) \equiv \exp(gz) \qquad (4.14)$$

where (by the scaling (3.32) and (4.4))

$$g = \lambda_3 \frac{4\pi\rho}{\lambda_w}\left(\frac{\gamma_r}{\langle\gamma\rangle_\circ}\right)^2 \qquad (4.14')$$

is the exponential or unsaturated gain per unit length. The instability condition is[17]:

$$27(1+\sigma_1)^2 - 4[\rho_1^3 + (1+\sigma_1)\delta_1^3] - \rho_1^2\delta_1^2 - 18(1+\sigma_1)\rho_1\delta_1 > 0$$

$$\rho_1 \equiv \sigma - 2\rho, \qquad \sigma_1 \equiv \sigma(\delta_1 + \rho^2) \tag{4.15}$$

4c. COMPTON AND RAMAN REGIMES

A deep insight into the physics of the system is gained by the analysis of the dependence of the imaginary part of λ, λ_3, on the detuning parameter δ, for different values of the basic FEL parameter ρ. Fig. 4.1 shows the behaviour of λ_3 vs. δ in four different cases, with ρ increasing by three orders of magnitude from $\rho = 10^{-3}$ to the (unrealistic) value $\rho = 1$. Only the positive part of the function, $\lambda_3 > 0$, is reported because λ_3 is symmetric with respect to the δ-axis.

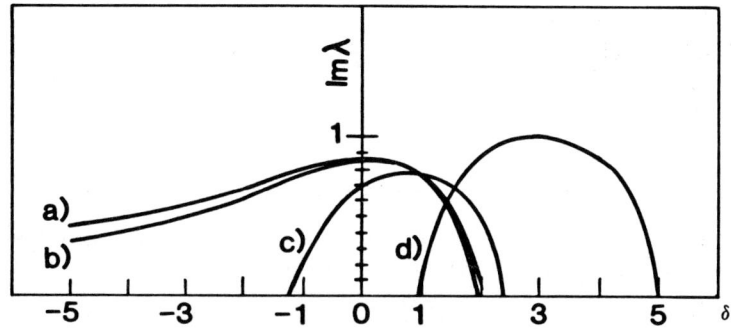

Fig. 4.1 Instability domain: imaginary part, λ_3, of the root of the dispersion relation (4.13) which rules the exponential amplification vs. detuning δ. a) $\rho = 0.001$, b) $\rho = 0.01$, c) $\rho = 0.1$, d) $\rho = 1$. Here $a_w = 1$.

Let us consider first the curve for $\rho = 10^{-3}$: we see that the imaginary part of λ i) has a maximum very close to $\delta = 0$, i.e., the maximum growth rate occurs very close to resonance, and ii) vanishes for $\delta = \delta_T \lesssim 2$, i.e., the system is unstable (stable) for all detunings $\delta \leq \delta_T$ ($\delta > \delta_T$). Really, in the limit $\rho \to 0$, the dispersion relation (4.16) reduces simply to

$$\lambda^3 - \delta\lambda^2 + 1 = 0 \tag{4.15'}$$

a cubic characteristic equation which is well-known since the theory of travelling-wave-tubes[22]. In this limit the maximum of λ_3 occurs exactly on resonance, $\delta = 0$, and its value is $\lambda_3 = \sqrt{3}/2$; the critical detuning, or the threshold value for the instability is $\delta = \delta_T = (27/4)^{1/3} \simeq 1.89$. For a given wiggler, the limit $\rho \to 0$ is approached for high electron energy and low current ($\rho \propto \gamma^{-1} n^{1/3}$); this is the <u>Compton regime</u>, in which high-density effects are quite negligible and the basic FEL process is just the stimulated Compton backscattering described in the previous sections. For $\rho = 10^{-2}$ the curve λ_3 vs. δ is very close

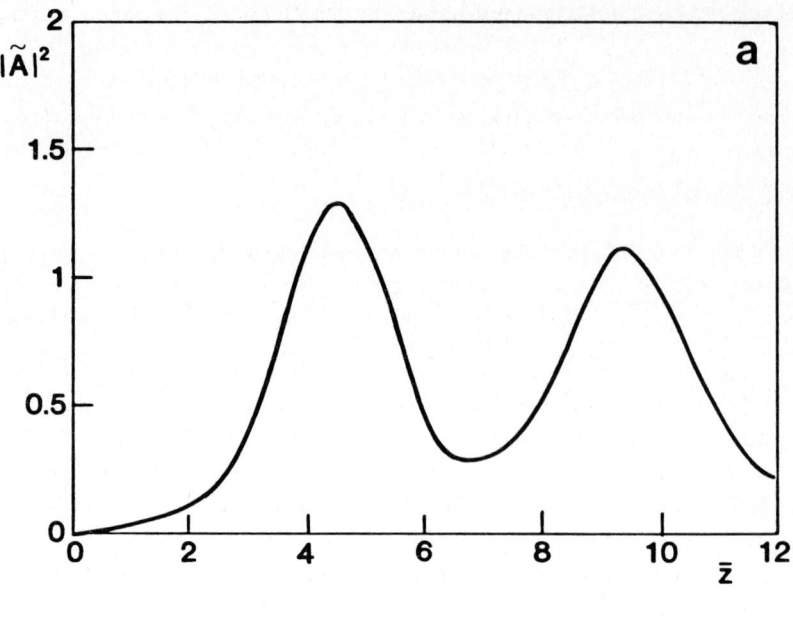

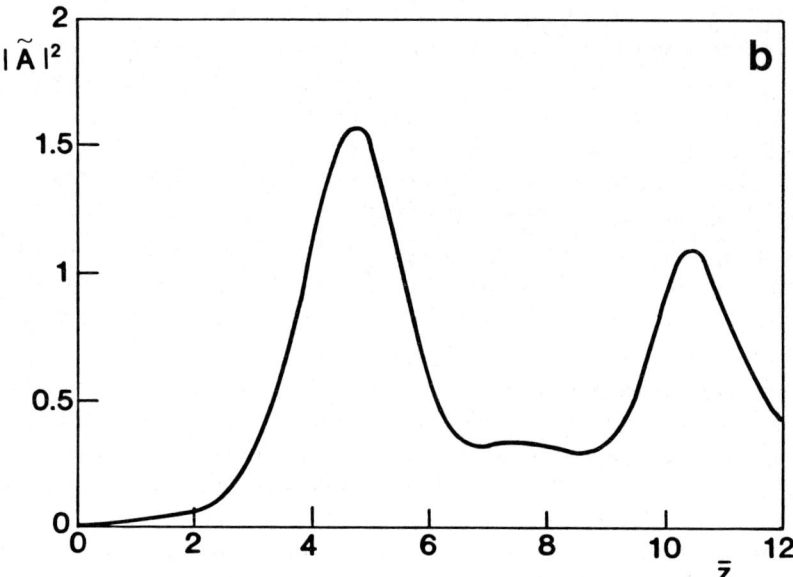

Fig. 4.2 Normalized field intensity $|\tilde{A}|^2$ vs. longitudinal coordinate $\bar{z}$ from eqs. (4.4) with initial conditions $\tilde{A}_\circ = 0$, $(\Gamma_j)_\circ = 1/\rho$, $\langle e^{-i\tilde{\theta}} \rangle \simeq 0.15$. a) Compton regime. The parameters are: $\rho = 0.01$, $\delta = 0.09$, $a_w = 1$. b) Raman regime. $\rho = 0.1$, $\delta = 0.78$, $a_w = 1$.

to that for $\rho = 10^{-3}$, i.e., the system is still fully in the Compton regime (or "strong-pump", "high-gain" Compton regime, in the unstable situation).

The picture changes drastically when $\rho = 10^{-1}$, namely, for lower electron energy and higher current. The instability domain becomes limited and shrinks as ρ increases; the maximum moves from $\delta \gtrsim 0$ to higher values. In this case, the exact dispersion relation (4.15) can be approximated to first order in ρ by

$$\lambda^3 - \delta\lambda^2 - (\sigma - 2\rho)\lambda + 1 + \sigma\delta = 0 \qquad (4.15'')$$

From (4.15"), the maximum growth rate turns out to occur at $\delta \simeq \sqrt{\sigma}$, whereas for $\rho \to 1$ the on-resonance gain vanishes. Hence space-charge definitely plays a role, and the FEL operates in a different regime, the Raman regime[7], in which the electrons interact appreciably not only via the common radiation field, but also directly due to space-charge effects. The electron system can exhibit collective plasma oscillations, i.e., the instability is collective in a stronger sense with respect to the Compton regime. Accordingly, in the electron longitudinal rest frame, the resonance relation is no longer the two-wave Compton relation, $\omega' = \omega'_w$ (see the relativistic mirror in Sec.2), but is a three-wave relation, $\omega' = \omega'_w - \omega'_p$, which describes a stimulated Raman backscattering with the radiation (signal) frequency equal to the difference between the wiggler (pump) and the plasma (idler) frequencies.

The scaling (4.3) allows for a description of the passage of the FEL dynamics from the Compton to the Raman regime only by increasing the FEL parameter ρ. Examples of radiated intensity as a function of the longitudinal coordinate $\bar{z}$ in the two regimes are reported in Fig. 4.2. These results are obtained from eqs. (4.4) with the same initial conditions but different values of ρ and for values of the detuning δ, obtained from Fig. 4.1, such that the growth rate is maximum in both cases.

We see that the scaled intensity $|A|^2$ grows from $|A|_o \ll 1$ to $|A|^2 = O(1)$ exhibiting the predicted exponential growth nearly up to the first peak, that is in saturation, where the linear analysis becomes completely invalid. This behaviour has been observed in several high-gain FEL experiments[13,15,16]. The result $|A|^2 = O(1)$ at saturation is relevant. Actually, since $|A|^2 \propto |E|^2/\rho n \propto |E|^2/n^{4/3}$ (see eq.(3.32)), where n is the electron density, it follows that at saturation $|E|^2 \propto n^{4/3}$, namely, the system exhibits a collective behaviour. However, the efficiency of the FEL process, $\eta \simeq \rho|A|^2 \simeq \rho$, is limited to a few percent. One way to avoid saturation and raise both the efficiency and the scaling of intensity with the electron density ($|E|^2 \propto n^{5/3}$, see ref. 9) is via a variable-parameter or "tapered" wiggler, as proposed in refs.20,23, and experimentally observed[15,24] (see the contributions 9,12). Another way is given by FEL superradiance, where $|E|^2 \propto n^2$, as will be discussed in Sec. 6.

Also, note from Fig. 4.2 that the exponential gain manifests itself after a lethargic stage. This stage is the longer, the closer the initial conditions are to the equilibrium state (4.8). It lasts as long as the three modes of the linear analysis interfere, until the divergent mode prevails over the two other modes. On the other hand, if the system is stable (or the wiggler is not long enough), the radiated intensity $|A|^2$ remains always close to its initial value $|A|_o^2$. In the following we shall focus on the Compton regime.

5. STEADY-STATE RESULTS IN THE COMPTON REGIME

5a. COMPTON FEL EQUATIONS AND HIGH-GAIN REGIME

As discussed in the previous section, for values of $\rho \lesssim 0.01$ the system operates in the Compton regime. In this case we can neglect both the space-charge contribution in eq.(4.4b) and the radiative corrections in eqs.(4.4a) and (4.4c); furthermore, we can surely perform the approximation of small relative variations of the electron energy in eqs.(4.4),

$$\left|\frac{\gamma_j - \langle\gamma\rangle_\circ}{\langle\gamma\rangle_\circ}\right| \ll 1 \tag{5.1}$$

Since $\Gamma_j = \gamma_j/\rho\langle\gamma\rangle_\circ$, this condition implies that $\rho\Gamma_j \simeq 1$. Hence, if we define for the jth electron the variable $\tilde{p}_j$ proportional to the relative energy variation,

$$\tilde{p}_j = \Gamma_j - \frac{1}{\rho} = \frac{1}{\rho}\frac{\gamma_j - \langle\gamma\rangle_\circ}{\langle\gamma\rangle_\circ} \tag{5.2}$$

eqs.(4a–c) reduce to

$$\frac{d\tilde{\theta}_j}{d\bar{z}} = \tilde{p}_j$$
$$\frac{d\tilde{p}_j}{d\bar{z}} = -(\tilde{A}e^{i\tilde{\theta}_j} + c.c.) \tag{5.3}$$
$$\frac{d\tilde{A}}{d\bar{z}} = \langle e^{-i\tilde{\theta}}\rangle + i\delta\tilde{A}$$

The linear stability analysis around the state (4.8) leads from eqs. (5.3) to just the cubic equation (4.15').

In order to better discuss the physics of the Compton regime, and also for future convenience, we can reabsorb the parameter δ from eqs. (5.3). In practice, we go back from the present scaling (4.3) to the original scaling (3.32). The steady-state Compton FEL equations then read:

$$\frac{d\theta_j}{d\bar{z}} = p_j \tag{5.4a}$$

$$\frac{dp_j}{d\bar{z}} = -(Ae^{i\theta_j} + c.c.) \tag{5.4b}$$

$$\frac{dA}{d\bar{z}} = \langle e^{-i\theta}\rangle \equiv b \tag{5.4c}$$

where

$$p_j = \frac{1}{\rho}\frac{\gamma_j - \gamma_r}{\gamma_r} \tag{5.5}$$

with $(\gamma_j - \gamma_r)/\gamma_r \ll 1$ and θ, A, ρ defined as in (3.32). Clearly, the detuning can be introduced at the level of the initial conditions, and we already know that on resonance the system is unstable in the Compton regime if it starts close to the equilibrium condition (4.8) (which now reads $A_\circ = 0$, $\langle e^{-i\theta}\rangle_\circ = (p_j)_\circ = 0$). Several remarks are in order. First of all, the universal scaling is here at its best: just no parameters are left in eqs.(5.4). Furthermore, these equations can be derived as Hamilton equations from a ("universal") Hamiltonian

$$H = \sum_{j=1}^{N} \left[\frac{p_j^2}{2} + i\left(A^* e^{-i\theta_j} - \text{c.c.}\right)\right] \tag{5.6}$$

using the Poisson brackets $\{\theta_i, p_j\} = \delta_{ij}$ and $\{A, A^*\} = -i/N$ [17]. Likewise, eqs.(5.3) could be derived from a one-parameter (the detuning δ) Hamiltonian. The constant of motion (4.5) simply reads here

$$\langle p \rangle + |A|^2 = \text{const.} \tag{5.7}$$

and still describes the energy exchange between the particles and the radiation field. From $|\langle p \rangle| = |\Delta\gamma|/\rho\gamma = O(1)$ in the high-gain regime, it follows that, like the efficiency η, also the energy spread is on the order of ρ.

Giving a closer look at eqs.(5.4), we see that if one neglects the variation of the field in both modulus and phase ($A \simeq A_\circ$, constant and real) the electron equations (5.4a, b) can be combined to give

$$\frac{d^2\theta_j}{d\bar{z}^2} = -2|A|\cos\theta_j \tag{5.8}$$

Namely, the electron system is described as an ensemble of N weakly-coupled pendula. We shall derive the results for this regime below, in a suitable limit of our treatment.

On the other hand, let us consider the field equation (5.4c). We see that the source term is the bunching parameter b, introduced in (3.29) as a measure of the degree of phase modulation in the electron beam. Thus, a strong bunching, $|b| \simeq 1$, is essential to drive the radiation field so to reach a high efficiency of the FEL process. This implies a strong coupling between the particles and the field, with the possibility for the electrons to communicate via the common radiation field. Indeed, in the exponential gain as described in the previous Section, $|b|$ (like $|A|$) can evolve from very low initial values to peak values approaching one, before oscillating in saturation.

If one integrates eq.(5.4c) in $\bar{z}$ and assumes b constant with respect to $\bar{z}$, forgetting the electron equations, and then goes back to the original variables, the result is that $|E|^2 \propto |b|^2 n^2$, that is, for appreciable bunching, the superradiant scaling of intensity $|E|^2 \propto n^2$. However, as we know, this scaling is never reached in the steady-state regime (see also Sec. 6).

5b. EXPONENTIAL GAIN RESULTS

The Compton FEL equations can be easily solved in the linear approximation. The solutions of the linearized system are particularly simple in the relevant case of the system on resonance, $\delta = 0$, in which the field has the form $A(\bar{z}) = \sum_{j=1}^{3} c_j e^{i\lambda_j \bar{z}}$, where λ_j ($j = 1, 2, 3$) are the solutions of the cubic eq. (4.15'). With initial conditions close to the unstable state (4.8), e.g.

$$A_\circ \equiv (\delta A)_\circ \neq 0, \qquad (p_j)_\circ = \langle e^{-i\theta} \rangle_\circ = 0 \tag{5.9}$$

where the only perturbation to the equilibrium condition (4.8) is a small injected signal, one finds[25] when looking at the divergent mode:

$$|A|^2(\bar{z}) \simeq \frac{1}{9} \left[4 \cosh^2\left(\frac{\sqrt{3}}{2}\bar{z}\right) + 4 \cos\left(\frac{3}{2}\bar{z}\right) \cosh\left(\frac{\sqrt{3}}{2}\bar{z}\right) + 1 \right] |A|_\circ^2 \tag{5.10}$$

The resonance gain $\mathcal{G}_{res}(\bar{z}) = \left(|A|^2(\bar{z}) - |A|_\circ^2\right)/|A|_\circ^2$ obtained from eq.(5.10) is plotted in Fig.5.1.

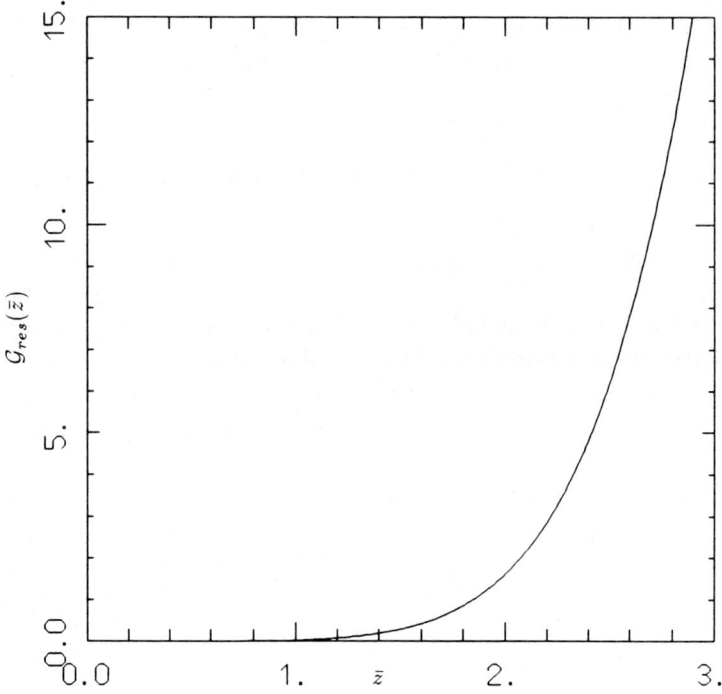

Fig. 5.1 Linear regime: Gain at resonance $\mathcal{G}_{res}(\bar{z})$ from eq. (5.10)

Note that for $\bar{z} \ll 1$, the behaviour of the output intensity is[25]

$$|A|^2(\bar{z}) \simeq (1 + \bar{z}^6/40)|A|_\circ^2 \qquad (5.10')$$

that is, it remains very close to the initial value $|A|_\circ^2$ (lethargy). On the contrary, for $\bar{z} \gtrsim 1$, $|A|^2(\bar{z})$ diverges exponentially as

$$|A|^2(\bar{z}) \simeq \frac{1}{9} e^{\sqrt{3}\bar{z}} |A|_\circ^2 \qquad (5.10'')$$

in agreement with (4.14) (recall that for $\delta = 0$, $\lambda_3 = \sqrt{3}/2$ and $\langle\gamma\rangle_\circ = \gamma_r$). The factor 1/9, or 1/3 for $|A|$, also referred to as "launching loss" term, is due to the initial power distribution among the three modes of the linearized dynamics. Going back to the original variable z, in the exponential gain regime

$$|A|^2(z) \propto e^{gz} |A|_\circ^2 \qquad (5.11)$$

where

$$g = \sqrt{3}\, \frac{4\pi\rho}{\lambda_w} \qquad (5.11')$$

is the exponential or unsatured gain per unit length, in agreement with (4.14'). If the peak power is near the wiggler end, one can define a total unsatured gain

$$G = gL_w = \sqrt{3}\, 4\pi\rho N_w \qquad (5.12)$$

The high-gain regime is such that $G > 1$. Also, from (5.12) it follows that the gain per wiggler period is on the order of $4\pi\rho$.

5c. MADEY'S SMALL SIGNAL GAIN

With still the initial conditions (5.9), in the stable region, $\delta > \delta_T$, and in the limit $\bar{z}/\sqrt{\delta} \ll 1$, the linear calculation gives for the radiated intensity

$$|A|^2(\bar{z}) \simeq \left[1 + \frac{4}{\delta^3}\left(1 - \cos\delta\bar{z} - \frac{\delta\bar{z}}{2}\sin\delta\bar{z}\right)\right]|A|_\circ^2 \qquad (5.13)$$

In eq.(5.13), $|A|^2(\bar{z})$ is an oscillating function: the hyperbolic functions of eq.(5.10) have become simple trigonometric functions. From eq.(5.13) we can define the small-signal or interference gain

$$\mathcal{G}(\bar{z},\delta) = \frac{|A|^2(\bar{z}) - |A|_\circ^2}{|A|_\circ^2} = \frac{4}{\delta^3}\left(1 - \cos\delta\bar{z} - \frac{\delta\bar{z}}{2}\sin\delta\bar{z}\right) \qquad (5.14)$$

where in the original variables (setting $\gamma_\circ \equiv \langle\gamma\rangle_\circ$)

$$\delta\bar{z} = \frac{4\pi}{\lambda_w}\, \frac{\gamma_\circ - \gamma_r}{\gamma_r}\, z \qquad (5.15)$$

At the wiggler end ($z = L_w = N_w \lambda_w$, $\bar{z} = 4\pi\rho N_w$)

$$\delta\bar{z} \longrightarrow 4\pi N_w \frac{\gamma_\circ - \gamma_r}{\gamma_r} \equiv \Delta \tag{5.15'}$$

which is a function of the injection energy $\gamma_\circ$ for fixed N_w and γ_r. Thus we can write

$$\mathcal{G}(\Delta) = 4(4\pi\rho N_w)^3 f(\Delta)$$
$$f(\Delta) \equiv \frac{1}{\Delta^3}\left(1 - \cos\Delta - \frac{\Delta}{2}\sin\Delta\right) \tag{5.16}$$

The small-signal-gain, proportional to the function $f(\Delta)$, is plotted in Fig. 5.2a. It is an asymmetric curve, quite peculiar of a dispersive process. In particular, the small signal gain is zero on resonance, just the opposite result with respect to the high-gain regime, where the growth rate is maximum at $\delta = 0$.

On the contrary, here the gain is maximum at $\Delta \simeq 2.5$; i.e., a necessary condition to have gain, is to inject electrons with $\langle\gamma\rangle_\circ > \gamma_r$, as anticipated in Sec. 2. A phase-space analysis would show that the injection of electrons with $\langle\gamma\rangle_\circ > \gamma_r$ gives an optimum gain, for a given undulation length, if it corresponds to the electrons executing on average one half-"synchrotron" oscillation in the ponderomotive bucket. On the other hand, for increasing $\bar{z}$ and thus relaxing the limit $\bar{z}/\sqrt{\delta} \ll 1$, the peak gain not only becomes remarkably higher, but moves towards smaller detuning values, in agreement with the previous discussion. In conclusion, the Madey regime is valid even in the unstable region $\delta \leq \delta_T$ up to $\bar{z} \overset{<}{\sim} 1$, whereas increasing $\bar{z}$ the simmetry is broken and the exponential gain at resonance increases (see Fig. 5.2b,c and Fig. 5.1).

In order to understand the apparent conflict between the zero small-signal gain and the fastest exponential rate both occurring on resonance in the two regimes, one could simply refer to the resonance condition in terms of the equality of the longitudinal electron velocity, $\bar{v}_\parallel$, with the ponderomotive phase velocity, v_p. In the self-consistent scheme of Sec. 3, eq.(3.14), v_p was evaluated as $v_p = \omega/(k + k_w)$ for a given radiation (and wiggler) field. In that case, in a resonant and randomly phased electron beam, nearly one half electrons absorb energy and one half lose energy, with no net gain; in fact, the particles (slightly) bunch around a phase for which there is no coupling with the radiation. However, if the field varies appreciably as in the high-gain regime, the ponderomotive phase velocity is modified as

$$v_p = \frac{\omega - \dot{\phi}}{k + k_w} \tag{3.14'}$$

where we set $A = |A|\exp(i\phi)$. Since $\dot{\phi} > 0$[25], the electron beam acts as a dielectric medium which slows down the phase velocity of the ponderomotive field. Hence, resonant electrons get a longitudinal velocity $\bar{v}_\parallel > v_p$ and bunch around a phase corresponding to gain.

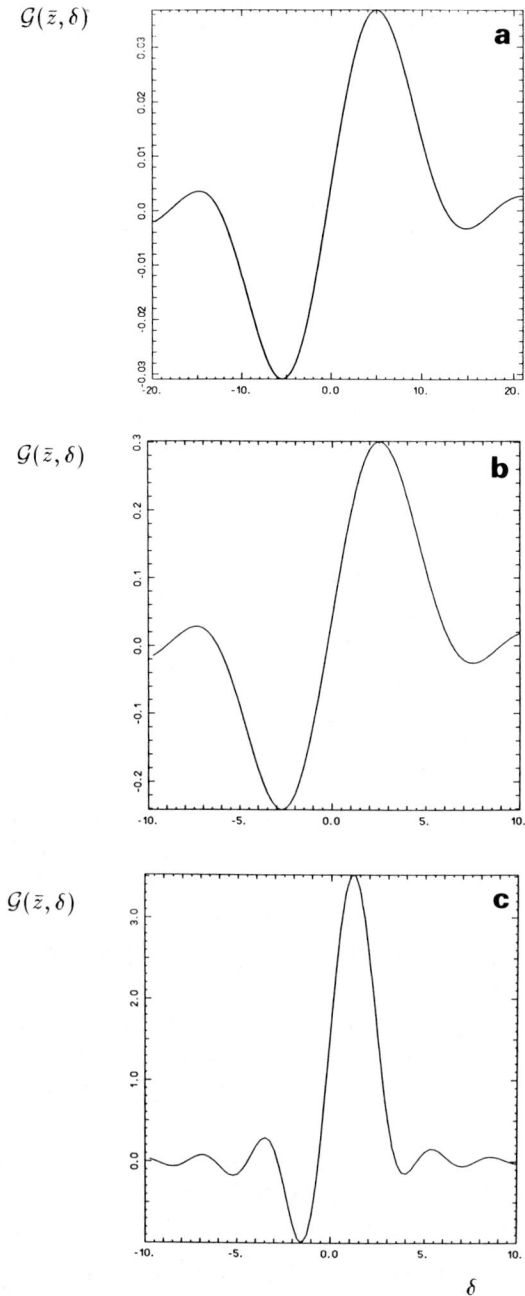

Fig. 5.2 Gain $\mathcal{G}(\bar{z},\delta)$ vs. detuning δ from the Compton FEL eqs. (5.4): transition from the odd Madey's gain to high-gain with increasing $\bar{z}$. a) $\bar{z} = 0.5$, b) $\bar{z} = 1$, c) $\bar{z} = 2$. Here $|A|_\circ = 10^{-4}$, $(p_j)_\circ = \langle \exp -i\theta \rangle_\circ = 0$.

Going back to the small-signal, low-gain analysis, the gain function (5.14) can be rewritten in the form

$$\mathcal{G}(\bar{z},\delta) = -\frac{\bar{z}^3}{2}\frac{\mathrm{d}}{\mathrm{d}(\delta\bar{z}/2)}\mathrm{sinc}^2(\delta\bar{z}/2) \tag{5.14'}$$

or, at the wiggler end,

$$\mathcal{G}(\Delta) = -\frac{(4\pi\rho N_w)^3}{2}\frac{\mathrm{d}}{\mathrm{d}(\Delta/2)}\mathrm{sinc}^2(\Delta/2) \tag{5.16''}$$

Now, from (5.15') and the resonance relation (3.18),

$$\frac{\Delta}{2} = 2\pi N_w \frac{\Delta\gamma}{\gamma_r} = \pi N_w \frac{\Delta\omega}{\omega} \tag{5.17}$$

so that

$$\mathcal{G}(x) \propto \frac{\mathrm{d}}{\mathrm{d}x}\mathrm{sinc}^2 x$$
$$x \equiv \pi N_w \frac{\Delta\omega}{\omega} \tag{5.18}$$

By recalling eq.(2.3a), we see that eq.(5.18) expresses the basic result that the small-signal gain is proportional to the derivative of the spontaneous spectrum. In fact, in this regime and for quite general undulators, two Madey's theorems hold[26]:
i) $\langle\gamma - \gamma_\circ\rangle = (1/2)\mathrm{d}/\mathrm{d}\gamma_\circ\left(\langle(\gamma-\gamma_\circ)^2\rangle\right)$, namely, there is no gain without energy spread;
ii) $\langle(\gamma-\gamma_\circ)^2\rangle \propto \mathrm{d}^2 I(\omega)/\mathrm{d}\omega\mathrm{d}\Omega$, in agreement with the previous derivation.

Also, it follows that in this regime the "gain" linewidth is on the order of the spontaneous linewidth $\simeq 1/N_w$ (sec. 2); hence, the electron energy spread must be less than $\simeq 1/N_w$, and the energy transfer from the electrons to the field, that is the FEL efficiency, is limited to within $\simeq 1/N_w$. This is valid only in the low gain situation $\bar{z} \ll 1$. In the high gain region $\bar{z} > 1$ the linewidth is given by $\Delta\omega/\omega \simeq \rho$. This can be easily inferred as follows:

$$\delta \simeq \frac{\gamma_j - \gamma_r}{\rho\gamma_r} < \delta_T \simeq 2 \quad\text{that is}\quad \frac{\Delta\gamma}{\gamma_r} \lesssim 2\rho$$

From the resonance relation (3.18) $\Delta\gamma/\gamma_r = (1/2)(\Delta\omega/\omega)$ so that $\Delta\omega/\omega \lesssim 4\rho$, i.e., the linewidth is on the order of ρ.

We conclude this section by recalling that FEL dynamics in the Compton regime can be described by means of only three (complex) electron and field collective variables[27]. These variables are the complex field amplitude A, the bunching parameter b, and another electron variable which describes electron energy modulation. The evolution equations for these quantities, obtained in suitable approximations, turn out to nicely reproduce the numerical results from the full $(2N+2)$ eqs. (5.3) or (5.4) even in the saturation regime.

6. SLIPPAGE EFFECTS AND SUPERRADIANCE IN THE COMPTON REGIME OF A HIGH-GAIN FEL*

6a. INTRODUCTION

In this section, our concern will be the investigation of the effects which arise i) from the velocity difference between electrons and light, ii) from electron bunches of finite length ℓ_b. The evolution of the radiation and electron pulses for a wide range of electron bunch length is examined, allowing us to investigate the FEL output as a function of the bunch length and of the gain of the system.

Up to now, like in most theoretical treatments, we have assumed an infinitely long and uniformly dense electron beam, with all sections of it evolving identically as they pass through the wiggler. Here we make no such assumptions and present numerical results obtained by a computer code in which we follow the evolution in the FEL process of a finite length electron bunch and a radiation pulse, taking into account the slippage between the two pulses, which move with different velocities along the wiggler.

An analysis of the characteristic lengths of the FEL process allows us to define two basic limits, of "long bunch" and of "short bunch". In the first case (long bunch) we find that only the leading region of the radiation (and electron) pulse exhibits the steady-state behaviour described in the previous sections, with the radiation peak power proportional to $n^{4/3}$. The rest of the radiation pulse exhibits a spiking behaviour, due to superradiant emission from the electrons in the trailing edge of the bunch. In the short bunch case no steady-state evolution is seen to take place, but we observe an emitted peak power which is proportional to the squared electron number density (n^2), typical of a superradiant behaviour. In this case, where the bunch length is smaller than a properly defined "cooperation length", the slippage is unneglibible, and we have a continuous radiation escape from the bunch.

6b. EQUATIONS

Let us now start from a generalized version of the 1D equations (5.4) for the electrons and the radiation field in the Compton Regime

$$\left(\frac{\partial}{\partial \bar{z}} + \frac{1}{\bar{v}_\parallel}\frac{\partial}{\partial \bar{t}}\right)\theta_j = p_j \tag{6.1a}$$

$$\left(\frac{\partial}{\partial \bar{z}} + \frac{1}{\bar{v}_\parallel}\frac{\partial}{\partial \bar{t}}\right)p_j = -\left(Ae^{i\theta_j} + c.c.\right) \tag{6.1b}$$

$$\left(\frac{\partial}{\partial \bar{z}} + \frac{1}{c}\frac{\partial}{\partial \bar{t}}\right)A = \chi_e(\bar{z} - \bar{v}_\parallel \bar{t})\langle e^{-i\theta}\rangle \tag{6.1c}$$

---

* Most of the results presented in this section were obtained in collaboration with Dr. B. McNeil

In eqs.(6.1) we consider the evolution of the electron and radiation beams in the vacuum, and we have slightly modified the definition of ρ to take into account non uniform electron bunches. In fact

$$\rho \equiv \frac{1}{\gamma_r}\left(\frac{a_0}{4}\frac{\tilde{\omega}_p}{ck_w}\right)^{2/3}, \qquad \tilde{\omega}_p \equiv \sqrt{\frac{4\pi e^2 \tilde{n}}{m}} \qquad (6.2)$$

and $\tilde{n}$ is the average electron density along the bunch. The function $\chi_e(\bar{z}-\bar{v}_\|\bar{t})$ in eq.(6.1c) is the macroscopic current shape of the bunch, and is defined through the relation $n(\bar{z},\bar{t}) = \chi_e(\bar{z}-\bar{v}_\|\bar{t})\,\tilde{n}$. We do not consider longitudinal spread of the electron bunch, so that the macroscopic density profile evolves unmodified with velocity $\bar{v}_\|$, that is, the mean initial velocity ($\bar{v}_\| \equiv (v_\|)_\circ$)

The other quantities appearing in eqs.(6.1) are the same ones introduced in the steady-state analysis.

For the sake of simplicity we assume a density profile defined as

$$\chi_e = \begin{cases} 1 & \text{inside the electron bunch } (0 \leq z - \bar{v}_\| t \leq \ell_b) \\ 0 & \text{elsewhere} \end{cases} \qquad (6.3)$$

With this choice our system reduces to

$$\begin{cases} \left(\dfrac{\partial}{\partial \bar{z}} + \dfrac{1}{\bar{v}_\|}\dfrac{\partial}{\partial \bar{t}}\right)^2 \theta_j = -\left(Ae^{i\theta_j} + c.c.\right) \\ \left(\dfrac{\partial}{\partial \bar{z}} + \dfrac{1}{c}\dfrac{\partial}{\partial \bar{t}}\right) A = \langle e^{-i\theta}\rangle \end{cases} \quad \text{inside the bunch}$$

$$\left(\frac{\partial}{\partial \bar{z}} + \frac{1}{c}\frac{\partial}{\partial \bar{t}}\right) A = 0 \quad \text{elsewhere (radiation evolution in vacuum)}$$

(6.1')

A linear analysis of this system is presented elsewhere[28].

Equations (6.1) assume a very simple form (very convenient for a numerical analysis) if we transform to the dimensionless characteristics

$$\begin{cases} z_1 = \dfrac{\bar{z} - \bar{v}_\|\bar{t}}{1 - \bar{\beta}_\|} \\ z_2 = \dfrac{c\bar{t} - \bar{z}}{1 - \bar{\beta}_\|}\bar{\beta}_\| \end{cases} \qquad (6.4)$$

By doing this we obtain

$$\frac{\partial \theta_j}{\partial z_2} = p_j \qquad (6.5a)$$

$$\frac{\partial p_j}{\partial z_2} = -\left(Ae^{i\theta_j} + c.c.\right) \qquad (6.5b)$$

$$\frac{\partial A}{\partial z_1} = \chi_e \langle e^{-i\theta}\rangle \qquad (6.5c)$$

These are the equations whose results will be discussed in the following paragraphs, in which we will present the code output for different values of the bunch length ℓ_b and of the coupling parameter ρ.

6c. FUNDAMENTAL PARAMETERS

As the radiation and the electron bunch travel with different velocities (c and $\bar{v}_{\|}$, respectively), the slippage between photons and electrons is given by $\ell_s = (c - \bar{v}_{\|})\Delta z / \bar{v}_{\|}$, where Δz is the distance travelled by the electrons.

The quantity

$$\ell_s = \frac{1 - \bar{\beta}_{\|}}{\bar{\beta}_{\|}} \Delta z \tag{6.6}$$

is called the "Slippage Length" after Δz.

If $\Delta z = \lambda_w$, the wiggler period, $\ell_s = \lambda_w(1 - \bar{\beta}_{\|})/\bar{\beta}_{\|} \equiv \lambda$, where we have used the resonant relation (3.17) between $\bar{\beta}_{\|}$, λ_w and λ. Indeed the exact resonant condition is obtained by imposing that, after one wiggler period, the radiation slips over an electron by one wavelength, so to maintain the same relative phase.

After N_p wiggler periods the slippage length is given by

$$\ell_s^{N_p} = N_p \lambda \tag{6.7}$$

We recall from the steady-state linear analysis of the high-gain FEL that the field grows exponentially with respect to z, the distance along the wiggler. From the expression (5.11') of the exponential gain per unit length (neglecting a factor $\sqrt{3}$), we can define the "Gain Length" as

$$\ell_g \equiv g^{-1} = \frac{\lambda_w}{4\pi\rho} \tag{6.8}$$

whereas we shall call "Gain" the quantity (5.12) (again neglecting a factor $\sqrt{3}$)

$$G \equiv g L_w \equiv 4\pi\rho \frac{L_w}{\lambda_w} \equiv 4\pi\rho N_w \tag{6.9}$$

The High-Gain regime requires that $G > 1$.

We can therefore evaluate the slippage in a gain length by substituting (6.8) for Δz in (6.6). This quantity is the "Cooperation Length" ℓ_c of ref.29, and is given by

$$\ell_c = \frac{1 - \bar{\beta}_{\|}}{\bar{\beta}_{\|}} \frac{\lambda_w}{4\pi\rho} = \frac{\lambda}{4\pi\rho} \tag{6.10}$$

where again we have used the resonance relation.

We have now a whole set of characteristic lengths by which we can define different dynamic regimes of a High-Gain FEL.

It is convenient to define the "Slippage Parameter" S as the ratio between the slippage parameter at the end of the wiggler and the bunch length ℓ_b.

$$S \equiv \frac{\ell_s}{\ell_b} = \frac{N_w \lambda}{\ell_b} \equiv \frac{N_w}{N_b} \tag{6.11}$$

where we have introduced $N_b \equiv \ell_b/\lambda$, the bunch length in units λ.

This parameter depends only on the relative length of the electron bunch with respect to the wiggler length, therefore we can define two opposite cases:

$$\begin{aligned} S \gtrsim 1 & \quad \text{Long Wigglers} & (N_w \gtrsim N_b) \\ S \ll 1 & \quad \text{Short Wigglers} & (N_w \ll N_b) \end{aligned} \tag{6.12}$$

However, this parameter is not useful to distinguish different dynamic regimes, because it does not take into account the interplay between gain and slippage in a FEL driven by finite-length bunches. Hence, we introduce the "Superradiant Parameter", K,[29] which is the ratio between the cooperation length and the bunch length,

$$K \equiv \frac{\ell_c}{\ell_b} = \frac{\lambda}{4\pi\rho\ell_b} = \frac{1}{4\pi\rho N_b} = \frac{S}{G} \tag{6.13}$$

This parameter, K, describes the interplay between gain and slippage.

Hence we define two cases:

$$\begin{aligned} K \ll 1 & \quad \text{"Long Bunch"} & (\ell_b \gg \ell_c \text{ or } N_b \gg \frac{1}{4\pi\rho}) \\ K \gtrsim 1 & \quad \text{"Short Bunch"} & (\ell_b \lesssim \ell_c \text{ or } N_b \lesssim \frac{1}{4\pi\rho}) \end{aligned} \tag{6.14}$$

K is equal to S in the case of unitary gain $G = 1$.

From this analysis it follows that only in the long bunch case we can expect to see a steady-state like evolution for the pulses, whereas in the short bunch case the steady-state theory is completely inadequate to model the pulse dynamics, due to high radiation losses from the bunch. However, even in the long bunch limit, we will show that the steady-state theory does not hold if the gain is high enough, due to superradiant trailing edge effects[30].

After introducing these fundamental quantities we can understand why the z_1–z_2 scaling is suitable for the analysis of the slippage problem.

By recalling the definitions (6.4) and (6.10) we can write

$$\begin{cases} z_1 = \dfrac{z - \bar{v}_\parallel t}{\bar{\beta}_\parallel \ell_c} \\ z_2 = \dfrac{ct - z}{\ell_c} \end{cases} \tag{6.15}$$

that is, z_1 and z_2 are the characteristics of the electrons and the light, respectively, normalized to the cooperation length. Furthermore, from the definitions (6.4) we have

$$z_1 + z_2 = \bar{z} \equiv \frac{4\pi\rho}{\lambda_w} z \qquad (6.16)$$

i.e., the dimensionless distance along the wiggler introduced by the scaling (3.32).

As the electron bunch travels for a wiggler period λ_w, the variation of $\bar{z}$ is equal to $4\pi\rho$, namely, the gain in a wiggler period. In fact

$$\Delta\bar{z} = \frac{4\pi\rho}{\lambda_w}\Delta z = 4\pi\rho \qquad (6.17)$$

But even z_1 and z_2 change by steps of $4\pi\rho$ per wiggler period along the characteristics $z_2 = $ const. and $z_1 = $ const., respectively. Indeed, if $z_2 = $ const., i.e., $\Delta z = c\Delta t$, then $\Delta z_1 = (4\pi\rho/\lambda_w)\Delta z$. Likewise, if $z_1 = $ const. ($\Delta z = \bar{v}_{\parallel}\Delta t$), $\Delta z_2 = (4\pi\rho/\lambda_w)\Delta z$.

Note that for $\Delta z = \lambda_w$ we have $\Delta z_1 = \Delta z_2 = 4\pi\rho$.

At a given time t, $\Delta z_1 = \Delta z_2 = 4\pi\rho$ corresponds to $\Delta z \simeq \lambda$, the radiation wavelength. Hence the natural length by which we can discretise the radiation and electron pulse evolution is λ. We have integrated the system (6.5), with a step $\Delta z_1 = \Delta z_2 = 4\pi\rho$, at each wiggler period for all the N_w periods, and every period we let the radiation slip over the electron pulse by one wavelength λ. In this way the equations are integrated for a total interval $4\pi\rho N_w$, that is the gain G (6.9), corresponding to a total slippage length $N_w\lambda$, as required by the resonance condition.

In the next paragraphs we will present numerical results for the three relevant limit cases:
i) long bunch, short wiggler ($K \ll 1$, $S \ll 1$)
ii) long bunch, long wiggler ($K \ll 1$, $S \gtrsim 1$)
iii) short bunch ($K \gtrsim 1$)

We note that, since $G > 1$ in high-gain systems and $K = S/G$, for a short bunch one has always $S \gtrsim 1$, so there is no short wiggler subcase.

Now we can explain why the superradiant parameter K describes the interplay between gain and slippage, and so it is the "true" parameter which permits us to distinguish two really different dynamic regimes, the short bunch and the long bunch one.

By the definition (6.13) we can write $K^{-1} = 4\pi\rho N_b$, that is the gain experienced by a photon emitted by the trailing edge of the electron bunch in passing through the whole bunch of length $\ell_b = N_b\lambda$. If $K \ll 1$, then $4\pi\rho N_b \gg 1$ and saturation can occur before the radiation escapes from the electron bunch, so there is enough time for the interplay between photon emission and reabsorption from the electrons, a process which leads to a steady-state like evolution of (some part of) the electron and radiation bunches. In the opposite case, $K \gtrsim 1$ we have $4\pi\rho N_b \lesssim 1$, and the radiation has not enough time to saturate while slipping over the electron bunch. In this case the slippage interferes with the basic mechanism of the steady-state gain and no stationary evolution can be observed. A completely new dynamic is seen to take place all over the radiation and the electron pulse.

6d. NUMERICAL RESULTS

We now use our computational model to examine the three previously mentioned cases.

The code output is plotted in a simple format, which shows the pulses at various positions through the wiggler. We plot the radiation output intensity $|A|^2$ after N_p wiggler periods in a window travelling at velocity c; positions within the window are referred in units of the wavelength λ from its trailing edge. The length of the window in these units is $N_b + N_w$, that is, the bunch length plus the total slippage. Above the radiation pulse another window of length N_b shows the electron energy variation parameter

$$\langle \dot{\theta} \rangle = \langle p \rangle = \frac{1}{\rho} \langle \frac{\gamma - \gamma_r}{\gamma_r} \rangle$$

along the electron bunch.

As the electrons travel with a velocity smaller than that of light, this window is seen to slip behind the radiation pulse window by one wavelength per wiggler period, as required by the resonance condition.

At the beginning of the wiggler the two windows have their leading edge in the same position, the radiation pulse is initialised with a small uniform field (simulating the resonant signal to amplify) and a perfectly tuned ($p_j = 0$) and nearly unbunched ($b \simeq 0$) electron beam is created.

6d.1. The Long Bunch, Short Wiggler case ($K \ll 1$, $S \ll 1$)

Let us consider now a typical long bunch, short wiggler case: an electron bunch 400 wavelength long is injected in a 100 period long wiggler, the coupling parameter is $\rho \simeq 0.02$.

Here $S = 1/4$ and $K = 0.01$, so that we are in the previously defined long bunch short wiggler case.

Parameters for this run are resumed in table 6.1.

Table 6.1.

Long bunch short wiggler case						
$N_\circ$	N_b	ρ	$\Delta \bar{z} = 4\pi\rho$	G	S	K
100	400	0.02	0.24	24	1/4	0.01

In figure 6.1a)–6.1d) we show the radiation and electron pulses at different positions along the wiggler, respectively after 20, 40, 60 and 100 periods. In Fig. 6.1a) we can see that the electron bunch, passing through 20 periods of the wiggler, slipped back on the radiation pulse by 20 radiation wavelengths, as required by the resonance condition.

The evolution of the pulses in the interacting region (that is, the region occupied by the electrons) takes place in a uniform way, except a small region around the trailing edges. Indeed we see an almost uniform value of the energy detuning parameter $\langle\dot{\theta}\rangle$ and of the radiation intensity $|A|^2$ all over the electron bunch, denoting a steady-state-like process. Here it is worth trying to imagine the results of a pure steady state analysis (with $v_\parallel = c$, so considering eqs. 5.4) when plotted in a similar way. Because the slippage is neglected in a steady state model, the two windows will remain with the same relative position in travelling through the wiggler at the same velocity. In fact we would always see a perfectly uniform level of $\langle\dot{\theta}\rangle$ and $|A|^2$ all over the pulses, whose height would obey the steady state evolution with respect to the distance z along the wiggler.

The deviations of our model from this steady state picture are seen here in two respects. First of all, as we have already noted, radiation is escaping from the leading edge of the electron bunch, and hence is no more interacting with electrons, but propagating in vacuum. The uniform radiation level in the electron bunch "traces" the shape of the radiation escaping from it. The second point is that there exists a region in the back of the pulses (up to 20 radiation wavelengths from the trailing edge for Fig. 6.1a)) which is evolving in a different way than the rest of the pulses. Here the energy detuning $\langle\dot{\theta}\rangle$ and the radiation intensity are no longer uniform. This is an effect due to the slippage and to the finite length of the electron bunch.

Indeed the electrons in the trailing edge (TE) are not experiencing radiation emitted from electrons behind them (since there are none), while the other electrons in the bunch do. These trailing electrons are always interacting with the pump field injected in the wiggler, that is, generally a weak radiation field. The radiation emitted by them propagates forward along the electron pulse, interacting with other electrons. The dynamics of the trailing edge electrons cannot be described by the steady state model. At the 20th wiggler period (Fig. 6.1a) the radiation emitted by the trailing edge electrons entering the wiggler traveled by 20 radiation wavelengths, so determining the width of the slippage region, in which propagation effects are observed. Electrons in the leading part of the pulse evolve accordingly to the steady state model only up to the moment in which the trailing edge radiation reaches them.

The instantaneous slippage parameter $S_i = N_p/N_b$ gives the fraction of the electron pulse which did not evolve with a steady state behaviour.

From Fig. 6.1 we can follow the evolution of the radiation along the slippage region of the electron pulse passing through the wiggler. As we can see, in this region the field and electron parameters are always non-uniform and radiation spikes of high-peak intensity are seen to evolve, reaching values one order of magnitude greater than the steady state saturation intensity $|A|^2_{sat} \simeq 1.4$. This spiking behaviour is yet still unreported (to the authors' knowledge) and is a direct consequence of the inclusion of the slippage in our treatment.

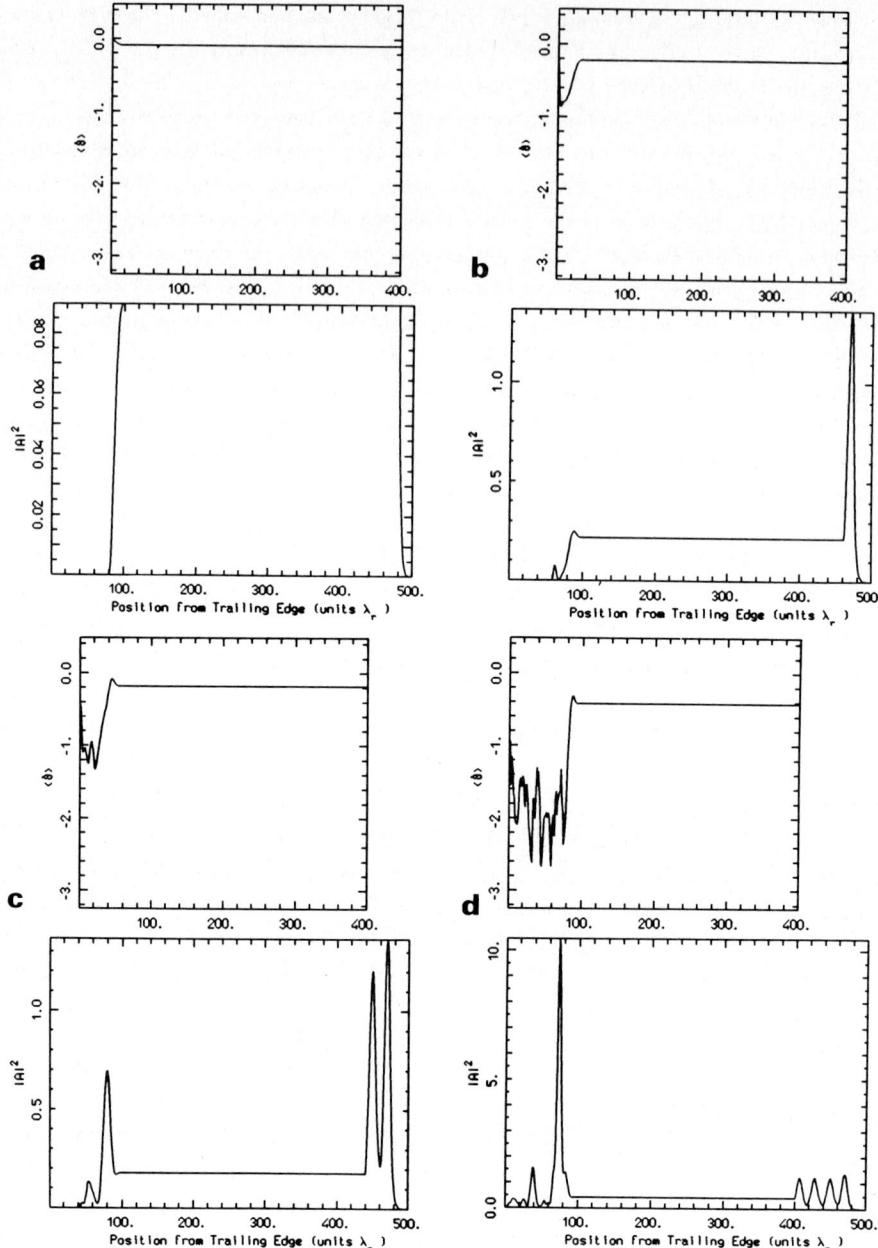

Fig. 6.1 Long bunch, short wiggler case. Radiation and electron pulses after N_P wiggler periods are shown from a) to d) for N_P =20, 40, 60, 100, respectively. Parameters for this case are listed in table 6.1. For figure description refer to the text.

We note that, from our analysis, the spikes begin to grow only after the steady state (leading) region of the radiation pulse reached the saturation intensity of $\simeq 1.4$ (see fig 6.1a–6.1b). Moreover the spikes seem to grow in a roughly exponential way with respect to the steady state "plateau" value, and no saturation phenomena were ever observed.

For the pulse evolution in the so called steady state region we can observe a perfect agreement with the results of the steady state model. We can see this by the shape of the radiation escaped from the electron pulse, which closely corresponds to the steady state plots of $|A|^2$ vs. $\bar{z}$, the only difference being the reference frame and the length unit.

From this analysis we can conclude that, even for a very long electron bunch and a short wiggler, the steady state model cannot describe the dynamics of the whole pulses. However, if the wiggler is sufficiently short ($S \ll 1$), the newly reported phenomena due to the slippage between the pulses, take place in a small part of the pulses, so they contribute little to the average pulse behaviour.

Indeed we can define a quantity which measures the average energy extracted from the electron beam as

$$E_L = \frac{\text{Total energy gain of radiation pulse (units } |A|^2)}{N_b} \qquad (6.19)$$

E_L is related to the usual FEL efficiency η_{eff} (defined as the ratio between the radiation energy gain and the initial electron energy) via the following formula:

$$E_L = \frac{1}{\rho}\eta_{eff}$$

In fig 6.2 we plot E_L as a function of the dimensionless position $\bar{z}$ through the wiggler.

Its behaviour closely resembles that of $|A|^2$ for the steady state model, with an initial lethargy followed by an exponential growth which reaches the saturation value of $\simeq 1.4$. After saturation we can see the average radiated intensity oscillating with peaks around the saturated value. The spiking behaviour has little effect on the effective average energy extracted from the electron pulse. Only at the end of the wiggler, for high values of $\bar{z}$, we can see a departure from the behaviour of $|A|^2$ predicted by the steady state theory. It is only after this averaging process that we can say that there is an agreement between our pulsed model and the steady state one, because the latter is completely wrong when trying to study the trailing edge dynamic.

In order to understand that there are two different kinds of evolution in this process we can look at the pulse properties at a given position on the electron bunch as a function of $\bar{z}$. In Fig. 6.3 we plot $|A|^2$, $|b|^2$, $\langle \dot{\theta} \rangle$ at the position $z_1 = 18$ along the bunch (i.e. 60 radiation wavelengths from the trailing edge). Due to the slippage, the radiation emitted from the trailing edge at the start of the wiggler will reach the electrons in our fixed z_1 at the 60th wiggler period ($\bar{z} = 18$). Indeed up to $\bar{z} = 18$ we see that the electrons are experiencing a stationary evolution, obeying the local steady state conservation law $|A|^2 + \langle \dot{\theta} \rangle =$const.

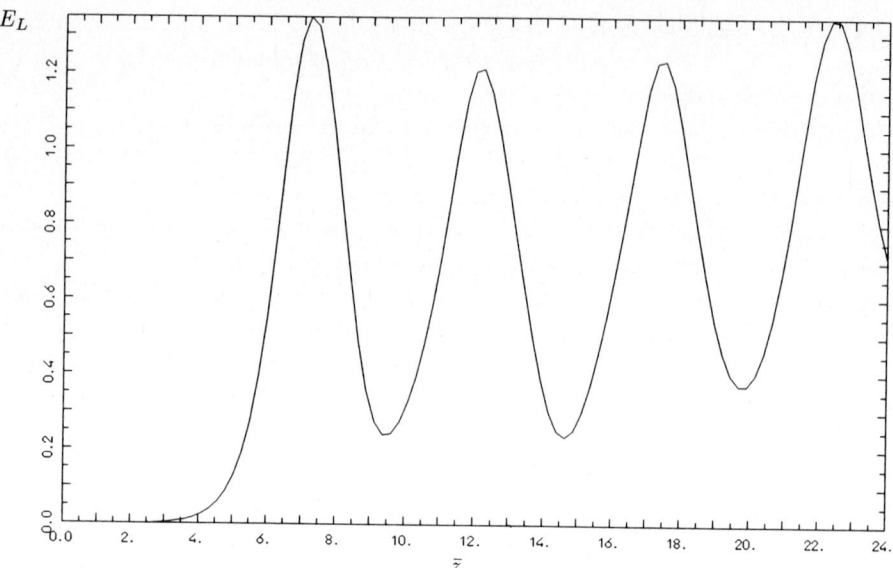

Fig. 6.2 Long bunch, short wiggler case. Average energy extraction E_L (eq. (6.19)) as a function of the dimensionless position $\bar{z}$ along the wiggler.

As the trailing radiation interacts with the electrons, the slippage effects are clearly seen: the steady state behaviour vanishes and the electrons loose much of their energy to the spike, which rapidly escapes from them, leaving them nearly unbunched.

6d.2. The Short Bunch Case ($K \gtrsim 1$)

It is now worth investigating the dynamic of the high gain FEL operating with short electron pulses ($\ell_b \lesssim \ell_c$, that is $K \gtrsim 1$). We recall that, since the relation $S = KG$ holds, for high gain systems (where $G > 1$) we have $S \gtrsim 1$, i.e. every wiggler is to be considered long.

As $S > 1$ we cannot expect a steady state like evolution on any part of the pulse, because the slippage region extends all over the pulse. In this case the radiation is quickly escaping from the electrons, due to the slippage between the pulses.

In Fig. 6.4 we show the output for a 100 period wiggler and a 4 wavelength long electron bunch, coupled via $\rho = 0.02$. Hence here $S = 25$, $K \simeq 1$.

Fig. 6.4a corresponds to the output at the 30th period, and we can clearly see that there is no flat region of the pulse, denoting no steady state evolution in any part of the pulse. The slippage between the pulses is now much greater than the electron bunch length ℓ_b. Here K is almost equal to one, but the gain experienced by a photon traversing the whole electron bunch ($K^{-1} = 4\pi\rho N_b$) is too small for the radiation to saturate at the steady state value $|A|^2_{eff} \simeq 1.4$ before escaping the bunch.

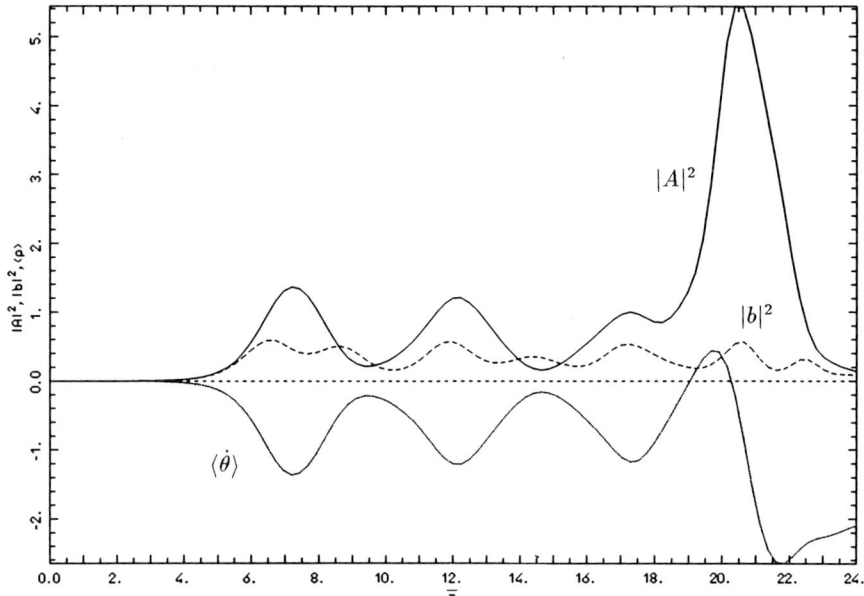

Fig. 6.3 Long bunch, short wiggler case. $|A|^2$ (solid line), $|b|^2$ (dashed line) and $\langle \dot\theta \rangle = \langle p \rangle$ (dotted line) as a function of $\bar z$ at a fixed position along the electron bunch $z_1 = 14.4$, i.e. 60 radiation wavelengths from the trailing edge.

Indeed in Fig. 6.4b, where we show the output at the 50th period, we see that the intensity of the radiation pulse escaping from the electrons is much smaller than in the steady state case.

The last two pictures, Fig. 6.4c–6.4d, show the output from the 80th and the 100th wiggler period, respectively. The radiation pulse, due to the slippage, continues to grow in length, but never reaches high peak values for the radiated intensity. Electron always loose energy to the radiation, which goes rapidly away from them. The stationary conservation law for $|A|^2 + \langle \dot\theta \rangle$ no longer holds locally within the whole electron pulse, due to the escape of radiation from the bunch.

From Fig. 6.4 we see that the radiation escaping from the bunch is still experiencing an exponential growth roughly up to the first peak. The height of this peak, I_p, however, turns out to be dependent on the gain parameter ρ. More precisely, executing several runs for different values of ρ, we found a linear dependence of I_p on ρ^2 for values of ρ such that $K \gtrsim 1$ (Fig. 6.5). This is in full agreement with the model discussed elsewhere[29], where a ρ^2 dependence of $|A|^2$ was found to occur analitically. We remember that such a behaviour indicates the occurrence of a superradiant emission of radiation, with output power scaling as n^2, the squared electron density. In fact, as $|A|^2 \propto |E|^2/\rho n$, and $\rho \propto n^{1/3}$, from $|A|^2 \propto \rho^2$ we have $|E|^2 \propto n^2$.

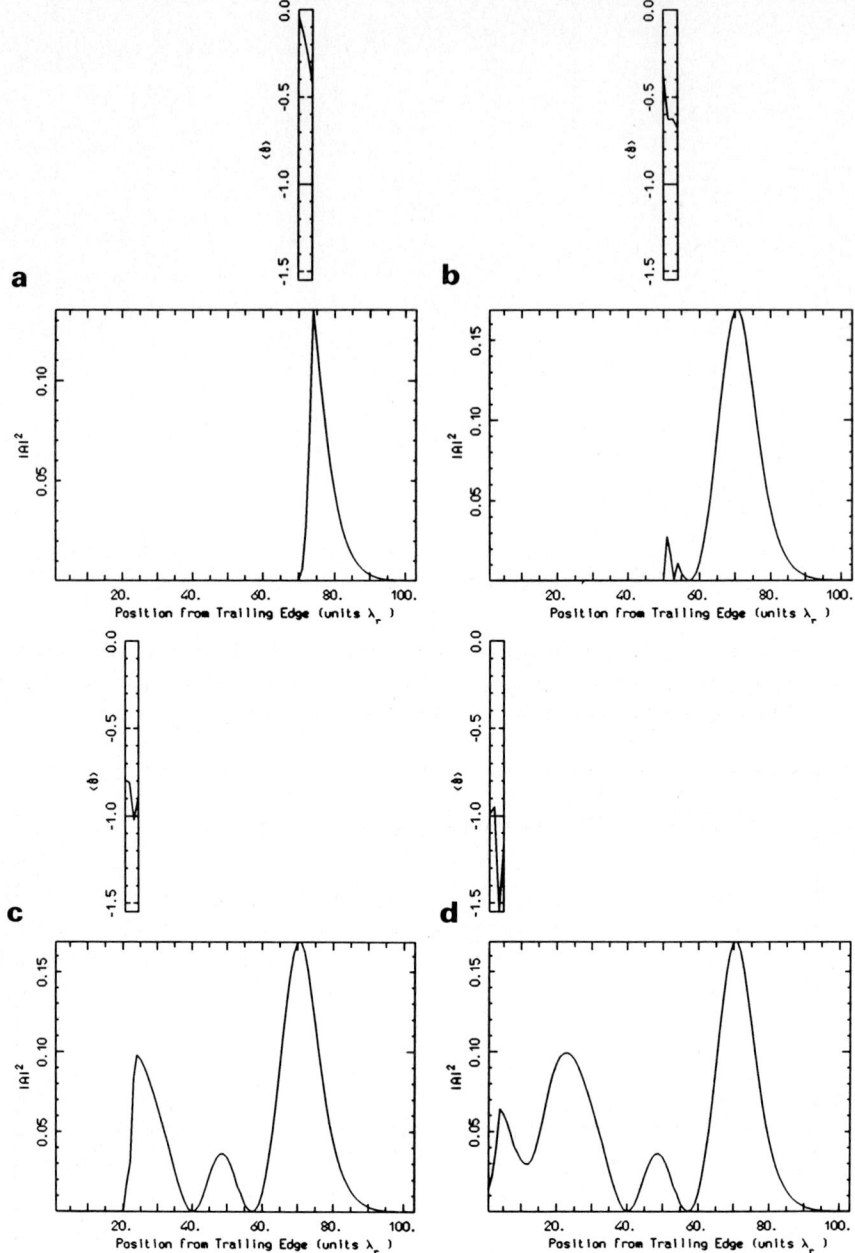

Fig. 6.4 Short bunch case. Radiation and electron pulses after N_P wiggler periods are shown from a) to d) for N_P =30, 50, 80, 100, respectively. Parameters for this case are listed in table 6.2.

One-dimensional theory of a FEL amplifier 75

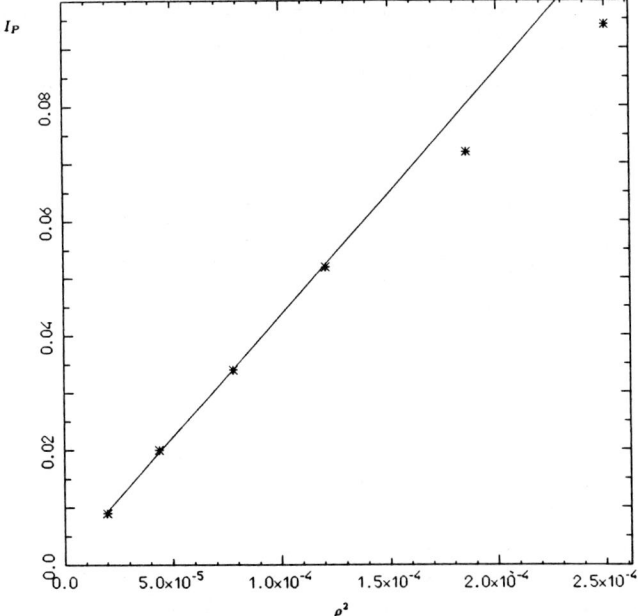

Fig. 6.5 Intensity of the first radiation peak emitted by the electrons (I_P) as a function of ρ^2 for 6 different values of ρ corresponding to $K \gtrsim 1$. The solid line shows the linear behaviour for small ρ (high K).

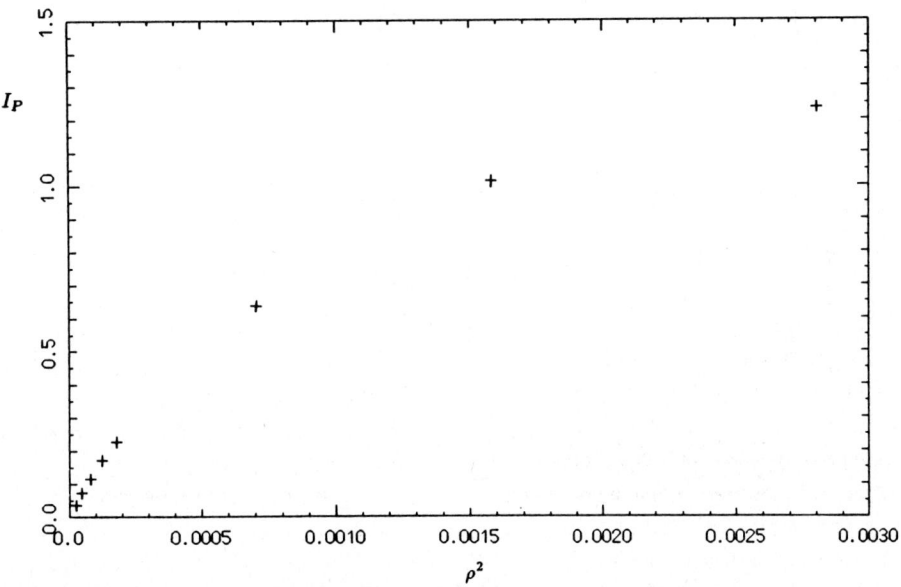

Fig. 6.6 As above but for a wider range of K, showing I_P flattening to 1.4 as K decreases (ρ increases).

In Fig. 6.6 we have plotted I_p for a wider range of values for ρ, including values corresponding to $K \ll 1$. As we can see, for values of K close to or smaller than one the linear dependence of I_p with respect to ρ^2 vanishes, and I_p tends to "flatten out" to the stationary value of $|A|^2_{sat} \simeq 1.4$, no longer dependent on ρ.

Another noticeable difference between the short and the long bunch regimes is seen comparing the behaviour of E_L as a function of $\bar{z}$.

In Fig. 6.7 we see the E_L plot corresponding to the parameters of Fig. 6.4, except that a longer wiggler (300 periods long) was chosen. As we can see E_L grows almost monotonically along the wiggler, never executing the quasi-periodic steady state oscillations. Moreover, energy is continuously extracted from electrons and the stationary limiting value of 1.4 is trespassed.

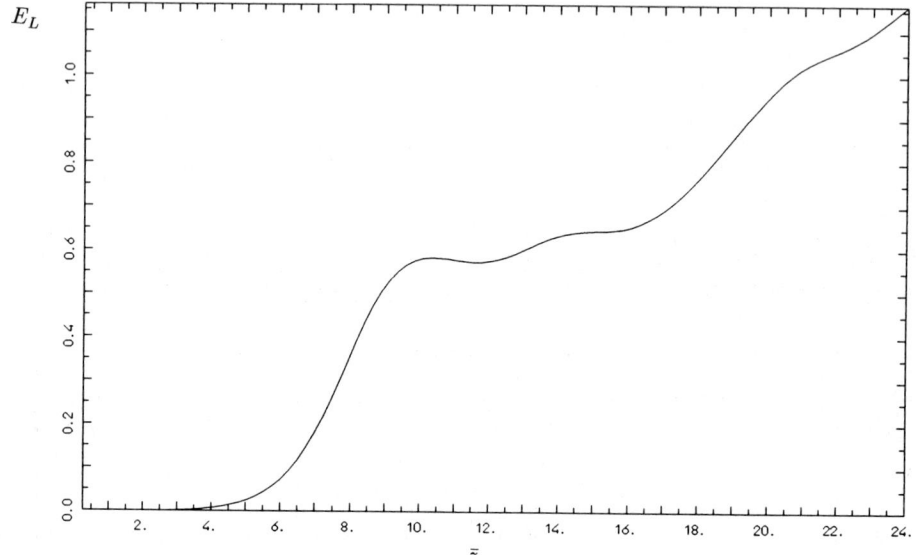

Fig. 6.7 Short bunch case. Average energy extraction E_L vs. $\bar{z}$.

In this process, where the radiation pulse has little peak efficiency, the longitudinal radiation pulse spread introduced by the slippage is responsible for the high electron energy extraction. We also remark that it is not necessary to have very short electron bunches ($K \gg 1$) to obtain these high values of E_L. This is clearly showed by Fig. 6.8, where, with a wiggler of a reasonable length, the remarkable value of $E_L \simeq 4$ was reached. The parameters used for this simulation were $N_w = 150$, $N_b = 4$, $4\pi\rho = 0.48$, giving $K = 0.52$, $S = 37.5$.

In Fig. 6.9 we plot the $|A|^2$, $|b|^2$, $\langle\dot{\theta}\rangle$ values in the leading edge of the pulse for parameters in table 6.2: we clearly see the difference between the short pulse and long pulse dynamic. No steady state like conservation law is followed here, electrons continue to loose their energy

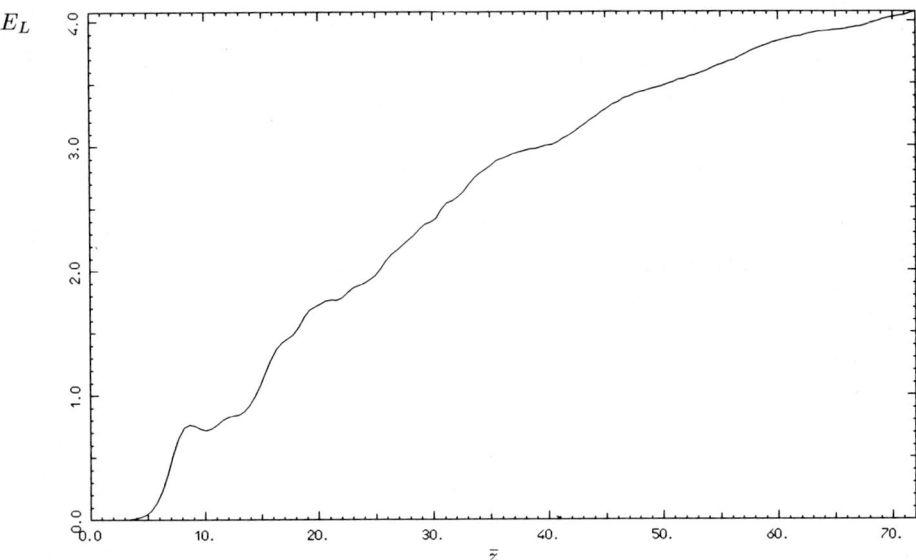

Fig. 6.8 Intermediate case. E_L vs. $\bar{z}$. Here $N_w = 150$, $N_b = 4$, $4\pi\rho = 0.48$, $K = 0.52$ and $S = 37.5$.

almost monotonically, while the radiation intensity oscillates around values smaller than the stationary value of 1.4.

Table 6.2.

Short bunch case						
$N_\circ$	N_b	ρ	$\Delta\bar{z} = 4\pi\rho$	G	S	K
100	4	0.02	0.24	24	25	1.01

6d.3. The Long Bunch, Long Wiggler Case ($K \ll 1$, $S \gtrsim 1$)

We are now going to show the behaviour of a long electron pulse ($K \ll 1$) injected in a long wiggler ($S \gtrsim 1$).

Comparing to the Long Bunch, Short Wiggler case, now the two pulses have more time to interact. The radiation emitted by the trailing electrons at the beginning of the wiggler has enough time to travel through the electron pulse and escape from the leading edge.

In Fig.6.10, for the set of parameters given in table 6.3, we show the output at $N_p = 30$, 50, 80, 150. Here, at the 50th period, the slippage region extends over the entire bunch length ℓ_b. We do not see anymore the steady state flat region, and the spike occurring after the emission of the first saturated peak takes place on a large portion of the bunch, as we can see in Fig. 6.10b.

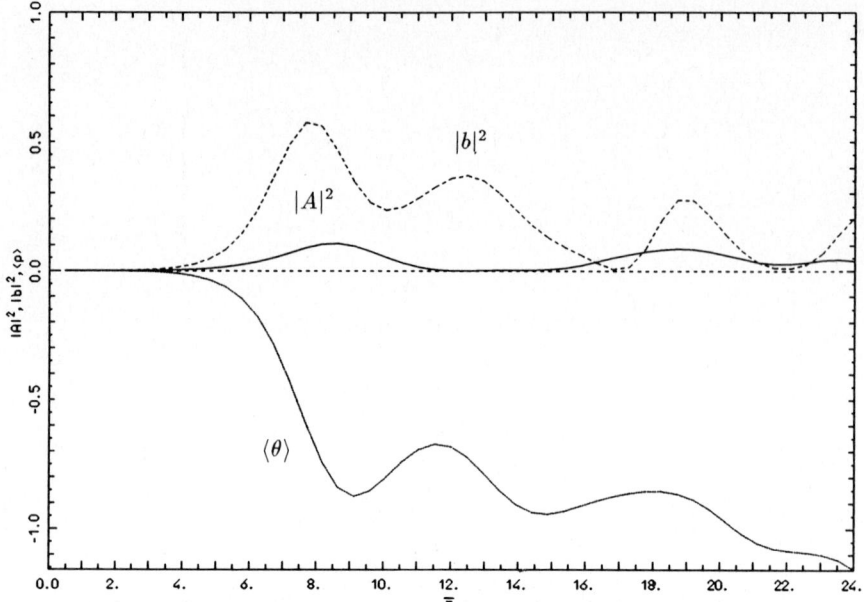

Fig. 6.9 Short bunch case. Same as fig. 6.3 but with $z_1 = 0.96$, i.e. 4 radiation wavelengths from the trailing edge.

Table 6.3.

Long bunch long wiggler case						
N_o	N_b	ρ	$\Delta \bar{z} = 4\pi\rho$	G	S	K
150	50	0.02	0.24	36	3	1/12

At the 80th period (Fig. 6.10c) the large spike is already escaped from the leading edge of the electron bunch, and hence will not grow further, propagating in the vacuum. At the end of the wiggler (Fig. 6.10d) we see many spikes escaped from the electron pulse. In this case the spikes do not reach the high peak value observed in the Long Bunch, Short Wiggler case previously reported, but their relative width (with respect to the bunch length) is considerably greater, so that they contribute much more to the average energy extracted from the electrons. This is clearly seen in Fig. 6.11, where we show the E_L plot for this simulation. Here, a steady state behaviour is observed only at the beginning of the wiggler (small values of $\bar{z}$). Then it grows in an almost continuous way, reaching higher values than 1.4 and showing a superradiant energy extraction behaviour.

6e. DISCUSSION

From the results showed in the previous section we conclude that there exist three possible dynamic regimes for a high gain single pass FEL amplifier. Only one of these can be effectively

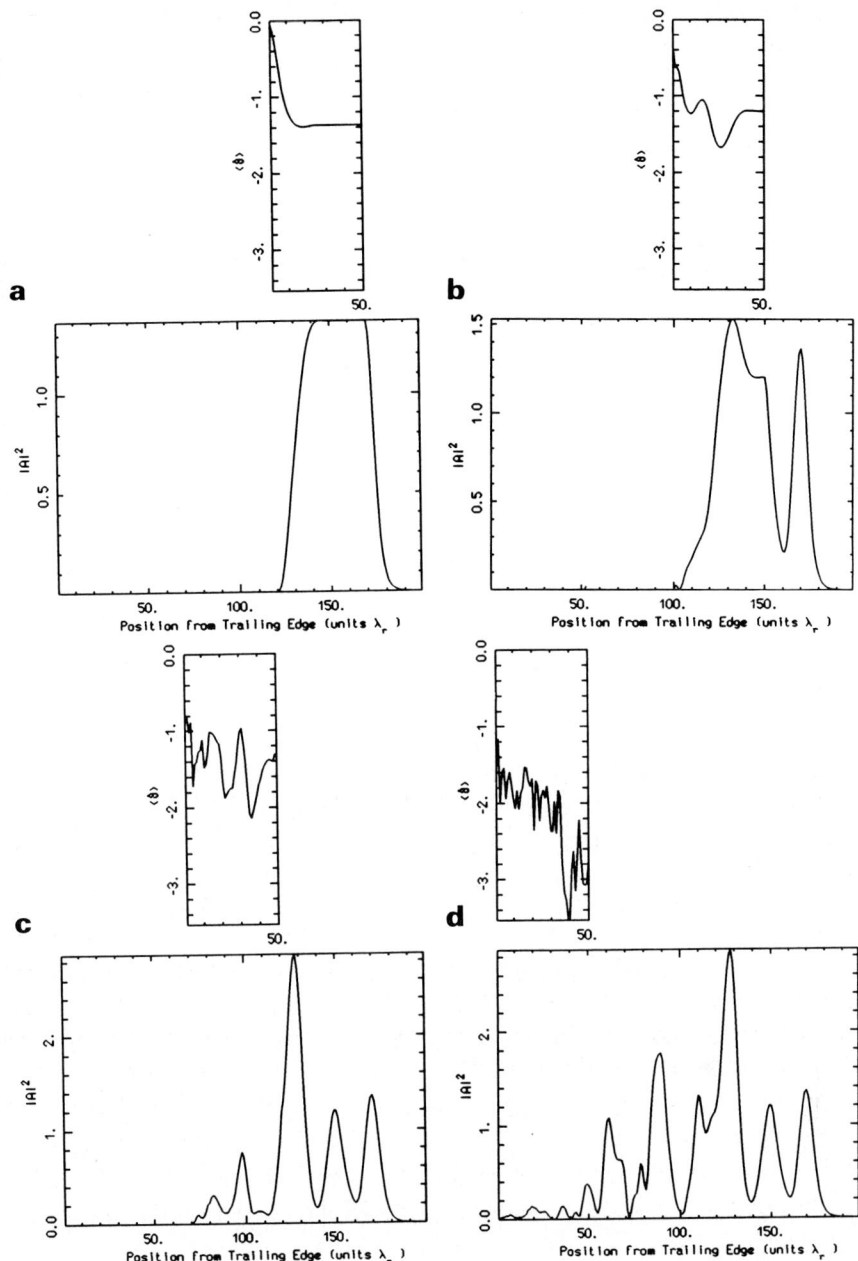

Fig. 6.10 Long bunch, long wiggler case. Radiation and electron pulses after N_P wiggler periods are shown from a) to d) for N_P =30, 50, 80, 150, respectively. Parameters for this case are listed in table 6.3.

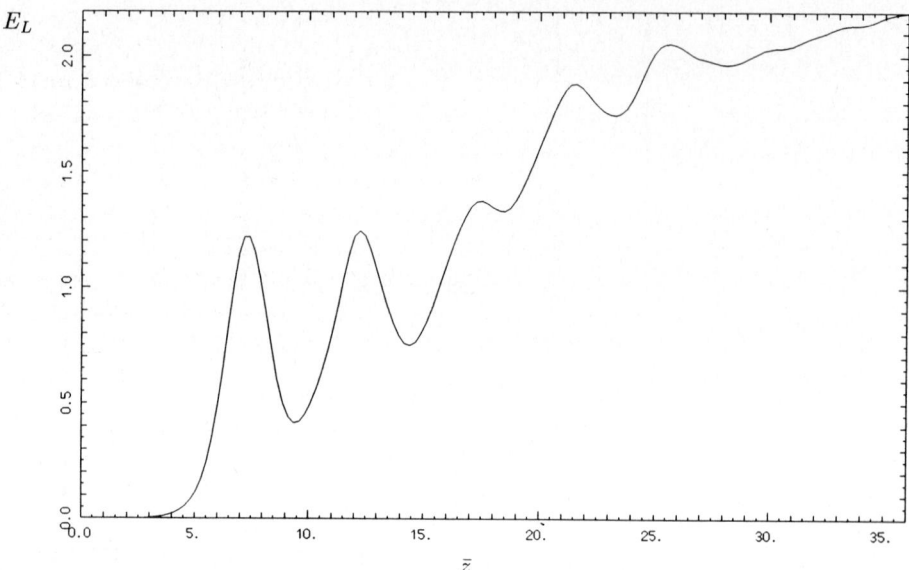

Fig. 6.11 Long bunch, long wiggler case. Average energy extraction E_L vs. $\bar{z}$.

modeled by a steady state theory, as far as average pulse properties are considered. It is the case of a Long Bunch, Short Wiggler FEL, the only case experimentally tested, where effects related to the slippage take place in a small region around the trailing edge of the pulses, contributing little to the average behaviour of the pulses.

But, if we allow the interaction to continue, using a longer wiggler, the "superradiant" trailing effects begin to rule the behaviour of the whole radiation pulse, permitting to reach high values of the average electron energy extraction (i.e. of the efficiency). This is the case of a FEL operated in the Long Bunch, Long Wiggler regime.

Finally, with our model, we were able to describe a FEL driven by Short Bunches, case in which we found the occurrence of the theoretically predicted phenomenon of superradiance (i.e. spontaneous radiation scaling with n^2), and of continuous electron energy extraction.

It is worth here reporting that a similar long bunch analysis performed by Dr. W. Sharp of the Lawrence Livermore National Laboratory confirmed our results and found out that the same spiking behaviour occurs even with a non-resonant (or detuned) electron beam. In this case, if the detuning corresponds to a stable steady state solution, only the spiking signal in the slippage region experiences high-gain, whereas in the steady-state region the radiation intensity oscillates around the initial value. The spikes are seen growing more than the resonant case, continuously reducing their width[31]. We checked these results with our code, obtaining almost identical results.

7. HIGH-GAIN FEL AMPLIFIERS IN A WAVEGUIDE

In this section we consider the FEL process occurring in a waveguide, which is the case, e.g., of amplifiers operating in the microwave range[13,15,32]. For sake of simplicity we shall assume a rectangular waveguide, with only one (transverse electric) mode which couples to the electron beam. Hence we shall generalize the FEL equations and the universal scaling to include off-axis propagation in a rectangular waveguide and with a planar wiggler. As an application, we shall discuss the optimization of gain, tuning and slippage with the waveguide[33]. In the final subsection we discuss the possibility of experimental test of all high-gain FEL regimes by means of the waveguide.

7a. PLANAR WIGGLER

Unlike in the previous sections, we shall consider a planar wiggler, with a geometry such that the electron beam, injected along the z-axis, is subjected to a transverse, periodic magnetostatic field in the y-direction, thus wiggling in the x-z plane (see Sec. 2). The modifications to be introduced in the previous treatment are well-known[12,34]. First of all, the undulator parameter (2.5), which can be written as $a_w = eB_w/mc^2 k_w$ for a helical wiggler, where B_w is the peak value of the wiggler field, becomes for a planar wiggler

$$a_w = \frac{eB_w}{\sqrt{2}mc^2 k_w} \tag{7.1}$$

Note that the definitions (2.5) and (7.1) can be unified if one introduces the root-mean-square wiggler field, $(B_w)_{rms}$, so that $a_w = e(B_w)_{rms}/mc^2 k_w$ in both cases. Second, in the FEL equations and in the scaling (3.32) one must perform the substitution

$$a_w \to a_w d_B \equiv \bar{a}_w \tag{7.2a}$$

where d_B is the following difference of Bessel functions[35]

$$d_B = J_0(\xi) - J_1(\xi)$$
$$\xi = \frac{a_w^2}{2(1 + a_w^2)} \tag{7.2b}$$

The quantity d_B is on the order of one and accounts for a reduction in the electron-field coupling in the presence of a planar wiggler instead of a helical one. Furthermore, in a planar wiggler the motion gets a longitudinal, fast-oscillating component superimposed on the wiggle motion in the (x-z) plane. As a consequence of averaging out this jitter effect, the relative phase of the electron in the ponderomotive field, θ (eq.(3.13)), must be referred to the average longitudinal electron motion:

$$\theta = (k + k_w)z - \omega \bar{t}$$

$$\bar{t} = \frac{z}{c\bar{\beta}_\|} \simeq \frac{z}{c}\left(1 + \frac{1 + a_w^2}{2\gamma^2}\right) \tag{7.3}$$

With a_w and θ defined as in (7.1), (7.3), as we shall understand from now on, the FEL dynamic equations both in the original and in the scaled form remain unchanged with respect to the case of a helical wiggler except for the substitution (7.2a). E.g., the FEL parameter ρ (eq.(3.32)) with a planar wiggler becomes

$$\rho = \frac{1}{\gamma_r}\left(\frac{\bar{a}_w}{4}\frac{\omega_p}{ck_w}\right)^{2/3} \tag{7.4}$$

7b. FEL EQUATIONS WITH A RECTANGULAR WAVEGUIDE

Let the FEL process occur within a rectangular waveguide, whose short (long) dimension is b (a), parallel (orthogonal) to the wiggler field ($\vec{B}_w \parallel \hat{y}$). We assume that only one guided mode, the transverse electric TE_{01} mode, couples with the electron beam. We must start from a full three-dimensional treatment of the field. However, we introduce the following ansatz on the radiation vector potential (compare eq.(3.4)):

$$\vec{\mathcal{A}}(\vec{x},t) = -\hat{x}|\tilde{a}|(z,t)\sin(k_\| z - \omega t + \phi(z,t))\sin(\vec{k}_\perp \cdot \vec{x}_\perp) \tag{7.5}$$

where $|\tilde{a}|(z,t)$ and $\phi(z,t)$ are the slowly varying real amplitude and phase of the TE_{01} mode; the dispersion relation is

$$\frac{\omega}{c} = (k_\|^2 + k_\perp^2)^{1/2} \equiv k, \tag{7.6}$$

and with our geometry the transverse component of the field wave-vector, $\vec{k}_\perp$, which describes off-axis propagation, is such that

$$\vec{k}_\perp \cdot \vec{x}_\perp = k_\perp \cdot y = \left(\frac{\pi}{b}\right)y \tag{7.7}$$

The phase and group velocities are then

$$\begin{aligned} v_{ph} &= \frac{\omega}{k_\|} = \left(\frac{k}{k_\|}\right)c > c \\ v_g &= \frac{d\omega}{dk_\|} = \left(\frac{k_\|}{k}\right)c < c \end{aligned} \tag{7.8}$$

We shall neglect dispersive effects treating the group velocity v_g as a constant.

Next, we assume an oversized waveguide, such that the electron beam radius, r, satisfies the inequality

$$\frac{\pi}{b}r \ll 1 \tag{7.9}$$

and such that we can limit our analysis to the electron motion in the x-z plane. From now on we shall assume $k_\perp/k_\parallel \ll 1$, i.e., $\lambda \ll b$

Let us consider first the electron equations. In this case the assumption (7.9) allows us reducing at once the $3D$ problem to a $1D$ problem again, but now generalized to off-axis propagation. Actually, if one retraces the same steps of Sec. 3 leading to the electron energy eq.(3.15), just the same equation is obtained, with only the definition of the phase θ generalized as follows (compare eq.(7.3)):

$$\theta = (k_\parallel + k_w)z - \omega \bar{t}, \qquad (7.10)$$

where ω and $k_\parallel$ are linked via the dispersion relation (7.6). Actually, if one derives eq.(7.10) with respect to z, using eq.(7.6) and approximation (3.19), and averaging over a wiggler period, one obtains:

$$\frac{d\theta}{dz} = k_\parallel + k_w - \frac{\omega}{c}\left(1 + \frac{1 + a_w^2}{2\gamma^2}\right) \qquad (7.10')$$

Imposing $\frac{d\theta}{dz} = 0$, we obtain the generalized expression of the resonant electron energy

$$\bar{\gamma}_r = \left\{\frac{k_\parallel}{2\bar{k}_w}(1 + a_w^2)\right\}^{1/2} \qquad (7.11)$$

where

$$\bar{k}_w \equiv k_w + k_\parallel - \frac{\omega}{c} \simeq k_w\left(1 - \frac{X}{2}\right), \qquad (7.12)$$

$$X \equiv \frac{k_\perp^2}{k_w k_\parallel} = \frac{\lambda \lambda_w}{4b^2} \qquad (7.13)$$

where $0 < X < 2$. In the limit $X \to 0$, $\bar{\gamma}_r$ reduces to the familiar expression (2.6). Again from $\frac{d\theta}{dz} = 0$, recalling that $\lambda = 2\pi/k_\parallel$, we derive the resonance relation generalized to off-axis propagation in a waveguide (valid only for $X \ll 1$)

$$\lambda = \frac{\lambda_w}{2\gamma_r^2}\left(1 + a_w^2 + \gamma_r^2 \varphi^2\right) \qquad (7.14)$$

where $\varphi = \arctan k_\perp/k_\parallel$, and $\gamma_r^2 = k_\parallel/2k_w(1 + a_w^2)$.

It is easily verified that, if one drops the approximation $\vec{a}_{tot} \simeq \vec{a}_w$ and introduces an effective wiggler wavenumber $\bar{k}_w$ defined in (7.12), one still recovers the phase equation (3.21). Thus, going from one to many electrons, one obtains the $1D$ electron equations in the paraxial approximation:

$$\left(\frac{\partial}{\partial z} + \frac{1}{\bar{v}_\parallel}\frac{\partial}{\partial t}\right)\theta_j = \bar{k}_w\left(1 - \frac{\bar{\gamma}_r^2}{\gamma_j^2}\right) + \frac{k_\parallel}{2\gamma_j^2}\left[i\bar{a}_w(ae^{i\theta_j} - c.c.) - |a|^2\right] \qquad (7.15a)$$

$$\left(\frac{\partial}{\partial z} + \frac{1}{\bar{v}_\parallel}\frac{\partial}{\partial t}\right)\gamma_j = -\frac{k\bar{a}_w}{2\gamma_j}(ae^{i\theta_j} + c.c.) \qquad (7.15b)$$

Eqs.(7.15) resemble eqs.(3.20), (3.21) very closely, with space-charge neglected and for a planar wiggler. However, we recall that θ_j, γ_r and k_w have been redefined due to the waveguide (in eqs.(7.10), (7.11), (7.12), respectively). Furthermore, the dimensionless complex field amplitude a is assumed to coincide with that of the r.m.s. TE_{01} cavity mode, i.e., $a \equiv a_{01}/\sqrt{2}$.

Now let us consider the field equation. One must start from the full $3D$ wave equation, i.e., including the second-order derivatives with respect to the transverse coordinates x, y (compare eq.(3.22)). However, on using the ansatz (7.5) and the approximation (7.9), by projecting the wave equation on the guided mode TE_{01} and integrating over the transverse mode section, one derives the following $1D$ field equation in the paraxial approximation:

$$\left[\frac{\partial}{\partial z} + \frac{1}{v_g}\frac{\partial}{\partial t}\right] a = \frac{\bar{\omega}_p^2}{2k_\parallel c^2}\chi\left[\bar{a}_w\left\langle\frac{e^{-i\theta}}{\gamma}\right\rangle - ia\left\langle\frac{1}{\gamma}\right\rangle\right] \qquad (7.16)$$

Eq.(7.16) generalizes eq.(3.27) in many respects. In the l.h.s., v_g is the group velocity (7.6), and the SVEA has been used so that second-order derivatives in z and t are neglected; the second-order derivatives with respect to x and y cancelled exactly due to the ansatz (7.5) and the dispersion relation (7.6). In the r.h.s., where an electron bunch profile, $\chi(z,t)$, has been introduced as in Sec. 6 to discuss propagation effects, the electron number density refers to the area of the mode transverse section, $S_{mode} = ab/2$, rather than to that of the electron beam transverse section[34]; that is,

$$\begin{aligned}\bar{\omega}_p^2 &= \frac{4\pi e^2 \bar{n}}{m} \\ \bar{n} &= \frac{I_{01}}{ec}\frac{1}{ab/2} = \frac{J_{01}}{ec}\end{aligned} \qquad (7.17)$$

where J_{01} is the electron current density with respect to the TE_{01}-mode cross section.

7c. UNIVERSAL SCALING WITH A WAVEGUIDE

The universal scaling of subsec. 3c can be generalized as follows:

$$\begin{aligned}\Gamma_j &= \frac{1}{\bar{\rho}}\frac{\gamma_j}{\gamma_r} \\ A &= \frac{ck}{\bar{\omega}_p\sqrt{\bar{\rho}\gamma_r}}\sqrt{\frac{k_\parallel}{k}}\,a \\ \bar{z} = 2\bar{k}_w\bar{\rho}z, \quad \bar{t} &= 2\bar{k}_w\bar{\rho}t \\ \bar{\rho} &= \frac{1}{\gamma_r}\left(\frac{\bar{a}_w}{4}\frac{\bar{\omega}_p}{c\bar{k}_w}\right)^{2/3}\left(\frac{k}{k_\parallel}\right)^{1/3} \\ \bar{\sigma} &= \frac{2}{\xi}\bar{\rho}\frac{k_\parallel}{k}\end{aligned} \qquad (7.18)$$

with $\bar{a}_w$, ξ, $\bar{\gamma}_r$, $\bar{k}_w$, $\bar{\omega}_p$ defined in eqs. (7.2), (7.11), (7.12), (7.17), respectively. Letting the variables θ_j, Γ_j, A keep the same symbols as in the previous sections, without waveguide and with a helical wiggler, the dimensionless form of the FEL eqs.(7.15a, b), (7.16) is

$$\left(\frac{\partial}{\partial \bar{z}} + \frac{1}{\bar{v}_\|}\frac{\partial}{\partial \bar{t}}\right)\theta_j = \frac{1}{2\bar{\rho}}\left(1 - \frac{1}{\bar{\rho}^2 \Gamma_j^2}\right)$$
$$+ \frac{k_\|}{k}\left[i\frac{1}{\bar{\rho}}\left(A\frac{e^{i\theta_j}}{\Gamma_j^2} - c.c.\right) - \frac{\bar{\sigma}}{2}\frac{|A|^2}{\Gamma_j^2}\right] \quad (7.19a)$$

$$\left(\frac{\partial}{\partial \bar{z}} + \frac{1}{\bar{v}_\|}\frac{\partial}{\partial \bar{t}}\right)\Gamma_j = -\frac{1}{\bar{\rho}}\left(A\frac{e^{i\theta_j}}{\Gamma_j} + c.c.\right) \quad (7.19b)$$

$$\left[\frac{\partial}{\partial \bar{z}} + \frac{1}{v_g}\frac{\partial}{\partial \bar{t}} - i\bar{\rho}\frac{\bar{k}_w}{\bar{k}_\|}\left(\frac{\partial^2}{\partial \bar{z}^2} - \frac{1}{c^2}\frac{\partial^2}{\partial \bar{t}^2}\right)\right]A = \chi\left[\frac{1}{\bar{\rho}}\left\langle\frac{e^{-i\theta}}{\Gamma}\right\rangle - i\frac{\bar{\sigma}}{2}A\left\langle\frac{1}{\Gamma}\right\rangle\right] \quad (7.19c)$$

A remarkable feature of the universal scaling with the waveguide (7.18) is that the second-order derivatives in the field eq.(7.19c) are multiplied times a coefficient $\bar{\rho}(\bar{k}_w/k_\|) \ll 1$. In particular, in the Compton limit ($\rho \leq 10^{-2}$) these contributions can be dropped without invoking the usual SVEA, whose validity may become questionable in some propagation problems.

Furthermore, in a waveguide one can arrange the group velocity, v_g, to be equal to the average longitudinal electron velocity, $\bar{v}_\|$, so that the slippage can be neglected. This technique was used in recent experiments[32] in order to eliminate the sideband instability which may develop in the presence of slippage. On using the expression of the resonant velocity with the waveguide,

$$v_r = \frac{ck}{k_\| + k_w} \quad (7.20)$$

the condition of no-slippage, $v_g = v_r$, can be written as

$$X = 1 \quad (7.21)$$

where X is the waveguide correction parameter introduced in eq. (7.13).

7d. TUNING, SLIPPAGE AND GAIN WITH A WAVEGUIDE

Relevant results concerning tuning, slippage and gain can be derived in a simple way[33], without consideration of the FEL equations.

First of all, from the dispersion relation (7.6) and the resonance relation (7.10), we obtain

$$\frac{1 + a_w^2}{\gamma^2} \simeq \frac{1}{\gamma_\|^2} = 2\alpha X \frac{1 - X/2(1-\alpha)}{(1 + \alpha X)^2} \quad (7.22)$$

where the parameter

$$\alpha \equiv \frac{\lambda}{\lambda_w} \frac{1}{X} = \frac{4b^2}{\lambda_w^2} \qquad (7.23)$$

Note that $\alpha X = \lambda/\lambda_w$. Now, up to the millimeter range $\alpha X \ll 1$ and (7.22) can be well approximated by

$$\frac{1+a_w^2}{\gamma^2} \simeq 2\alpha X \left(1 - \frac{X}{2}\right) \equiv y(X) \qquad (7.22')$$

where α has been taken constant, which means that the variable parameter is $\lambda \propto X$ with fixed wiggler and waveguide, so that $y(X)$ is a tuning curve. In Fig. 7.1 we see that $y(X)$ has a maximum equal to α for $X = 1$, which is the condition of zero slippage (7.22).

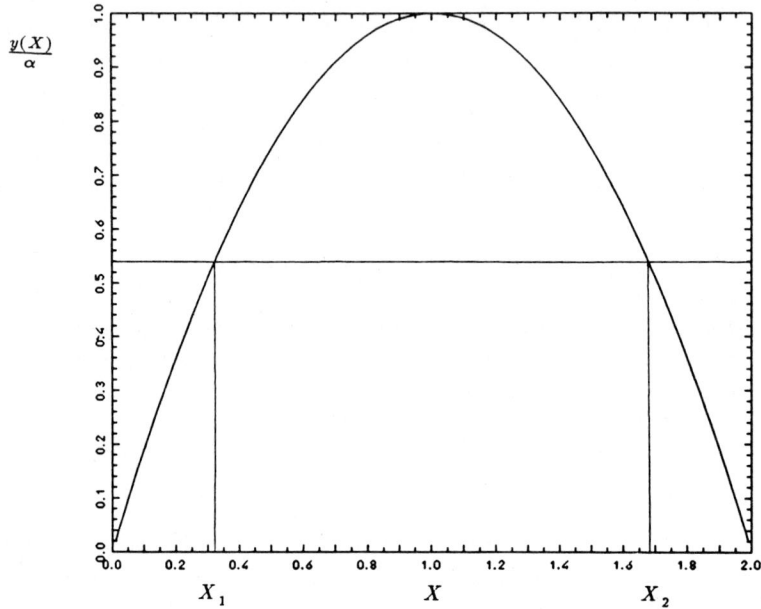

Fig. 7.1 Function $\frac{y(X)}{\alpha}$ (eq. (7.22')). For $0 < (1 + a_w^2)/\gamma^2 < \alpha$ (see text) there are two resonant operating points with either positive ($0 < X_1 < 1$) or negative ($1 < X_2 < 2$) slippage.

The resonance condition (7.22') is satisfied at the intersections of the line $(1 + a_w^2)/\gamma^2$ with the curve $y(X)$. Hence, while with no waveguide ($X \ll 1$, $y \simeq 2\alpha X$) there is always one intersection, with the waveguide there are no intersections for $(1 + a_w^2)/\gamma^2 > \alpha$ and two intersections for $0 < (1 + a_w^2)/\gamma^2 < \alpha$, one for positive and another for negative slippage. Also, from eq.(7.22') it follows that the radiation wavelength λ is always greater than without waveguide, and this effect is maximum just at zero slippage ($X = 1$), where λ is doubled with respect to the no-guide case.

The total slippage length (see eqs.(6.6), (6.11)) becomes with the waveguide

$$\ell_s = \left(\frac{v_g}{\bar{v}_\parallel} - 1\right) L_w \qquad (7.24)$$

By means of eqs.(7.6) and (7.20) one easily finds

$$\ell_s = L_w \alpha X(1 - X), \qquad (7.24')$$

which generalizes the usual result $\ell_s = L_w \alpha X = N_w \lambda$. Fig. 7.1 shows that also the slippage length has a maximum as a function of X with α constant, (at $X = 1/2$), i.e., with respect to the wavelength λ.

Most important, also the FEL gain can be optimized with respect to the wavelength λ for fixed wiggler and waveguide, e.g., by varying the electron energy. Let us consider the steady-state unsaturated gain per unit length g (eq.(6.8)), here generalized with the waveguide as

$$\bar{g} = 2\bar{k}_w \bar{\rho} \qquad (7.25)$$

with $\bar{k}_w$ and $\bar{\rho}$ defined in (7.12) and (7.18). Now, we can write the latter parameter in the form

$$\bar{\rho} = \rho \left(1 - \frac{X}{2}\right)^{-1/6} \left(\frac{k}{k_\parallel} \frac{S_{beam}}{S_{mode}}\right)^{1/3} \qquad (7.26)$$

where ρ is the FEL parameter (7.4) without waveguide (with a planar wiggler).

If one sets the waveguide geometrical factor $(k/k_\parallel)(S_{beam}/S_{mode})$ equal to one, as we shall assume from now on, the ratio of the gains with and without waveguide is

$$\frac{\bar{g}}{g} = \frac{\bar{k}_w \bar{\rho}}{k_w \rho} = \left(1 - \frac{X}{2}\right)^{5/6}, \qquad (7.27)$$

showing a gain reduction due to the waveguide, which is a monotonic decreasing function of X. However, if one is interested in the global dependence of the gain $\bar{g}$ on the wavelength λ (i.e., with $X \propto \lambda$ variable and α constant as above for tuning and slippage) then, since $\rho \propto \sqrt{\lambda/\lambda_w} = \sqrt{\alpha X}$,

$$\bar{g} \propto \sqrt{X} \left(1 - \frac{X}{2}\right)^{5/6} \equiv W(X) \qquad (7.28)$$

The curve $W(X)$ is reported in Fig. 7.2. A remarkable result is the existence of a maximum of $W(X)$, which occurs for $X = \lambda \lambda_w / 4b^2 = 0.75$. Namely, for a fixed ratio b/λ_w, the gain can be optimized with respect to the radiation wavelength by the choice $\lambda \simeq 3b^2/\lambda_w$.

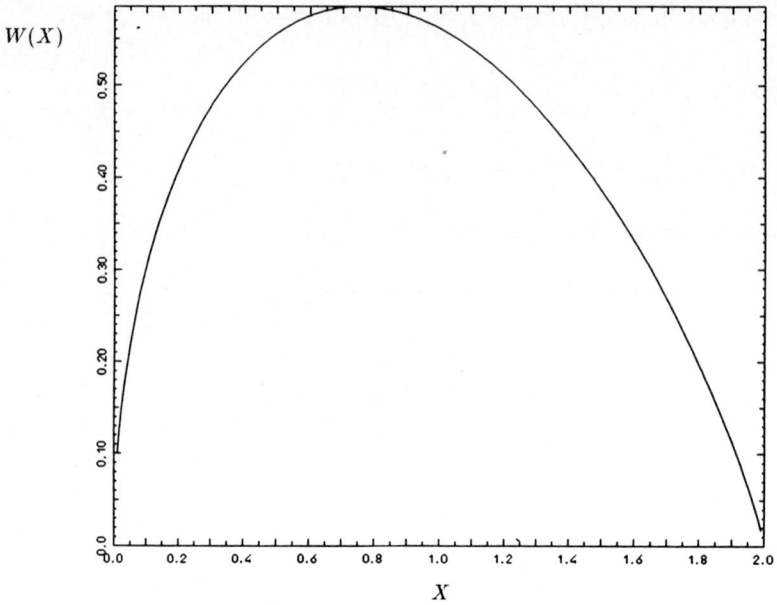

Fig. 7.2 Function $w(X)$ (eq. (7.28)) describing the gain per unit length in the waveguide as a function of the radiation wavelength.

7e. ON THE POSSIBILITY OF EXPERIMENTAL OBSERVATION OF THE THREE HIGH-GAIN FEL REGIMES

Let us now briefly discuss the fascinating possibility, offered by the waveguide, of observing the high-gain FEL regimes described in the previous Section, i.e., steady-state, weak and strong Superradiance, with basically the same experimental set-up. First of all we remark that the FEL equation (7.19) in a waveguide, neglecting second order derivatives, are formally equal to the propagation equations in free space provided that definitions (7.18) are taken into account. Hence, we can generalize the fundamental parameters S, G, K introduced in sec.6 to describe the three high gain regimes. From eq.(7.27) the gain parameter in a waveguide $\bar{G} = 2\bar{k}_w \, \bar{\rho} L_w$ can be written as

$$\bar{G} = G \left(1 - \frac{X}{2}\right)^{5/6} \tag{7.29}$$

where G is the gain parameter introduced in eq. (6.9). From eq. (7.24), imposing that the bulk velocity $\bar{v}_\parallel$ is equal to the resonant velocity v_r (eq. (7.20)), and using the expression (7.8) for v_g we have:

$$\bar{\ell}_s = \left(\frac{k_w k_\parallel - k_\perp^2}{k_\perp^2 + k_\parallel^2}\right) L_w \simeq \frac{\lambda}{\lambda_w} L_w (1 - X)$$

where again we have assumed $k_\perp^2 \ll k_\parallel^2$. To compare with eq. (6.7) we can write:

$$\bar{\ell}_s = \ell_s(1 - X) \qquad (7.30)$$

Hence we can generalize the slippage parameter introduced in eq. (6.11) as follows:

$$\bar{S} \equiv \frac{\bar{\ell}_s}{\ell_b} = S(1 - X) \qquad (7.31)$$

Therefore, according to eq. (6.13), the superradiant parameter K becomes:

$$\bar{K} \equiv \frac{\bar{S}}{\bar{G}} = K \frac{1 - X}{\left(1 - \frac{X}{2}\right)^{5/6}} \qquad (7.32)$$

Expression (7.29), (7.31) and (7.32) have a very remarkable implication. Namely, one can change all the three FEL parameters S, G, K without varying the electron beam and wiggler parameters, but only changing the waveguide height b, since $k_\perp = \pi/b$. More precisely, for $X = 1$ one recovers the steady state regime, whereas for $X \lesssim 1$ and $X \ll 1$ one can match the conditions to observe strong and weak superradiance, respectively, as discussed in detail in ref. 3.

Acknowledgments

We thank Bill Barletta, Brian Mc Neil, Ted Scharlemann, and Bill Sharp for helpful discussions, and all components of the ELFA group in Milano for their assistance in the preparation of the manuscript.

LIST OF SYMBOLS

$\vec{A}_w$	$\equiv$	Wiggler magnetostatic field vector potential
a_w	$\equiv$	Dimensionless wiggler parameter or undulator parameter (helical wiggler)
$\bar{a}_w$	$\equiv$	Dimensionless wiggler parameter or undulator parameter (planar wiggler)
$\vec{A}$	$\equiv$	Radiation vector potential
$\tilde{a}$	$\equiv$	Radiation complex field amplitude
a	$\equiv$	Dimensionless radiation complex amplitude (scaled e/mc^2)
$\vec{A}_{tot}$	$\equiv$	Total (radiation+wiggler) vector potential
$\vec{a}_{tot}$	$\equiv$	Dimesionless total vector potential (scaled e/mc^2)
A	$\equiv$	Dimensionless field amplitude (universal scaling) (eq.(3.32))
$\tilde{A}$	$\equiv$	Dimensionless field amplitude (universal scaling) (eq.(4.3a))
a	$\equiv$	Long waveguide dimension (along x)
$\vec{B}_w$	$\equiv$	Wiggler magnetostatic field
$\vec{B}$	$\equiv$	Radiation magnetic field
b	$\equiv$	Bunching parameter
b	$\equiv$	Short waveguide dimension (along y)
c	$\equiv$	Light velocity
d_B	$\equiv$	Difference of Bessel functions
e	$\equiv$	Electron charge
$\vec{E}$	$\equiv$	Radiation electric field
$\hat{e}$	$\equiv$	Circular polarization unit vector
E_L	$\equiv$	Average energy extracted from the electron beam
g	$\equiv$	Gain per unit length
$\bar{g}$	$\equiv$	Gain per unit length in a waveguide
G	$\equiv$	Unsaturated gain parameter
$\mathcal{G}$	$\equiv$	Gain
H	$\equiv$	Hamiltonian (4.6)
$\tilde{H}$	$\equiv$	Hamiltonian (4.6')
k_w	$\equiv$	Wiggler wavenumber
k	$\equiv$	Radiation wavenumber
K	$\equiv$	Superradiant Parameter
I	$\equiv$	Electron current
I_{01}	$\equiv$	Electron current flowing in the waveguide (TE_{01} mode)
$\vec{J}_\perp$	$\equiv$	Transverse electron current density
J_0	$\equiv$	Zero order Bessel function of the first kind
J_1	$\equiv$	First order Bessel function of the first kind
J_{01}	$\equiv$	Transverse electron current density in a waveguide (TE_{01} mode)
L_w	$\equiv$	Wiggler length
ℓ_b	$\equiv$	Bunch length

ℓ_s	$\equiv$	Total slippage length
$\ell_s^{N_p}$	$\equiv$	Slippage length after N_p wiggler periods
ℓ_g	$\equiv$	Gain length
ℓ_c	$\equiv$	Cooperation length
n	$\equiv$	Electron density
$\bar{n}$	$\equiv$	Electron density in the waveguide
$n_\perp$	$\equiv$	Transverse electron number density
$\tilde{n}$	$\equiv$	Average electron density along the bunch
N	$\equiv$	Total number of electrons
N_b	$\equiv$	Bunch length in units of the radiation wavelength
N_w	$\equiv$	Total number of wiggler periods
N_p	$\equiv$	Number of wiggler periods (p=1,2,...)
p_j	$\equiv$	Canonical momentum of the jth electron (universal scaling) (eq.(5.5))
$\tilde{p}_j$	$\equiv$	Canonical momentum of the jth electron (universal scaling) (eq.(5.2))
r	$\equiv$	Electron beam radius
S_{mode}	$\equiv$	TE$_{01}$ mode area
t	$\equiv$	Time variable
$\bar{t}$	$\equiv$	Scaled time variable (Universal scaling) (eq.(3.32))
	$\equiv$	Also, average electron arriving time (in a planar wiggler)
$\bar{t}'$	$\equiv$	Transformed time variable (eq.(4.1))
v	$\equiv$	Electron velocity
$v_\parallel$	$\equiv$	Longitudinal electron velocity
$\bar{v}_\parallel$	$\equiv$	Average longitudinal electron velocity
$v_\perp$	$\equiv$	Transverse electron velocity
v_p	$\equiv$	Ponderomotive phase velocity
v_{ph}	$\equiv$	Radiation phase velocity
v_g	$\equiv$	Radiation group velocity
W	$\equiv$	Waveguide function (eq.(7.28))
z	$\equiv$	Longitudinal variable (along wiggler axis)
$\bar{z}$	$\equiv$	Dimensionless longitudinal variable (universal scaling)
z_1, z_2	$\equiv$	Dimensionless electron and radiation field characteristics
X	$\equiv$	Waveguide parameter
X, Y	$\equiv$	Collective electron variables (linear theory)
x	$\equiv$	Argument of the small–signal gain (eq.(5.18))
y	$\equiv$	Waveguide function (eq.(7.22'))
α	$\equiv$	Waveguide parameter (eq.(7.23))
β	$\equiv$	Dimensionless electron velocity
$\beta_\parallel$	$\equiv$	Longitudinal dimensionless electron velocity
$\bar{\beta}_\parallel$	$\equiv$	Average value of $\beta_\parallel$
$\beta_\perp$	$\equiv$	Transverse dimensionless electron velocity

γ	$\equiv$	Lorentz factor, dimensionless electron energy
$\gamma_\parallel$	$\equiv$	Longitudinal Lorentz factor
$\gamma_\perp$	$\equiv$	Transverse Lorentz factor
γ_r	$\equiv$	Resonant electron energy
γ_o	$\equiv$	Initial average electron energy
$\bar{\gamma}_r$	$\equiv$	Resonant electron energy in a waveguide (TE_{01} mode)
Γ	$\equiv$	Dimensionless electron energy (universal scaling)
δ	$\equiv$	Detuning parameter (universal scaling)
δ_1	$\equiv$	Effective detuning parameter (eq.(4.11))
Δ	$\equiv$	Argument of the small–signal gain (eq.(5.15'))
ϕ	$\equiv$	Radiation field phase
λ	$\equiv$	Radiation wavelength
λ_w	$\equiv$	Wiggler period
λ_s	$\equiv$	Slippage wavelength in a waveguide
$\lambda_{1,2,3}$	$\equiv$	Complex roots of the characteristic equation for the field (linear theory)
θ_j	$\equiv$	Phase of the jth–electron (eq.(3.13))
$\tilde{\theta}_j$	$\equiv$	Phase of the jth–electron (eq.(4.3a))
ρ	$\equiv$	Fundamental FEL parameter (universal scaling)
$\bar{\rho}$	$\equiv$	Fundamental FEL parameter in a waveguide (TE_{01} mode)
σ	$\equiv$	Space-Charge parameter (universal scaling)
$\bar{\sigma}$	$\equiv$	Transverse electron density
ω	$\equiv$	Radiation frequency
ω_s	$\equiv$	Spontaneous emission peak frequency
ω_p	$\equiv$	Plasma frequency
$\bar{\omega}_p$	$\equiv$	Plasma frequency in a waveguide (TE_{01} mode)
$\tilde{\omega}_p$	$\equiv$	Average plasma frequency along the bunch
ξ	$\equiv$	Argument of the Bessel functions
χ	$\equiv$	Macroscopic electron current profile

REFERENCES

1) J.M. Madey, *Jour. Appl. Phys.* **42** (1971) 1906
2a) L.R. Elias, W.M. Fairbank, J.M. Madey, H.A. Schwettman, T.I. Smith, *Phys.Rev.Lett.* **36** (1976) 717
2b) D.A.G. Deacon, L.R. Elias, J.M.J. Madey, G.J. Ramian, H.A. Schwettman, T.I. Smith, *Phys.Rev.Lett.* **38** (1977) 892
3) R. Bonifacio, I. Boscolo, F. Casagrande, G. Cerchioni, R. Corsini, L.De Salvo Souza, D. Fadini, M. Ferrario, C. Maroli, P. Pierini, N. Piovella, in print; and Contribution to these Proceedings (reduced version focussed on the theoretical aspects of the project)
4) S. Van Der Meer, Contribution to these Proceedings
5) P. Sprangle,C.M. Tang and J. Walsh (eds.), Free Electron Laser (North-Holland,Amsterdam,1988), and the previously published Proceedings of the FEL Conferences
6) Special Issue on the Free-Electron-Laser, *IEEE J. Quant.Electr.* **QE–23** (1987) and the previously published Special Issues
7) T.C. Marshall, *Free Electron Lasers* (Macmillan, New York,1985)
8) The review paper which is closest to ours is: C. Pellegrini and J.B. Murphy in: Proceedings of the Joint US-CERN Particle Acceleration School, eds. M. Month and S. Turner (Springer, Berlin, 1988)
9) R. Bonifacio, F. Casagrande, M. Ferrario, P. Pierini, N. Piovella, Contribution to these Proceedings
10) J.D. Jackson, *Classical Electrodynamics* (Wiley, New York,1975)
11) K.J. Kim, Contribution to these Proceedings
12) E.T. Scharlemann, Contribution to these Proceedings
13) T.J. Orzechowski, B.R. Anderson, W.M. Fawley, D. Prosnitz, E.T. Scharlemann, S.M. Yarema, D. Hopkins, A.C. Paul, A.M. Sessler, J.S. Wurtele, *Phys.Rev.Lett.* **54** (1985) 889
14) R. Bonifacio, C. Pellegrini, L. Narducci, *Opt. Comm.* **50** (1984) 373
15) T.J. Orzechowski, B.R. Anderson, J.C. Clark, W.M. Fawley, A.C. Paul, D. Prosnitz, E.T. Scharlemann, S.M. Yarema, D.B. Hopkins, A.M. Sessler, J.S. Wurtele, *Phys.Rev.Lett.* **57** (1986) 2172
16) D.B. McDermott, T.C. Marshall, S.P. Schlesinger, R.K. Parker and V.L. Granatstein, *Phys.Rev.Lett.* **41** (1978) 1368
R.K. Parker, R.H. Jackson, S.H. Gold, H.P. Freund, V.L. Granatstein, P.C. Efthimion, M. Herndon and A.K. Kinkead, *Phys.Rev.Lett.* **48** (1982) 238
S.H. Gold, D.L. Hardesty, A.K. Kinkead, L.R. Barret and V.L. Granatstein, *Phys.Rev.Lett.* **52** (1984) 1218
J.A. Pasour, R.F. Lucey and C.A. Kapetanakos, *Phys.Rev.Lett.* **53** (1984) 1728
17) R. Bonifacio, F. Casagrande, C. Pellegrini, *Opt. Comm.* **61** (1987) 55
18) W.B. Colson, *Phys.Lett.* A **64** (1977) 198
19) J.B. Murphy, C. Pellegrini, R. Bonifacio, *Opt. Comm.* **53** (1985) 197
20) N.M. Kroll, L.P. Morton, M.N. Rosenbluth, *IEEE J. Quant.Electr.* **QE– 17** (1981) 1436
21) D. Prosnitz, A. Szoke and V.K. McNeil, *Phys. Rev.* A **24** (1981) 1436
22) J.R. Pierce, *Traveling Wave Tubes* (Van Nostrand,Princeton,1950)
23) P. Sprangle, C.M. Tang and W.M. Manheimer, *Phys. Rev.* A **21** (1980) 302
24) J.A. Edighoffer, G.R. Neil, C.E. Hess, T.I. Smith, S.W. Fornaca and H.A. Schwettman, *Phys.Rev.Lett.* **52** (1984) 344

25) R. Bonifacio, F. Casagrande, L.De Salvo Souza, *Opt. Comm.* **58** (1986) 259
26) J.M.Madey, *Il Nuovo Cimento B* **50** (1979) 64
27) R. Bonifacio, F. Casagrande, L. De Salvo Souza, *Phys. Rev. A* **33** (1986) 2836
28) R. Bonifacio, C. Maroli, N. Piovella, *Opt. Comm.* **68** (1988) 369
    R. Bonifacio, C. Maroli, N. Piovella, Contribution to these Proceedings
29) R. Bonifacio, F. Casagrande, *Nucl.Instr.and Meth.A* **239** (1985) 36
30) R. Bonifacio, B.W.J. McNeil, *Nucl.Instr.and Meth.A* **272** (1988) 280
    R. Bonifacio, B.W.J. McNeil, P. Pierini, in print
31) W.M.Sharp,W.M.Fawley,S.S.Yu,A.M.Sessler,R.Bonifacio,L.De Salvo Souza, in print
32) J. Masud, T.C. Marshall, S.P. Schlesinger, F.G. Yee, W.M. Fawley, E.T. Scharlemann, S.S. Yu, A.M. Sessler and E.J. Sternbach, *Phys.Rev.Lett.* **58** (1987) 763
33) R.Bonifacio and L.De Salvo Souza, *Nucl.Instr.and Meth.A*, in print
34) T.J. Orzechowski, E.T. Scharlemann and D. Hopkins, *Phys. Rev. A* **35** (1987) 2184
35) W.B.Colson, *IEEE J. Quant.Electr.* QE- **17** (1981) 1417
36) F. T. Arecchi, R. Bonifacio, *IEEE J. Quant.Electr.* QE- **1** (1965) 169

SELECTED TOPICS IN FELS

E. T. SCHARLEMANN

Lawrence Livermore National Laboratory*, Livermore, CA USA

A tutorial introduction to wiggler tapering, electron beam dynamics in the wiggler — with particular emphasis on focusing — and optical guiding is presented.

1. INTRODUCTION

In this chapter I will discuss three different and generally unrelated topics of free-electron laser physics, wiggler tapering, electron beam dynamics in the wiggler, and optical guiding. The common theme is the physics that becomes important in long wigglers, which require tapering to prevent saturation, focusing to hold the electron beam together, and optical guiding to keep the signal near the electron beam. Most of what I have to say about these topics has already been published in one form or another, but I hope that these notes will provide a clear tutorial on the topics and make the original literature somewhat more transparent.

I feel obliged to apologize for using only dimensional quantities throughout this chapter, in contrast to many of the other chapters in this book. My background in FELs is to a large extent as a computer code builder who has worked on codes for designing FELs, accounting for the engineering realities of wiggler errors, electron beam steering, misalignments, etc. When effects such as these are considered, dimensionless units become a genuine hindrance. Notation in this chapter also differs somewhat from other chapters, and is summarized in Table 1. The geometry assumed in this chapter is illustrated in Figure 1; the electron beam propagates in the positive z direction, the wiggler magnetic field (generally pictured as vertical) points in the $\pm y$ direction, and the wiggle motion of each electron occurs in the $\pm x$ plane.

2. TAPERING THE WIGGLER

The exponential gain regime of a high-gain FEL amplifier is described elsewhere in this volume[1] and also in Section 4 of this chapter (on optical

---

* Work performed jointly under the auspices of the U.S. Department of Energy by Lawrence Livermore National Laboratory under contract W-7405-ENG-48, for the U. S. Army Strategic Defense Command and the Department of Defense under Defense Advanced Research Projects Agency Order No. 5316 in support of SDIO/BMD-ATC MIPR No. W31-RPD-53-A127.

TABLE 1. Notation (based on Reference 2, cgs units)

Wiggler field -

B_w = peak field

λ_w = wiggler period $\qquad k_w = 2\pi/\lambda_w$

$$a_w = \frac{eB_w}{\sqrt{2}mc^2 k_w} \text{ (linear wiggler) or } \frac{eB_w}{mc^2 k_w} \text{ (helical)}$$

Signal field -

E_s = peak electric field

λ_s = wavelength $\qquad k_s = k = 2\pi/\lambda_s$

$$e_s = \frac{eE_s}{\sqrt{2}mc^2} \text{ (linear polarization) or } \frac{eE_s}{mc^2} \text{ (circular)}$$

$a_s = e_s/k$

ϕ = electric field phase with respect to vacuum plane wave

$\mathcal{E}$ = slowly varying complex amplitude of the electric field:

$$\vec{E} = \hat{x} \frac{i}{2} \mathcal{E}(x,y,z,t) e^{i(kz-\omega t)} + \text{c.c.} \quad \text{(linear polarization)}$$

$$\hat{e}_s = \frac{e\mathcal{E}}{\sqrt{2}mc^2} = |\hat{e}_s| e^{i\phi}$$

Electrons -

γ = Lorentz factor

$\vec{\beta} = \vec{v}/c$ = normalized velocity

$\beta_\parallel$ = axial (z) component of normalized velocity

$\beta_\perp$ = transverse component, including both wiggle and betatron motion

$$\gamma_\parallel = (1 - \beta_\parallel^2)^{-1/2} \approx \gamma/(1 + \gamma^2 \beta_\perp^2)^{1/2}$$

FEL interaction -

$\psi = (k+k_w)z - \omega t + \phi$

$\vartheta = (k+k_w)z - \omega t = \psi - \phi$

$f_B = [J_0(\xi) - J_1(\xi)]$ with

$$\xi = \frac{a_w^2}{2(1+a_w^2)}$$

Particle equations of motion -

$$\frac{d\gamma}{dz} = - \frac{a_w f_B e_s}{\gamma} \sin \psi$$

$$\frac{d\psi}{dz} = k_w - \frac{\omega/c}{2\gamma^2}(1 + a_w^2 - 2a_w f_B a_s \cos \psi) + \frac{d\phi}{dz}$$

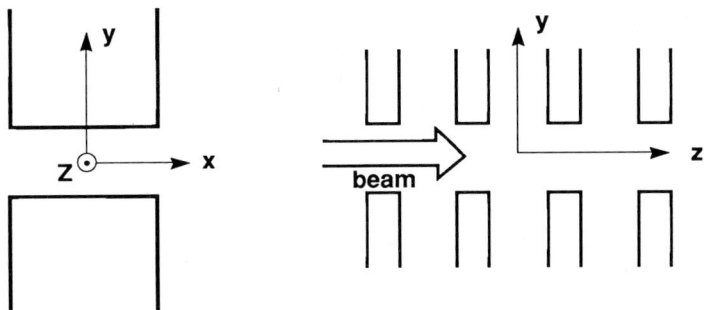

FIGURE 1
The coordinate system used throughout this chapter. The electron beam propagates in the positive z direction, the wiggler field is in the vertical (y) direction.

guiding). In exponential gain, the electron beam physically bunches on the scale of a signal wavelength, and gain occurs as long as the $<\sin \psi>$ (the angle brackets denote an average over the electron distribution) for the bunch remains positive. Exponential gain saturates when electrons lose enough energy that the electron bunches fall back — because they decelerate as they lose energy — to $<\sin \psi> < 0$. Saturation corresponds to an extraction efficiency, defined as the fractional electron beam power converted to signal power, of only several percent. Beyond saturation, the electron bunches can reabsorb the signal power, producing a *decrease* in signal power as a function of wiggler length. As long as the electron bunches stay together and nonuniformly distributed in the ponderomotive potential well formed by the signal and the wiggler fields, the bunches can alternately absorb and emit signal power and produce oscillations in the measured power. The reabsorption and subsequent oscillations are observed in most numerical simulations and have also been seen in the microwave FEL experiment, ELF, at LLNL.

Figure 2 illustrates, from a numerical simulation by FRED (one of several LLNL FEL simulation codes), the saturation of exponential gain and subsequent power oscillations in a 21 μm FEL driven by a 38 MeV, 1 kA electron beam. Table 2 lists the other parameters of the simulation. The figure shows exponential gain saturating at about 3.8 m of wiggler, with power oscillations (~2-m period) beyond saturation.

Figure 3 illustrates an electron bunch, pictured in longitudinal phase space (γ and ψ), just at saturation of exponential gain. The electron beam is assumed to be locally periodic in ψ (an assumption valid as long as slippage effects are not important) and so ψ only extends between $-\pi$ and $+\pi$. The

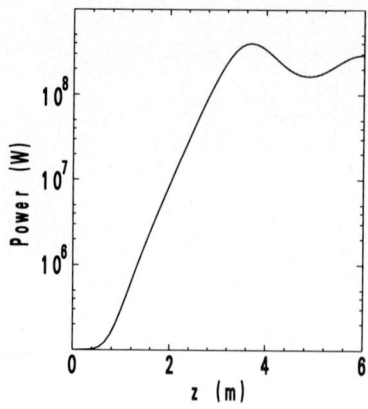

FIGURE 2
Power vs wiggler length for an untapered wiggler amplifier with the electron beam parameters of Table 2.

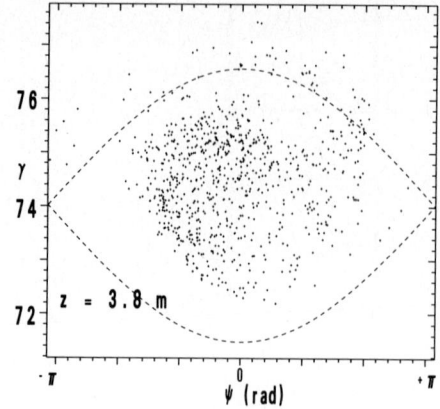

FIGURE 3
The longitudinal phase space of an electron bunch within a single ponderomotive potential well, at saturation of an untapered wiggler.

TABLE 2. Simulation parameters

38 MeV
1 kA
10^3 mm-mrad normalized edge emittance
0.26 cm beam radius
21 μm light
100 kW input power
5.5 cm wiggler period

average value of sin ψ for all electrons vanishes at this point, producing the power maximum at 3.8 m that is seen in Figure 2. The dotted line in Figure 3 outlines the "bucket", or ponderomotive potential well, within which electrons alternately lose and gain energy. Orbits of representative electrons within the bucket are shown in Figure 4; the orbits are only quasi-periodic because they are oscillating in a bucket that is itself oscillating in amplitude.

The limitation of extraction efficiency (hence power) to several percent can be circumvented by *tapering* the wiggler; that is, by decreasing the wiggler magnetic field B_w or wavelength λ_w with wiggler length. The effect of the tapering is to keep the electron bunches at positive <sin ψ> to permit continued extraction of power from the electrons.

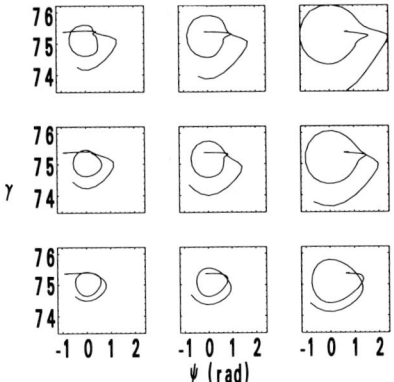

FIGURE 4
Orbits of nine representative electrons in the untapered wiggler amplifier of Figures 2 and 3. The electrons differ in initial pondermotive phase and in the properties of their transverse orbits.

The condition for FEL synchronism (for free-space optical propagation) is

$$\lambda_s = \frac{\lambda_w}{2\gamma_\parallel^2} = \frac{\lambda_w}{2\gamma^2}(1 + a_w^2) \ . \tag{1}$$

As electrons lose energy in an untapered wiggler, γ decreases and the synchronism condition is quickly violated. To keep the electron bunch near synchronism with $<\sin \psi> > 0$, one can decrease λ_w, decrease B_w, hence a_w, or any combination of the two that decreases the product $\lambda_w(1+a_w^2)$. Those electrons that continue to lose energy to the electromagnetic field are "trapped" in the bucket, which itself is now decelerating.

Tapering can be treated quantitatively by introducing the concept of a "resonant" electron at γ_R, ψ_R with γ_R decreasing monotonically in z. The equations for the γ,ψ position of the resonant electron follow from the single particle equations of motion of Table 1:

$$\frac{d\gamma_R}{dz} = - \frac{a_w(z) \ f_B \ e_s(z)}{\gamma_R} \sin \psi_R \ , \tag{2}$$

and

$$\frac{d\psi_R}{dz} = \text{something simple, such as 0.} \tag{3}$$

The standard references on tapering[2,3] use the concept of the resonant electron extensively. The physical idea is that a single electron is kept at a constant positive phase ψ_R in the ponderomotive potential well; as that electron decelerates, a_w is decreased (usually) to maintain the constant ψ_R.

Other trapped electrons undergo "synchrotron oscillations" around the resonant electron. The origin of the synchotron oscillations can be understood from Figure 3. Electrons with $\psi > \psi_R$ (= 0 for the untapered wiggler of Figure 3) and $\gamma > \gamma_R$ are losing energy faster than the resonant electron but they are moving faster, and so move downward and to the right in the γ-ψ plot of the figure. Electrons with $\psi > \psi_R$ but $\gamma < \gamma_R$ are still losing energy faster than the resonant electron, but are moving more slowly; hence they more downward and to the left. Once the electrons move to $\psi < \psi_R$, they lose energy more slowly than the resonant electron, and move upward in the figure.

Only trapped electrons are continuously decelerated in the ponderomotive potential well — untrapped electrons simply decouple from the electromagnetic field of the signal, because the untrapped electrons quickly move very far from synchronism. Electrons trapped in the bucket can become untrapped, if for example the bucket is decelerated too rapidly. One simple design tradeoff becomes immediately evident: rapid tapering (from large ψ_R) can decelerate trapped electrons quickly, but may also detrap electrons. The extraction efficiency is equal to the product of the fraction of electrons trapped and their fractional deceleration in energy. There is generally, therefore, an optimum value of ψ_R that maximizes the extraction efficiency.

Figure 5 illustrates the increase in power that can be obtained from a tapered wiggler amplifier. Saturation of exponential gain still occurs here at 3.8 m, but beyond 3.8 m the power continues to increase. The tapering is done (numerically) by fixing ψ_R = 0.35 radians and keeping λ_w constant, to yield a wiggler magnetic field profile that looks like Figure 6. The γ-z and γ-ψ orbits of the same representative electrons as in Figure 4 are pictured in Figures 7 and 8; the steady decrease of energy, with synchrotron oscillations superposed, is clearly seen.

Figure 9 shows the behavior of an electron bunch as it loses energy in a decelerating bucket. The bunch quickly smooths out within the bucket, but remains clumped around γ_R and ψ_R as γ_R decreases.

It is an important property of tapering that the <u>normalized</u> emittance (= $\gamma\beta$ × phase-space area/π) of the electron beam, both trapped and untrapped electrons, remains constant as γ_R decreases. This occurs because the transverse force on each electron due to the FEL coupling alone vanishes in an average over a wiggler period; hence, the transverse momentum is conserved while the transverse velocity increases — as γ decreases. The electron beam emerging from a tapered wiggler has two distinct electron populations with different energies but identical normalized emittances, and thus different unnormalized emittances.

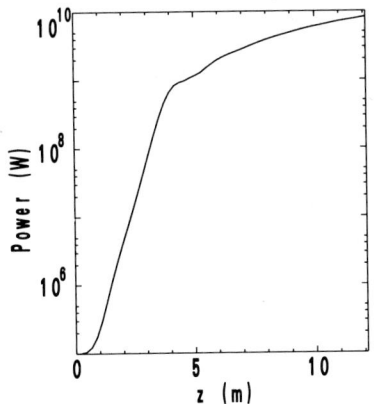

FIGURE 5
Power vs wiggler length in a tapered wiggler amplifier, again with the parameters of Table 2. Power continues to increase past the saturation observed in Figure 2.

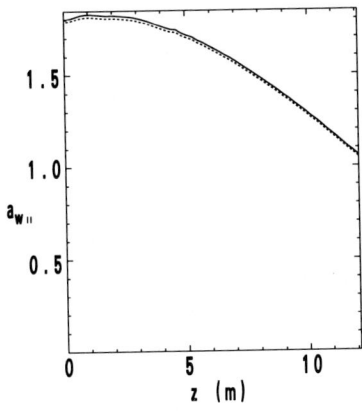

FIGURE 6
The magnetic field profile (plotted as a_w) that produced the power vs z plot of Figure 5.

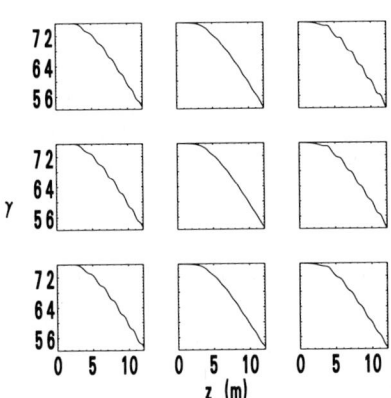

FIGURE 7
Electron energy (γ) vs z for the representative electrons of Figure 4 in a tapered wiggler amplifier.

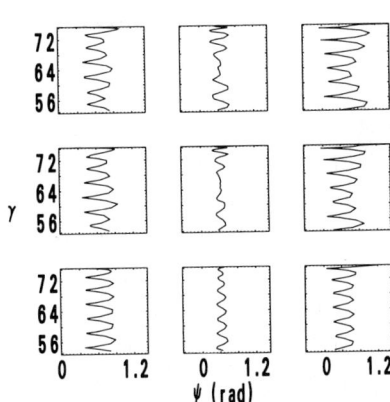

FIGURE 8
Longitudinal phase space (γ-ψ) orbits of the representative electrons of Figure 7.

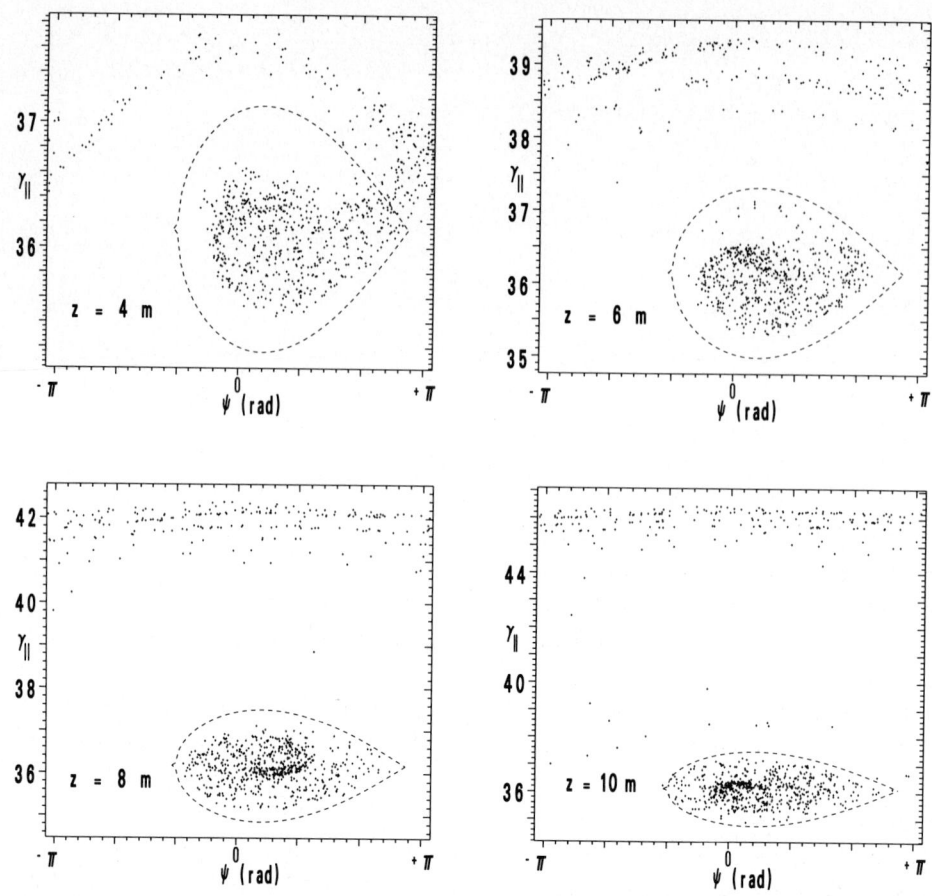

FIGURE 9
The bunched electrons at and beyond saturation of the exponential gain regime of a tapered wiggler amplifier. To include properly the effects of emittance (see Section 3 of this chapter) $\gamma_{\parallel}$ rather than γ is plotted vs ψ. The longitudinal Lorentz factor $\gamma_{\parallel}$ is reduced from γ by the wiggler field and the transverse velocities of the electrons, and remains constant for trapped electrons in a tapered wiggler amplifier with constant period. Steady deceleration of the electron bunch — the trapped electrons — appears here as a steady increase in $\gamma_{\parallel}$ of the untrapped electrons.

The bucket outline in some of the earlier figures is derived from an equation for the separatrix of the ponderomotive well obtained from the Hamiltonian for γ,ψ motion expanded to second order in $\delta\gamma = \gamma-\gamma_R$:

$$H(\delta\gamma,\psi) = \frac{k_w}{\gamma_R} \delta\gamma^2 - \frac{a_w f_B e_s}{\gamma_R} (\cos\psi + \psi \sin\psi_R). \tag{5}$$

The bucket is described by

$$\delta\gamma^2 = \frac{a_w f_B e_s}{k_w} [\cos\psi + \cos\psi_R - (\pi-\psi-\psi_R)\sin\psi_R]. \tag{6}$$

Unfortunately for easy analytical analysis, the parameters in $H(\delta\gamma,\psi)$ — for example, e_s and a_w — can change non-adiabatically, so that the notion of a precisely defined ponderomotive well is useful but not exact.

An example to illustrate this last remark is provided by the exponential gain regime. Saturation necessarily occurs at approximately half a bounce (synchrotron) period, and thus all the trapping done in exponential gain is by definition non-adiabatic. Nonetheless, simulations indicate that trapping in exponential gain, starting from an infinitesimal bucket and growing non-adiabatically, is often better than trapping in an initially imposed, large amplitude bucket.

The great efficiency enhancement permitted by wiggler tapering was dramatically demonstrated in the ELF experiment[4] at LLNL. ELF was a microwave FEL, operating at 35 GHz with an 850 A, 3.5 MeV electron beam provided by the Experimental Test Accelerator at LLNL. Table 3 lists the other important experimental parameters; the experimental conditions were known well enough to leave no free parameters for numerical fits. ELF was a high-gain, strong-pump FEL, with only small (~20%) space-charge corrections to the gain. The longitudinal energy spread of the electron beam in the wiggler was ~6.5% (full width), due almost entirely to beam emittance. Details of the experiment itself are described in Ref. 4, but the important result is shown in Figure 10, a plot of power vs wiggler length for an untapered and a tapered wiggler in ELF. With no tapering, the FEL saturated at about 1.4 m at a power level of 180 MW, or 6% extraction. The wiggler was an air-core electromagnet, with individual power supplies controlling the coils for each two wiggler periods. Tapering was accomplished by adjusting the power supplies sequentially along the wiggler length, past saturation, to maximize the power at each achievable wiggler length. The procedure yielded a power enhancement of more than five, leading to 1 GW of peak microwave power or 35% extraction.

One point perhaps obscured by the use of ψ for the longitudinal coordinate is that the electron bunches of Figures 3 and 9 are in fact physical bunches in the electron beam, radiating coherent synchrotron radiation. The bunches

TABLE 3. ELF parameters

850 A
3.5 MeV
700 mm-mrad emittance
9.8 cm wiggler period
30 periods
3.8 kG filed
60 G/cm horizontal focusing
34.6 Ghz
50 kW input power
3×10 cm waveguide
TE_{01} design mode

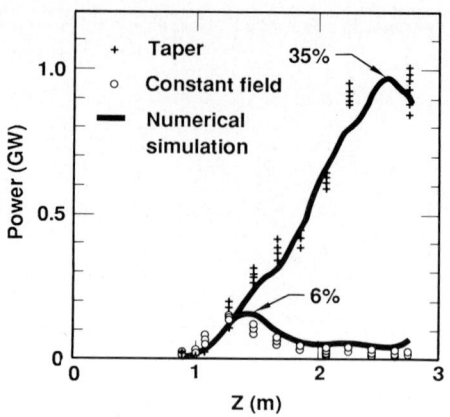

FIGURE 10
Power vs wiggler length as measured experimentally at ELF, with comparison with numerical simulations. Tapering enhanced the extraction efficiency from 6% to more than 35%.

are some fraction (~1/2) of a signal wavelength thick (axially), are separated by a signal wavelength, and are the electron beam diameter wide. For ELF, this means bunches about 8 mm thick and 1 cm in diameter. For short wavelength FELs, however, the relative scales can be very different. In PALADIN (the 10.6 μm FEL at the Advanced Test Accelerator of LLNL), the bunches are less than 10.6 μm thick but nearly a centimeter across — much thinner than any cookable pancake. These crepes are formed of electrons which individually are not moving in straight lines, but rather in transverse trajectories across the electron beam. If the FEL is to work well, with the bunches staying together and not dispersing, the individual trajectories must permit the crepes to stay together as the beam moves down the wiggler. That issue brings this chapter to its second part.

3. ELECTRON BEAM DYNAMICS IN THE WIGGLER

The potential beam transport problem is illustrated in Figure 11, in which an attempt is made to represent the electron bunches and the individual electron trajectories. The electron beam in the figure (and in any FEL with a long wiggler) is maintained at approximately constant radius by some form of focusing, while the individual electrons bounce back and forth inside the beam envelope. With no focusing at all, any electron beam with non-zero emittance would fly apart simply from the random transverse velocities. For long wigglers, focusing in both transverse planes is required to confine the beam.

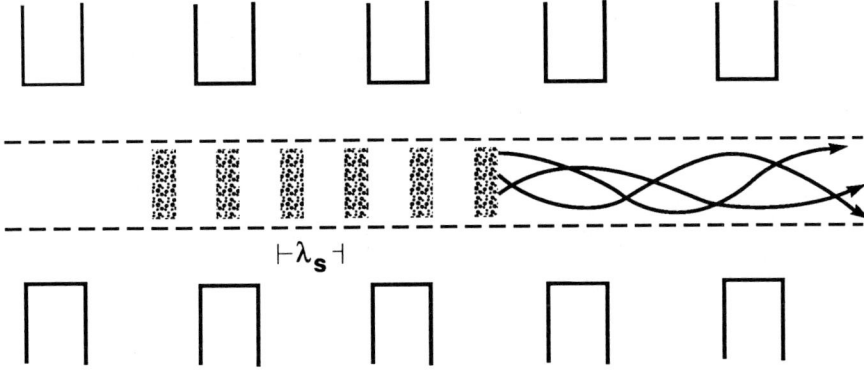

FIGURE 11
The electron bunches transported through a wiggler are roughly an optical wavelength in thickness but a beam diameter wide. Within the bunches, electrons move back and forth across the beam. Focusing in the wiggler must preserve the bunches as the electrons move back and forth.

Standard linear wigglers focus the beam in the vertical direction (the direction in which the wiggler field points) spontaneously. The focusing force is provided by the cross-product of the wiggle motion (in x) with the z-component of magnetic field that accompanies the periodic wiggler field. The most common form for representing the wiggler field,

$$\vec{B}_w = \hat{y} B_0 \sin k_w z \quad , \tag{7}$$

is only valid very near the $y = 0$ plane; the field of eq. (7) does not satisfy Maxwell's equations for a vacuum magnetic field [the curl of eq. (7) does not vanish]. An expression for a linear wiggler field that exactly solves Maxwell's equations is

$$\vec{B}_w = \hat{y} B_0 \sin k_w z \cosh k_w y + \hat{z} \cos k_w z \sinh k_w y \quad . \tag{8}$$

The periodic behavior of the y-component of eq. (8) requires a z component of the field, and the z component, as we shall now see, focuses the electron beam in y.

The wiggle motion from the field of eq. (8) is

$$\frac{\vec{v}_w}{c} = -\hat{x} \frac{eB_w}{\gamma mc^2 k_w} \cos k_w z \cosh k_w y \quad ; \tag{9}$$

this motion arises from the force $F_x = e(v_z/c)B_y \simeq eB_y$, where the latter approximation arises from the assumption that $v_z \simeq c$. Note that the wiggle motion increases away from the midplane (y = 0); that increase will become important later. The motion of eq. (9) interacts with B_z to focus the beam:

$$\frac{d^2y}{dz^2} = \frac{e}{\gamma mc^2} \frac{v_x}{c} B_z \qquad (10)$$

$$= -\left(\frac{eB_0}{\gamma mc^2}\right)^2 \frac{1}{k_w} \cos^2 k_w z \; \sinh k_w y \; \cosh k_w y \; .$$

An average over a wiggler period converts $\cos^2 k_w z$ to a factor of 1/2, and an expansion of the sinh and cosh to first order in $k_w y$ yields

$$\frac{d^2y}{dz^2} = -k_{\beta y}^2 y \; , \qquad (11)$$

with

$$k_{\beta y}^2 = \frac{1}{2}\left(\frac{eB_0}{\gamma mc^2}\right)^2 = \frac{a_{w0}^2 k_w^2}{\gamma^2} \; . \qquad (12)$$

Eq. (11) can be recognized as the equation of motion for a particle in a harmonic potential well, centered at y = 0. The solution is straightforward:

$$y(z) = y_\beta \cos(k_{\beta y} z + \phi_y) \; , \qquad (13)$$

where ϕ_y is an arbitrary phase determined by initial conditions. The motion described by eq. (13) is referred to as "betatron motion", and arises from the focusing force described by eq. (11). A standard linear wiggler focuses only in y, not in x, leaving the problem of horizontal plane focusing unsolved.

The technique used above for finding the focusing in y implicitly assumes that the focusing force is much weaker than the force that drives the wiggle motion, and occurs on a much different spatial scale. The procedure is clearly a perturbation expansion in a_w/γ, with the wiggle motion [eq. (9)] the first order term, and the focusing [eq. (11)] the second order term.

Horizontal focusing can be, and often is, provided by a quadrupolar magnetic field — either from external quadrupoles as in ELF or by canting the magnet pole faces as in the wiggler built by Spectra Technology, Inc.[5] for the Boeing FEL experiment. The quadrupole magnetic field can be written

$$\vec{B}_Q = -Q_0 (\hat{x}y + \hat{y}x) \; . \qquad (14)$$

Using the approximation again that $v_z \simeq c$, the x and y forces on an electron come from

$$\frac{d^2x}{dz^2} \simeq \frac{eB_y}{\gamma mc^2} \; , \qquad (15)$$

$$\frac{d^2y}{dz^2} \simeq -\frac{eB_x}{\gamma mc^2} \; . \qquad (16)$$

Because B_x and B_y depend linearly on y and x, respectively, the focusing in x and y is again harmonic:

$$\frac{d^2x}{dz^2} = -\frac{eQ_0}{\gamma mc^2} x = -k_Q^2 x , \qquad (17)$$

$$\frac{d^2y}{dz^2} = \frac{eQ_0}{\gamma mc^2} y = k_Q^2 y , \qquad (18)$$

except that now for the quadupole field, the force in y is *de*focusing. The net effect of adding a quadrupole field that focuses an electron beam in the horizontal plane will be a reduction of the natural wiggler focusing in y. This reduction is not usually a problem, but must be remembered. With both wiggler and quadrupole focusing, the equations for betatron motion become

$$\frac{d^2x}{dz^2} = -k_Q^2 x , \qquad (19)$$

$$\frac{d^2y}{dz^2} = -(k_{\beta y}^2 - k_Q^2) y = -\hat{k}_{\beta y}^2 y . \qquad (20)$$

For equal focusing in both planes, to produce a round electron beam for equal x and y emittances, k_Q^2 must be $k_{\beta y}^2/2$ and the focusing force in y is reduced by a factor of 2. Note that the sum of the k_β^2 in the two planes is independent of k_Q^2.

A helical wiggler focuses in both x and y — no other focusing is required. The simplest helical wiggler field that satisfies Maxwell's equations can be obtained from the gradient of a scalar potential V, with $\vec{B} \equiv -\nabla V$, and

$$V = -\frac{2B_0}{k_w} I_1(k_w r) \sin(k_w z - \Phi), \qquad (21)$$

where I_1 is a modified Bessel function and Φ is the azimuthal angle. Near the axis,

$$B_x \simeq B_0 \sin k_w z , \qquad (22)$$
$$B_y \simeq B_0 \cos k_w z , \qquad (23)$$

and

$$B_z \simeq B_0 [k_w x \cos k_w z + k_w y \sin k_w z] . \qquad (24)$$

Focusing arises in both x and y from the same mechanism as the natural vertical focusing of a linear wiggler: $(v_x/c) \times B_z$ focuses in y, with v_x the x-directed wiggle motion, and $(v_y/c) \times B_z$ focuses in x. The difference from the linear wiggler occurs because in the helical wiggler, wiggle motion occurs in both x and y and so focusing can occur in both x and y. For the helical wiggler

$$\frac{d^2x}{dz^2} = -k_{\beta h}^2 x , \qquad (25)$$

$$\frac{d^2y}{dz^2} = -k_{\beta h}^2 y , \qquad (26)$$

with

$$k_{\beta h}^2 = \frac{a_{w0}^2 k_w^2}{2\gamma^2} . \qquad (27)$$

The focusing force, although in both planes, is weaker by a factor of two in either plane than for the linear wiggler. The sum of the focusing k_β^2 is the same as for a linear wiggler, almost as if equal two-plane focusing were provided by quadrupoles [cf. eqs. (19) and (20) and subsequent discussion].

Focusing in either plane establishes a unique equilibrium beam radius in that plane; the equilibrium radius is the radius of the beam envelope that remains constant as the beam propagates in z. Figure 12 illustrates how that equilibrium radius can be calculated from the emittance of the electron beam and the focusing. An electron whose turning point in its betatron trajectory is at the beam envelope will also have the largest transverse velocity of any electron in the beam, 1/4 betatron period later. The edge emittance of the beam is defined by

$$\epsilon_y(\text{edge}) = y_{max} y'_{max} \qquad (28)$$

and follows immediately from the consideration of the electron with turning point at the beam edge:

$$y_{max} = r_{beam} , \qquad (29)$$

$$y'_{max} = k_\beta r_{beam} , \qquad (30)$$

where the last expression follows from eq. (13) and k_β depends on the type of wiggler focusing employed. Combining eqs. (28) through (30), the equilibrium beam radius must be

$$r_{beam} = \left(\frac{\epsilon}{k_\beta}\right)^{1/2} \qquad (31)$$

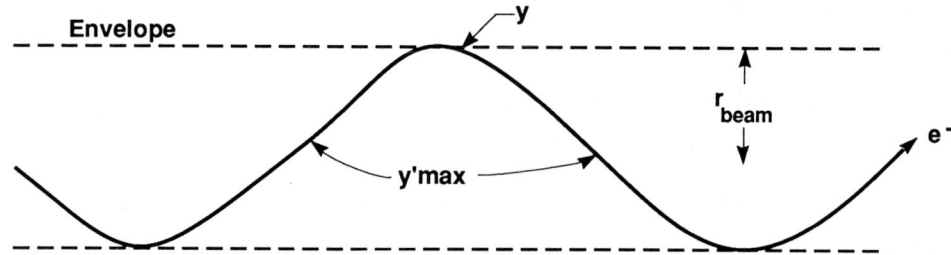

FIGURE 12
An nonzero radius electron beam in a focusing channel contains electrons moving at a range of angles y'. An electron that has a turning point in its transverse orbit at the beam edge also has maximum angle as it crosses the axis.

for either x or y. In practice, the emittance and the k_β can be different in x and y, and the equilibrium cross-section of the beam need not be round.

Both the betatron motion and the off-midplane (or off-axis) increase in the wiggler field — mentioned earlier, just below eq. (9) — affect FEL resonance. The important quantity for characterizing the shift from resonance is $\gamma\beta_\perp$, as we can see in the following derivation. The FEL synchronism condition can be written

$$\frac{\omega}{c} = (k + k_w)\beta_\parallel \tag{32}$$

obtained from setting $d\psi/dz = 0$ (see Table I). From the definition of γ,

$$\beta_\parallel^2 = 1 - \frac{1}{\gamma^2} - \beta_\perp^2 , \tag{33}$$

or, approximately,

$$\beta_\parallel \simeq 1 - \frac{1}{2\gamma^2}(1 + \gamma^2\beta_\perp^2) \tag{34}$$

for $\gamma \gg 1$. Here, $\gamma^2\beta_\perp^2$ includes both wiggle motion and betatron motion. The off-midplane increase in wiggle motion does not change γ (γ only changes through the FEL interaction on a scale much longer than we are concerned with here) so that changes in $\beta_\parallel$, hence synchronism in eq. (32), are controlled by changes in $\gamma^2\beta_\perp^2$.

The changes in $\gamma^2\beta_\perp^2$ are of two types: from the increase in wiggle motion away from the wiggler midplane or axis, and from the variation in transverse betatron velocity through the betatron orbit of an electron. For natural wiggler focusing (e.g., in a standard linear wiggler with no quadrupole focusing) these two components of $\gamma^2\beta_\perp^2$ combine to remain constant (averaged over a wiggler period) over the betatron orbit of any electron. From eq. (9),

$$\gamma^2\beta_\perp^2(\text{wiggle}) = 2 a_{w0}^2 \cos^2 k_w z \cosh^2 k_w y \tag{35}$$
$$\simeq a_{w0}^2 (1 + k_w^2 y^2) \text{ (averaged and expanded)},$$

with a_{w0} the value of a_w on the midplane. Eq. (13) can be used to rewrite this expression further:

$$\overline{\gamma^2\beta_\perp^2}(\text{wiggle}) \simeq a_{w0}^2[1 + k_w^2 y_\beta^2 \cos^2(k_{\beta y} z + \phi_y)] . \tag{36}$$

Also from eq. (13) it is clear that

$$\overline{\gamma^2\beta_\perp^2}(\text{betatron}) = \gamma^2 k_{\beta y}^2 y_\beta^2 \sin^2(k_{\beta y} z + \phi_y) . \tag{37}$$

Eq. (12) relates $k_{\beta y}^2$ to a_{w0} and γ, to permit writing eq. (37) as

$$\overline{\gamma^2\beta_\perp^2}(\text{betatron}) = a_{w0}^2 k_w^2 y_\beta^2 \sin^2(k_{\beta y} z + \phi_y) . \tag{38}$$

The sum of eq. (36) and (38) is the total $\gamma^2\beta_\perp^2$ (the wiggle motion and the betatron motion occur on widely separate scales [cm vs m], so that the cross term between wiggle and betatron motion averages to zero):

$$\overline{\gamma^2\beta_\perp^2}(\text{total}) = a_{w0}^2(1 + k_w^2 y_\beta^2) , \qquad (39)$$

which is a constant for each electron over its betatron trajectory — there is no longer a dependence of $\gamma^2\beta_\perp^2$ on z. As a consequence, the synchronism condition does not change for an electron over its betatron orbit in natural wiggler focusing. Figure 13 illustrates the origin of this constant of the (averaged) motion. The wiggle motion is largest, and the transverse betatron motion is least, at the turning point of each electron in its betatron motion. The wiggle motion is smallest and the betatron motion is largest when each electron crosses the midplane. The increase (or decrease) of the wiggle motion away from (toward) the midplane is always compensated by the decrease (increase) of the betatron motion.

If horizontal focusing is provided with quadrupoles, the constant of the motion vanishes. Eqs. (19) and (20) describe the betatron motion of an electron with quadrupole focusing, and yield

$$y = y_\beta \cos(\hat{k}_{\beta y} z + \phi_y) , \qquad (40)$$
$$x = x_\beta \sin(k_Q z + \phi_x) . \qquad (41)$$

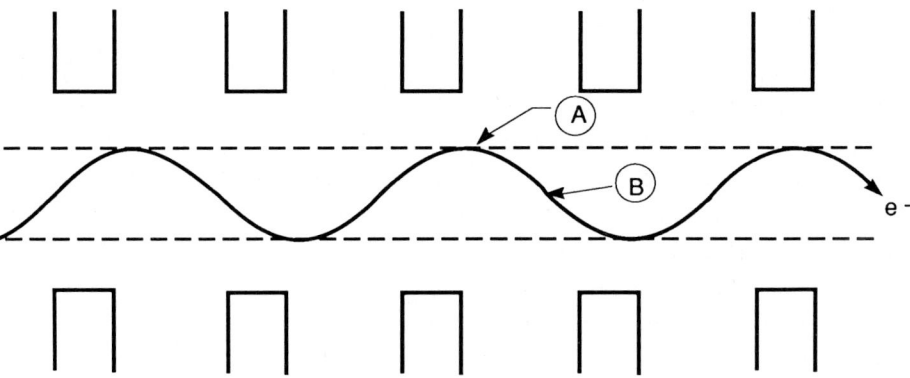

FIGURE 13
At point A in an electron's betatron orbit, the wiggle motion is largest (because of the off-axis increase of the wiggler field) but the transverse betatron motion vanishes. At point B, the wiggle motion is least, but the transverse betatron motion is largest. The two components of transverse motion add to a constant (averaged over a wiggler period) in natural wiggler focusing.

The phases ϕ_x and ϕ_y are random among the electrons, depending on the initial conditions of injection of each electron into the wiggler. Following through the argument that led to the constant of the motion, eq. (39), for natural wiggler focusing, one finds

$$\begin{aligned}\overline{\gamma^2\beta^2_\perp}(\text{total}) &= a_{w0}^2[1 + k_w^2 y_\beta^2 \cos^2(\hat{k}_{\beta y}z + \phi_y)] \\ &+ (a_{w0}^2 k_w^2 - \gamma^2 k_Q^2)y_\beta^2 \sin^2(\hat{k}_{\beta y}z + \phi_y) \\ &+ \gamma^2 k_Q^2 x_\beta^2 \sin^2(k_Q z + \phi_x) \\ &= a_{w0}^2(1+k_w^2 y_\beta^2) + \gamma^2 k_Q^2[x_\beta^2\sin^2(k_Q z+\phi_x) - y_\beta^2\sin^2(\hat{k}_{\beta y}z+\phi_y)]\end{aligned} \qquad (42)$$

Because of the random nature of x_β, y_β, ϕ_x, and ϕ_y, this final expression for $\gamma^2\beta^2_\perp$ is (except for a vanishingly small subset of electrons) not constant over a betatron orbit.

If $\gamma^2\beta^2_\perp$ varies over a betatron orbit, so does $\bar{\beta}_\parallel$; the varying $\bar{\beta}_\parallel$ couples betatron and synchrotron motion, modulating ψ with a randomly phased, periodic driving term at the betatron period. The driving can excite synchrotron motion and detrap electrons from ponderomotive potentials, or it can disperse bunching as it is forming in the exponential gain regime.[6] The importance of the detrapping or dispersal depends on the amplitude of the driving term, hence [from eq. (42)] on x_β and y_β. From an estimate of the variation in ψ of a single electron over its betatron orbit, with the condition that the variation remain much less than a radian, we can conclude that quadrupole focusing will be detrimental to the FEL interaction unless

$$k\epsilon \ll 1 \ . \qquad (43)$$

This condition looks very similar to the condition derived elsewhere in this volume that emittance not degrade FEL performance. FELs can operate in the regime in which inequality (43) is not satisfied; for example, infrared and optical FELs driven by induction linear accelerators, or ultraviolet FELs driven by rf linacs. For such FELs, the use of quadrupoles to focus the beam in the horizontal plane further degrades FEL performance. This particular form of degradation can be circumvented in two ways.

First, a helical wiggler provides natural wiggler focusing (as described above) in both x and y; using the off-axis approximations for the wiggler field of Ref. 7, one can show that helical wiggler focusing maintains constant $\bar{\beta}_\parallel$. Helical wigglers are unfortunately harder to build and harder to taper than linear wigglers, and perhaps worst of all, eliminate side access to the vacuum beampipe inside the wiggler. They also produce circularly polarized light, which at high power can be harder to handle than linearly polarized light.

An alternative that provides the focusing and resonance properties of the helical wiggler while retaining the advantages of a linear wiggler is described in Ref. 8. The problem produced by quadrupole focusing in a linear wiggler stems from the absence of an increase in wiggle motion with x to compensate for the decrease in betatron motion away from $x = 0$. An increase in wiggle motion with x can be provided by curving the magnet pole faces, as shown in Figure 14. The curvature also produces its own focusing, and the detailed analysis of Ref. 8 indicates that the focusing maintains a constant $\bar{\beta}_{\parallel}$.

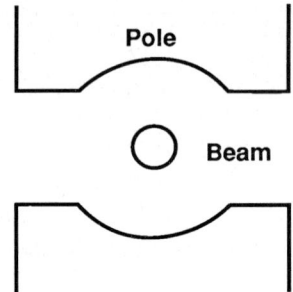

FIGURE 14
Magnet pole faces curved across the electron beam focus the beam in both transverse directions, with focusing properties identical to those of a helical wiggler.

An idealized form of the magnetic field produced by curved pole faces is

$$\vec{B}_w = \hat{y} B_w \cosh k_x x \cosh k_y y \sin k_w z$$
$$+ \hat{x} \frac{k_x}{k_y} B_w \sinh k_x x \sinh k_y y \sin k_w z \qquad (44)$$
$$+ \hat{z} \frac{k_w}{k_y} B_w \cosh k_x x \sinh k_y y \cos k_w z \ ,$$

with Maxwell's equations requiring that

$$k_x^2 + k_y^2 = k_w^2 \ . \qquad (45)$$

Because of the $\cosh k_x x \cosh k_y y$ dependence, B_y increases with both x and y away from $x = y = 0$. At the electron beam (near the axis, $x = y = 0$) the dominant field component is B_y, so the wiggle motion is predominantly in x. Vertical focusing in this magnetic field arises from the same effect as for a conventional linear wiggler: $(v_w/c) \times B_z$ produces an average force toward the $y = 0$ plane. Horizontal focusing arises from a totally different effect; because B_y is larger away from the $x = 0$ plane, each electron experiences a slightly larger force $(v_z/c) \times B_y$ at the outer edge of its wiggle trajectory than at the inner edge. The average effect is a force toward $x = 0$. Figure 15 clarifies the origin of the horizontal focusing force.

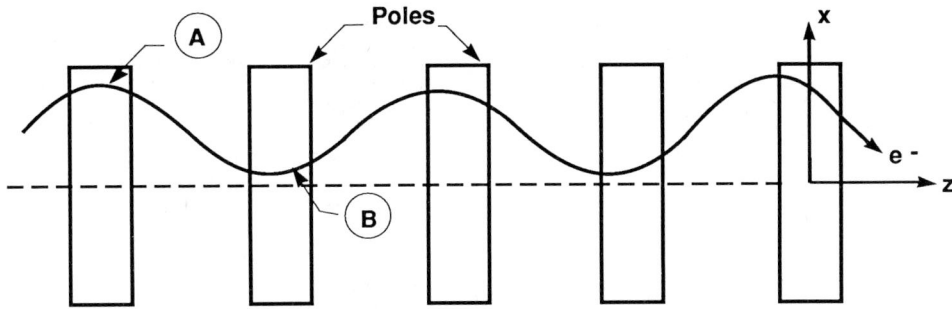

FIGURE 15
At point A in an electron's wiggle motion, the wiggler field is largest because of the off-axis increase of the field; the force toward $x = 0$ is therefore greatest. At point B, the force away from $x = 0$ is less, because the wiggler field is smaller. Averaged over a wiggler period, the net force on the electron is a focusing force toward $x = 0$.

Derivation of the horizontal and vertical focusing proceeds in a form very similar to the derivations above. The field of eq. (44) is expanded in power series in $(k_x x)$ and $(k_y y)$, retaining only the quadratic terms:

$$\vec{B}_w \simeq \hat{y} B_w (1 + k_x^2 x^2 + k_y^2 y^2) \sin k_w z$$
$$+ \hat{x} \frac{k_x}{k_y} B_w (k_x x)(k_y y) \sin k_w z \qquad (46)$$
$$+ \hat{z} \frac{k_w}{k_y} B_w (1 + k_x^2 x^2) k_y y \cos k_w z \; .$$

The resulting expressions for the x and y focusing forces are identical to the expressions for a helical wiggler, eqs. (25) through (27). More important, the evaluation of $\bar{\beta}_\parallel$ yields a constant over the betatron orbit, irrespective of k_x and k_y [although they must of course satisfy eq. (45)].

The field of eq. (44) is the exact solution of Maxwell's equations that provides the desired two-plane focusing in the wiggler. In practice, it is the coefficients in eq. (46) — an approximation to eq. (44) — that determine the focusing properties. For experimental confirmation that the appropriate pole face curvature has been achieved, it should be sufficient to measure the coefficients of the quadratic term in B_y to get k_x and k_y.

4. OPTICAL GUIDING

"Optical guiding" refers to a modification of the propagation of light in an FEL by the complex index of refraction of the electron beam. To understand the effect, it is helpful first to examine vacuum propagation of a coherent optical beam and then to look at the changes produced by the electron beam.

The calculation of optical propagation in an FEL (and in many other circumstances) is simplified by use of the "paraxial" or "slowly-varying envelope" approximation. The exact wave equation for the electric field $\vec{E}$,

$$\nabla^2 \vec{E} - \frac{1}{c^2} \frac{\partial^2 \vec{E}}{\partial t^2} = \frac{4\pi}{c^2} \frac{\partial \vec{J}}{\partial t} , \qquad (47)$$

simplifies by writing

$$\vec{E} = \frac{1}{2} \left[i \hat{p} \, \mathcal{E}(x,y,z,t) \, e^{i(kz - \omega t)} + c.c. \right] \qquad (48)$$

if the complex amplitude $\mathcal{E}$ is assumed to be a slowly varying function of z and t. The unit polarization vector $\hat{p}$ might be $\hat{x}$ for linear polarization or $(\hat{x} \pm i\hat{y})/\sqrt{2}$ for circular. The factor i inside the brackets of eq. (48) is an arbitrary phase factor introduced for consistency with other equations in this chapter. If $\mathcal{E}$ is slowly varying in z and t, then

$$\left| \frac{\partial^2 \mathcal{E}}{\partial z^2} \right| \ll k \left| \frac{\partial \mathcal{E}}{\partial z} \right| \qquad (49)$$

and

$$\left| \frac{\partial^2 \mathcal{E}}{\partial t^2} \right| \ll \omega \left| \frac{\partial \mathcal{E}}{\partial t} \right| . \qquad (50)$$

The small second derivatives of the complex amplitude $\mathcal{E}$ can be dropped, simplifying the left-hand side of the wave equation [eq. (47)] to

$$\nabla^2 \vec{E} - \frac{1}{c^2} \frac{\partial^2 \vec{E}}{\partial t^2} \approx \frac{1}{2} \{ i \hat{p} \, e^{i(kz-\omega t)} \times$$

$$[2ik \frac{\partial \mathcal{E}}{\partial z} + 2i \frac{\omega}{c^2} \frac{\partial \mathcal{E}}{\partial t} + \nabla_\perp^2 \mathcal{E} + (\frac{\omega^2}{c^2} - k^2)\mathcal{E}] + c.c. \}. \qquad (51)$$

The source term on the right-hand side of eq. (47) has been derived elsewhere in this volume in one-dimensional FEL theory. The new term introduced here is the $\nabla_\perp^2 \mathcal{E}$ term of eq. (51); that term is responsible for the most important 2D and 3D optical effect: diffraction. With some manipulation, the full 3D paraxial wave equation for an FEL becomes

$$\frac{\partial \hat{e}_s}{\partial z} + \frac{\omega}{c^2 k} \frac{\partial \hat{e}_s}{\partial t} - \frac{i}{2k} \left[\nabla_\perp^2 \hat{e}_s + (\frac{\omega^2}{c^2} - k^2) \hat{e}_s \right] \qquad (52)$$

$$= \frac{2\pi i e}{mc^3} (\frac{\omega}{ck}) a_w f_B \frac{I}{N} \sum_{j=1}^{N} \delta(x-x_j)\delta(y-y_j) \frac{e^{-i\vartheta_j}}{\gamma_j}$$

where the sum over the N electrons can be replaced, in a continuous beam limit, with a current density term:

$$\frac{I}{N} \sum_{j=1}^{N} \delta(x-x_j)\delta(y-y_j) \frac{e^{-i\vartheta_j}}{\gamma_j} \to J(x,y) \left\langle \frac{e^{-i\vartheta}}{\gamma} \right\rangle ; \tag{53}$$

the angle brackets refer to a local average over the electron distribution.

For time-independent, free-space ($\omega = ck$) propagation, the left-hand side of eq. (52) becomes

$$\frac{\partial \hat{e}_s}{\partial z} - \frac{i}{2k} \nabla_\perp^2 \hat{e}_s = (\text{r.h.s}) , \tag{54}$$

which is a complex diffusion equation for $\hat{e}_s$ with a source term provided by the particles [the right-hand side of eq. (52)]. The $\nabla_\perp^2$ term leads to a diffusion-like spreading of the coherent light as it propagates; the spreading is of course the well-known phenomenon of diffraction.

Steady-state vacuum propagation can be described by an arbitrary superposition of the orthonormal modes of eq. (54), with the right-hand side set to zero. The Gauss-Laguerre modes are one set:

$$\mathcal{E}_{nm}(r,\phi,z) = \frac{w_0}{w} L_n^m \left(\frac{2r^2}{w^2}\right) \left(\frac{\sqrt{2}r}{w}\right)^m e^{im\phi} \left[\frac{n!}{(n+m)!}\right]^{\frac{1}{2}} \times \tag{55}$$

$$e^{-r^2/w^2 + ikr^2/2R - i(2n+m+1)\tan^{-1}(z/z_R)} ,$$

where $L_n^m(z)$ is an associated Laguerre polynomial. The lowest mode, $\mathcal{E}_{00}$, is a Gaussian beam with spot size w (the radius to the $1/e^2$ intensity level) and a spherical phase front with a radius of curvature R. The spot size and radius of curvature evolve with z (in free-space propagation) as

$$w = w_0 \left(1 + \frac{z^2}{z_R^2}\right)^{\frac{1}{2}} \simeq w_0 \frac{z}{z_R} \quad (\text{at large } z) , \tag{56}$$

$$R = z + \frac{z_R^2}{z} \simeq z \quad (\text{at large } z) . \tag{57}$$

The Rayleigh range z_R is a characteristic diffraction length defined by

$$z_R = \frac{\pi w_0^2}{\lambda_s} , \tag{58}$$

w_0 is the spot size of the lowest order mode ($n = m = 0$) at a focus ($z = 0$), and z is the distance measured from the focus. The Rayleigh range z_R is the distance away from a focus at which the spot area of the lowest mode has increased by a factor of two. At a focus, $R \to \infty$, indicating that the phase fronts are flat. As $z \to \infty$, the phase fronts acquire a radius of curvature $R \simeq z$, indicating that the light appears to come from a point at the focus. At any z, the pair of coefficients (w,R) specifies a complete, orthonormal set of basis functions referred to as the TEM_{nm} modes.

Propagation of electromagnetic radiation in a waveguide can also be treated with eq. (52), but not as precisely. In a waveguide, ω is not equal to ck and the last term in parentheses of the left-hand side of eq. (52) does not vanish; however, for a single waveguide mode, with ω and k chosen for that mode, the entire term in brackets on the left-hand side of eq. (52) <u>does</u> vanish:

$$\nabla_\perp^2 \hat{e}_s + \left(\frac{\omega^2}{c^2} - k^2\right)\hat{e}_s = 0 \ . \tag{59}$$

The $\nabla_\perp^2$ term must nonetheless be retained if higher order waveguide modes are to be followed. Eq. (52) treats waveguide propagation only approximately because it neglects

- wall currents that generate E_y and E_z, the field components that do not directly couple to the electrons' wiggle motion,

- modifications to the $\vec{v}\cdot\vec{E}$ coupling from the E_z of the transverse magnetic (TM) modes, and perhaps most important,

- the dispersion of the electromagnetic signal in a waveguide that arises from those second-derivative terms in the wave equation that have been dropped in the paraxial wave equation. Nonetheless, considerable success has been achieved in modeling waveguide FEL experiments using the paraxial wave equation.[9]

This brief introduction to paraxial optics brings us, at last, to the subject of optical guiding. Optical guiding, as the term was originally defined,[10] actually refers to two physically distinct but conceptually similar phenomena, gain guiding and refractive guiding. Both variants of optical guiding can lead to a confinement of the coherent optical beam near the electron beam in an FEL. For gain guiding, the confinement is illusory, and arises from the transversely localized gain (in the electron beam) that balances diffraction; for refractive guiding, the confinement is real, and is precisely the confinement of light that occurs near and within an optical fiber.

Both forms of optical guiding counteract diffraction, and make possible long wigglers ($L_w \gg z_R$). Without optical guiding, wigglers would be constrained to be only a few Rayleigh ranges long, with the Rayleigh range determined by the electron beam size (it is the Rayleigh range of the light radiated by the electron beam — to provide gain — that is relevant). Because good quality electron beams have a very small cross section, the limitation to only a few Rayleigh ranges would mean a short physical length, hence a low gain, low extraction efficiency FEL.

The origin of gain guiding is very simple; light tends to remain where it is amplified until diffraction can carry it away. In a gain-guiding medium, the intrinsic gain length is shorter than or comparable to the Rayleigh length of the equilibrium gain-guided profile. Any transversely localized gain medium should be able to exhibit gain guiding as long as the gain is greater than diffraction.

Refractive guiding, on the other hand, is nearly unique to an FEL. It arises from the phase shift ($v_{phase} < c$) of light in an FEL. The phase shift and refractive guiding can be independent of the gain, and may occur even in the absence of gain (e.g., after saturation in an untapered wiggler). This phenomenon is somewhat more surprising than gain guiding, because in a conventional laser medium governed by the Kramers-Kronig relation, the real part of the index of refraction (hence the linear phase shift) vanishes at peak gain; at the design point of the laser medium, therefore, refractive guiding necessarily vanishes. The relevance of the Kramers-Kronig relations to refractive guiding in an FEL will be discussed below.

Refractive guiding does not arise from the usual index of refraction of a plasma; that index has the wrong sign for guiding — $Re(n_{plasma}) < 1$. Instead, guiding arises from the optical phase shift produced by the bunching of the electron beam in an FEL, as described below.

The significance of the optical phase shift (or the real part of an effective index of refraction) for guiding light was realized nearly simultaneously by Slater and Lowenthal,[11] Kroll, Morton, and Rosenbluth,[2] Prosnitz, Szöke and Neil,[3] and Sprangle and Tang.[12] The significant feature of refractive guiding — that wigglers much longer than a few Rayleigh ranges become possible — was first unambiguously pointed out in Ref. 2, in which reference is made to "optical beam trapping, which could be helpful for devices with $L > kr_e^2$." Their r_e is the electron beam radius, so that kr_e^2 is twice the Rayleigh range obtained by equating the optical mode size to the electron beam size, a reasonable approximation for a high gain amplifier. Very little analysis of refractive or gain guiding was done until Moore[13] realized that the exponential gain regime of an FEL, in the presence of arbitrarily strong diffraction, can be treated nearly exactly — with the main approximation that the light profile propagates self-similarly along with the electron beam. Similar conclusions were independently obtained from a useful analogy to optical fibers by others.[10] Since the appearance of these two papers there has been considerable theoretical analysis of optical guiding, primarily gain guiding, inspired by the possibility of actually building devices with long wigglers.

Analytical work in understanding optical guiding is still limited nearly completely to the exponential gain regime. The analysis can proceed through an analogy with a weakly guiding optical fiber, for which the propagation equation in the paraxial approximation can be written

$$\frac{\partial \hat{e}_s}{\partial z} - \frac{i}{2k} \nabla_\perp^2 \hat{e}_s - \frac{i}{2} k(n^2 - 1) \hat{e}_s \simeq ik \, \delta n \, \hat{e}_s , \qquad (60)$$

where n is the index of refraction (possibly complex), $\delta n = n - 1$ and $\delta n \ll 1$ has been assumed. The important feature of eq. (60) is that the source term on the right-hand side is linear in $\hat{e}_s$. The field equation for optical propagation in the exponential gain regime can be written to look like eq. (60) if the right-hand side of eq. (52) or (54) is linearized along with the single particle equations of motion — the linearization is of course an essential part of exponential gain theory even without optical guiding.

Linearization involves writing the equations for γ_j and ϑ_j of the jth electron as

$$\frac{d\gamma_j}{dz} = \frac{1}{2} \frac{a_w f_B}{\gamma_j} \left[i \, \hat{e}_s \, e^{i\vartheta_j} + \text{c.c.} \right] \qquad (61)$$

and

$$\frac{d\vartheta_j}{dz} = k_w - \frac{\omega/c}{2\gamma_j^2} (1 + a_w^2) + \frac{1}{2} \frac{a_w f_B}{\gamma_j^2} \left[\hat{e}_s \, e^{i\vartheta_j} + \text{c.c.} \right] \qquad (62)$$

and then expanding to first order in $\hat{e}_s$, with the assumption that $\hat{e}_s$ varies as $\exp(i\Gamma z)$ for complex Γ. Perturbation is around an equilibrium with $\hat{e}_s = 0$, $\gamma_j = \gamma_{j0}$ (a value perhaps varying with index j but constant in z) and $\vartheta_j = \vartheta_{j0} + z\Delta k_j$, where

$$\Delta k_j = k_w - \frac{\omega/c}{2\gamma_{j0}^2} (1 + a_w^2) ; \qquad (63)$$

the Δk_j term occurs because individual electrons need not be precisely at synchronism, and hence need not be (in equilibrium) at constant ϑ_j. The electrons are initially uniformly distributed in ϑ_j and have a spread in γ_{j0}, hence Δk_j.

The (real) perturbed quantities $\delta\gamma_j$ and $\delta\vartheta_j$ are written as

$$\delta\gamma_j = \frac{1}{2} \left[\delta\hat{\gamma}_j + \text{c.c.} \right] , \quad \delta\vartheta_j = \frac{1}{2} \left[\delta\hat{\vartheta}_j + \text{c.c.} \right] , \qquad (64)$$

with the caret denoting a complex perturbation amplitude.

From eq. (61), the equation for $\delta\hat{\gamma}_j$ becomes

$$\frac{d}{dz} \frac{\delta\hat{\gamma}_j}{\gamma_{j0}} = i(\Gamma + \Delta k_j) \frac{\delta\hat{\gamma}_j}{\gamma_{j0}} - \frac{a_w f_B}{\gamma_{j0}^2} i \, \hat{e}_s \, e^{i(\vartheta_{j0} + z\Delta k_j)} . \qquad (65)$$

so that

$$\frac{\delta\hat{\gamma}_j}{\gamma_{j0}} = \frac{a_w f_B}{\gamma_{j0}^2(\Gamma+\Delta k_j)} \hat{e}_s e^{i(\vartheta_{j0}+z\Delta k_j)} . \qquad (66)$$

For $\delta\vartheta_j$, the linearized equation becomes

$$\frac{d}{dz}\delta\hat{\vartheta}_j = i(\Gamma+\Delta k_j)\delta\hat{\vartheta}_j = \frac{\omega/c}{\gamma_{j0}^2}(1 + a_w^2)\frac{\delta\hat{\gamma}_j}{\gamma_{j0}} + \qquad (67)$$

$$\frac{a_w f_B}{\gamma_{j0}^2} \hat{e}_s e^{i(\vartheta_{j0}+z\Delta k_j)} .$$

Note that although $\hat{e}_s$ is proportional to $\exp(i\Gamma z)$, both $\delta\hat{\gamma}_j$ and $\delta\hat{\vartheta}_j$ are proportional to $\exp[i(\Gamma+\Delta k_j)z]$. From eqs. (66) and (67) we obtain

$$i\delta\hat{\vartheta}_j + \frac{\delta\hat{\gamma}_j}{\gamma_{j0}} = \frac{2a_w f_B}{\gamma_{j0}^2(\Gamma+\Delta k_j)^2}(\Gamma + k_w)\hat{e}_s e^{i(\vartheta_{j0}+z\Delta k_j)} . \qquad (68)$$

The linearized field equation, from eq. (52) becomes

$$\frac{\partial\hat{e}_s}{\partial z} - \frac{i}{2k}\nabla_\perp^2 \hat{e}_s = -\frac{2\pi i e}{mc^3} a_w f_B \frac{J}{N} \sum_j \frac{e^{i(\vartheta_{j0}+z\Delta k_j)}}{\gamma_{j0}}(i\delta\vartheta_j+\frac{\delta\gamma_j}{\gamma_{j0}}) . \qquad (69)$$

The perturbed particle quantities $\delta\vartheta_j$ and $\delta\gamma_j$ in eq. (69) are without carets; the real perturbation amplitudes must be replaced with the forms in eq. (64). The combination that appears in eq. (69) is of course the motivation for writing eq. (68).

Substituting eq. (68) into (69), we find

$$\frac{\partial\hat{e}_s}{\partial z} - \frac{i}{2k}\nabla_\perp^2 \hat{e}_s = -\frac{2\pi i e}{mc^3}\frac{J}{N} a_w^2 f_B^2 (\Gamma + k_w)\sum_j \frac{1}{\gamma_{j0}^3(\Gamma+\Delta k_j)^2}\hat{e}_s \qquad (70)$$

(plus other rapidly oscillating terms), which looks very much like eq. (60) but with an index of refraction δn that involves Γ and a sum over constant (in linear theory) particle quantities. The quantity Γ is in general complex, leading to a complex index of refraction: the imaginary part of the index of refraction produces gain and the real part produces a phase shift.

In one-dimensional theory, the $\nabla_\perp^2$ term does not appear, and eq. (70) becomes a straightforward equation for the complex growth rate Γ. If simple assumptions are made about the initial energy distribution of the electrons, the resulting equation for Γ even becomes solvable. For example, a square distribution function:

$$f(\gamma_{j0}) = \frac{1}{2\Delta\gamma} , \quad \gamma_0-\Delta\gamma < \gamma_{j0} < \gamma_0+\Delta\gamma , \qquad (71)$$

permits us to replace the sum with

$$\frac{1}{N}\sum_j \frac{1}{\gamma_{j0}^3(\Gamma+\Delta k_j)^2} = \frac{1}{\gamma_0^3[(\Gamma+\Delta k_0)^2 - 4k_w^2\,\Delta\gamma^2/\gamma_0^2]} \quad , \tag{72}$$

where

$$\Delta k_0 \equiv k_w - \frac{\omega/c}{\gamma_0^2}(1 + a_w^2) \tag{73}$$

and the approximations that $\Gamma \ll k_w$ and $\Delta\gamma \ll \gamma_0$ have been used. Eq. (70) in the one-dimensional approximation, with eq. (72), produces the standard cubic equation for Γ; considerable work is however required to demonstrate that the cubic is equivalent to other forms in which it appears.

Two- and three-dimensional effects, represented by the $\nabla_\perp^2$ term and the transverse variation of both J and the sum in eq. (70), complicate matters. If eq. (70) is still to be reduced to a cubic for Γ, the $\nabla_\perp^2$ term needs to be replaced by a constant times $\hat{e}_s$, and somehow the radial dependence of J and the bunching must be eliminated. The approach of Ref. 13 to solving eq. (70) involves simply assuming that J and the bunching sum in eq. (70) are radially constant out to the beam radius a, and that $\hat{e}_s$ varies self-similarly in z; that is,

$$\hat{e}_s(x,y,z) = \hat{e}_s(x,y)\,e^{i\Gamma z} \quad . \tag{74}$$

This assumption of self-similarity is in fact buried in the linearization performed above [cf. the line just past eq. (67)] but must be made explicit when diffraction effects are considered. Self-similar propagation in the exponential gain regime is illustrated in Figure 16.

With the two assumptions of self-similarity and radially constant source term, the standard theory of weakly guiding optical fibers[14] — trivially generalized to include a complex index of refraction — can be used. For axisymmetric fields (i.e., two-dimensional diffraction, as is relevant for an FEL with a circular electron beam) the paraxial wave equation becomes

$$\frac{1}{r}\frac{\partial}{\partial r}r\frac{\partial}{\partial r}\hat{e}_s - 2k(\Gamma - k\delta n)\hat{e}_s = 0 \quad , \tag{75}$$

with $\delta n = 0$ for $r > a$. The solutions to this equation are Bessel functions:

$$\hat{e}_s = e_0\,J_0(\kappa r) \quad , \qquad r < a \tag{76}$$

$$\hat{e}_s = Ae_0\,K_0(\chi r) \quad , \qquad r > a \tag{77}$$

where κ and χ are complex parameters that characterize the propagation, J_0 is a Bessel function and K_0 is a modified Bessel function. Ref. 14 uses the

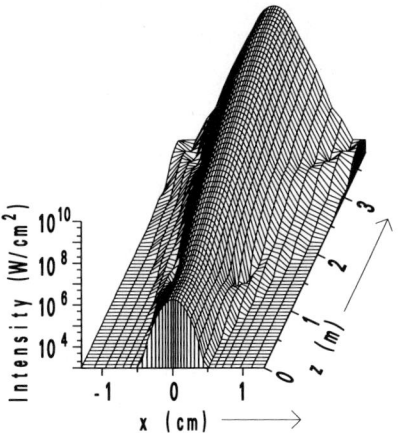

FIGURE 16
Optical intensity vs x and z in the exponential regime of the 21-μm amplifier of Figure 2. The transverse optical profile remains unchanged as the power increases, illustrating (at least by simulation) the approximation of self-similar propagation in the exponential gain regime. The bumps in the wings toward the end of the simulation arise from beampipe reflections.

symbol γ instead of χ; we switch to χ here to avoid confusion with the Lorentz factor of an electron. From the wave equation (75),

$$\kappa = [-2k(\Gamma - k\delta n)]^{\frac{1}{2}} , \tag{78}$$

$$\chi = [2k\Gamma]^{\frac{1}{2}} . \tag{79}$$

What has now been achieved with the assumptions of self-similarity and constant J is the replacement of the $\nabla^2_\perp$ term of eq. (70) by a constant:

$$\nabla^2_\perp \hat{e}_s = -\kappa^2 \hat{e}_s , \quad r < a , \tag{80}$$

$$\nabla^2_\perp \hat{e}_s = +\chi^2 \hat{e}_s , \quad r > a . \tag{81}$$

The choice of Bessel functions, J_0 and K_0, was determined by the boundary conditions (finite $\hat{e}_s$) at $r = 0$ and $r \to \infty$. Those boundary conditions also impose the constraint that $\text{Re}(\chi) > 0$, since $K_0(\chi r) \to \exp(-\chi r)/\sqrt{r}$ as $r \to \infty$. The condition that the phase velocity of the light be directed away from the fiber constrains $\text{Im}(\chi)$ to be < 0. The propagation constant Γ, and hence κ and χ, are determined by the boundary conditions on $\hat{e}_s$ at $r = a$:

$$\text{continuous } \hat{e}_s \to J_0(\kappa a) = A K_0(\chi a) , \tag{82}$$

$$\text{continuous } \frac{\partial \hat{e}_s}{\partial r} \to \kappa J_1(\kappa a) = A \chi K_1(\chi a) . \tag{83}$$

Eliminating A, we find a dispersion relation for κ and χ:

$$\frac{\kappa J_1(\kappa a)}{J_0(\kappa a)} = \frac{\chi K_1(\chi a)}{K_0(\chi a)} , \qquad (84)$$

with

$$\kappa^2 + \chi^2 = 2k^2 \delta n . \qquad (85)$$

Since κ and χ are complex, it is necessary to solve eqs. (84) and (85) numerically. Since κ and χ both depend on and determine Γ, the numerical solution must be iterative. Perhaps the best way to find a solution is to note that for $r > a$, eq. (75) yields a relation between Γ and χ^2:

$$\Gamma = \frac{\chi^2}{2k} ; \qquad (86)$$

this is the mathematical statement that the field gradients for $r > a$ are those required to diffract power outward, to maintain self-similar growth (at complex growth rate Γ) of the light profile.

Using eq. (86) to replace Γ by $\chi^2/2k$ everywhere in eq. (70) [as modified by eq. (71)], we obtain a cubic equation for χ^2:

$$\left(\frac{\chi^2+\kappa^2}{2k}\right)\left[\left(\frac{\chi^2}{2k} + \Delta k_0\right)^2 - 4k_w^2 \frac{\Delta\gamma^2}{\gamma_0^2}\right] + \frac{2\pi e}{mc^3} J \frac{a_w^2 f_B^2}{\gamma_0^3} \left(\frac{\chi^2}{2k} + k_w\right) = 0. \qquad (87)$$

The complex parameter κ^2 also appears in eq. (87), so that eq. (84) must be solved simultaneously. The solution is reasonably straightforward, but still requires numerical iteration. Several examples have been discussed by Moore.[13]

Several general statements can be made about optical guiding in the exponential gain regime. First, it is usually the case that the gain length obtained from eqs. (84) and (87) is shorter than the Rayleigh range defined by the electron beam size, at least with the parameters of presently feasible FELs. In these cases, gain guiding — provided by the imaginary part of the index of refraction — dominates. In other cases, with the gain length longer than the Rayleigh range (i.e., diffraction stronger than gain, for example if the electron beam is quite small), refractive guiding — from the real part of the index of refraction — can dominate.

In tapered wiggler amplifiers, refractive guiding usually dominates after saturation of exponential gain. It is possible (at least numerically) to construct an FEL amplifier that has neither gain nor loss well past saturation, yet guides the light quite strongly; clearly in such a case, refractive guiding is all that contributes to guiding the light.

To understand the origin of the index of refraction of the electron beam in an FEL, it is useful first to summarize what is usually meant by an index of refraction. In, for example, a glass or crystal with a real index of refraction n > 1, the index of refraction arises because light, propagating through the glass, excites atomic dipoles which reradiate 90° out of phase with the incident light. If the reradiated light is *precisely* 90° out of phase, it only shifts the phase of the incident light, with no loss or gain. The propagation of light in glass, or the guiding of light by a conventional optical fiber, can be completely described by summing the incident light and the reradiated light from each atomic dipole; since this would obviously be a tedious calculation, the effect of adding reradiated light is commonly treated by a phenomenological index of refraction n.

The important point in that brief description of a refractive index — at least for understanding the origin of the index in an FEL — is that the index can be used to represent the effect of light radiated by many microscopic dipoles. For understanding the propagation alone, the origin of the dipole radiation does not matter; the dipoles could be microscopic antennas wiggled by equally microscopic cockroaches without changing the description of propagation. In particular, for the FEL, the dipoles are individual bunches in the electron beam, moving in the magnetic field of the wiggler. As described above, the bunches are approximately the size of an optical wavelength, and are formed by the gain mechanism of the FEL — in a tapered wiggler, the bunches are the electrons trapped in ponderomotive potentials.

The possibility of refractive guiding in an FEL may at first glance be surprising; as mentioned earlier, the Kramers-Kronig relations usually (in a normal laser medium) require that the real part of the index of refraction of a gain medium vanish at peak gain.

The Kramers-Kronig relations relate the real and imaginary parts of the one-sided (in time) Fourier transform of a *linear* response, such as a conductivity or permittivity. Causality alone requires that the Fourier transform be one sided, and the one-sided nature of the transform ensures that the transformed function is analytic in the upper half complex ω plane. Then the real and imaginary parts of the transformed function form a Hilbert transform pair [see, for example, Ref. 15]. If then the gain is a symmetric function of frequency around a central ω_0, $\text{Re}[n(\omega_0)] = 0$.

A more detailed discussion here of the derivation of the Kramers-Kronig relations is not warranted, because the key word above is "linear." For an FEL beyond saturation of exponential gain, the response function is very nonlinear. For example, we can look at a short section of wiggler with a long, bunched electron beam moving through it. The field at the beginning of the short section of undulator is $\mathcal{E}_{in}$, and at the end is $\mathcal{E}_{out}$. Associated with the change from $\mathcal{E}_{in}$ to $\mathcal{E}_{out}$ is some gain and phase shift. Are the phase shift and gain related by the Kramers-Kronig relations? If one could write

$$\mathcal{E}_{out}(t) = \int_{-\infty}^{t} g(t - t') \mathcal{E}_{in}(t') \, dt' \tag{88}$$

where g is any function that depends only on $t - t'$, then indeed the gain and phase shift would be related by the Kramers-Kronig relations. In fact, though, we cannot write any equation like eq. (88), except in the exponential gain regime. In the opposite limit, in which the electron bunches do not respond to changes in $\mathcal{E}_{in}(t)$ but instead radiate a fixed $\Delta\mathcal{E}(t)$, we can write

$$\mathcal{E}_{out}(t) = \mathcal{E}_{in}(t) + \Delta\mathcal{E}(t) \tag{89}$$

The gain and phase shift are very nonlinear, and no general relation between them can be derived.

In the exponential gain regime the gain and phase shift <u>are</u> linear, and we would expect the Kramers-Kronig relations to apply. Solutions of the cubic equation for Γ indicate that indeed the relations do apply, but that the gain curve is asymmetric around peak gain. The asymmetry ensures that the phase shift does not vanish at peak gain, providing the possibility of refractive guiding even at peak exponential gain.

ACKNOWLEDGEMENT

I am grateful to many people for the remarks and suggestions in the past several years that have directly or indirectly contributed to the contents of this chapter: these people include, but are certainly not limited to, W. Fawley, D. Prosnitz, A. Sessler and J. Wurtele.

REFERENCES

1) R. Bonifacio, this volume

2) N. M. Kroll, P. L. Morton, M. R. Rosenbluth, IEEE Journ. Quantum Electronics QE-17 (1981) 1436.

3) D. Prosnitz, A. Szöke, and V. K. Neil, Phys. Rev. A24 (1981) 1436.

4) T. J. Orzechowski, B. R. Anderson, J. C. Clark, W. M. Fawley, A. C. Paul, D. Prosnitz, E. T. Scharlemann, S. Yarema, D. B. Hopkins, A. M. Sessler, and J. S. Wurtele, Phys. Rev. (Letters) 57 (1986) 2172.

5) J. Slater, T. Churchill, D. Quimby, K. Robinson, D. Shemwell, A. Valla, A. A. Vetter, J. Adamski, W. Gallagher, R. Kennedy, B. Robinson, D. Shoffstall, E. Tyson, A. Vetter, and A. Yeremian, in Free-Electron Lasers, Proc. 7th International Free-Electron Laser Conference, eds., E. T. Scharlemann and D. Prosnitz (North Holland, Amsterdam, 1986) p. 228.

6) W. M. Fawley, D. Prosnitz, and E. T. Scharlemann, Phys. Rev. A30 (1984) 2472.

7) J. P. Blewett and R. Chasman, Journ. Applied Phys. 48 (1977) 2692.

8) E. T. Scharlemann, Journ. Applied Phys. 58 (1985) 2154.

9) E. T. Scharlemann, W. M. Fawley, B. R. Anderson, and T. J. Orzechowski, in Free-Electron Lasers, Proc. 7th International Free-Electron Laser Conference, eds., E. T. Scharlemann and D. Prosnitz (North Holland, Amsterdam, 1986) p. 150.

10) E. T. Scharlemann, A. M. Sessler, and J. S. Wurtele, Phys. Rev. Lett. 54 (1985) 1925.

11) J. M. Slater and D. D. Lowenthal, Journ. Appl. Phys. 52 (1981) 44.

12) P. Sprangle and C.-M. Tang, Applied Physics Letters 39 (1981) 677; C.-M. Tang and P. Sprangle, in Physics of Quantum Electronics, vol. 9 (Addison-Wesley, Reading, Mass., 1982) 627.

13) G. T. Moore, Optics Communications 52 (1984) 46.

14) D. Marcuse, Theory of Dielectric Optical Waveguides, (Academic, New York, 1974), p. 60ff.

15) B. G. Levich, Theoretical Physics, (North Holland, Amsterdam, 1971) p. 540ff.

HIGH GRADIENT ACCELERATORS FOR LINEAR LIGHT SOURCES

William A. BARLETTA

University of California, Lawrence Livermore National Laboratory,
P.O. Box 808, Livermore, California* and Center for Advanced Accelerators,
Department of Physics, University of California at Los Angeles

Ultra-high gradient radio frequency linacs powered by relativistic klystrons appear to be able to provide compact sources of radiation at XUV and soft x-ray wavelengths with a duration of 1 picosecond or less. This paper provides a tutorial review of the physics applicable to scaling the present experience of the accelerator community to the regime applicable to compact linear light sources.

1. INTRODUCTION

Linear light sources capable of producing picosecond bursts of XUV and soft x radiation have gained ever increasing attention with the rapid advances both in free electron laser (FEL) physics and in the technology of ultra-high gradient linacs capable of producing high brightness electron beams at high energy. Compact accelerators for linear light sources are based upon conventional radio frequency structures. Desired average accelerating gradients ($\approx$ 200 MeV/m or more) are higher than those achievable in present-day accelerators (e.g., 17 MeV/m at SLAC). This goal is, however, being pursued actively at Lawrence Livermore National Laboratory (LLNL) in collaboration with the Stanford Linear Accelerator Center (SLAC) and the Lawrence Berkeley Laboratory (LBL) toward the design of a TeV electron-positron collider. A great deal is known about the basic scaling laws governing the operation of rf accelerators. They are presented here as a tutorial guide to those considering the design of compact linear light sources (or flavor factories).

A critical beam characteristic with respect to the performance of an x-ray FEL (XRFEL) is the emittance of the beam, because it strongly affects the gain of the

---

* Work performed under the auspices of the U.S. Department of Energy by Lawrence Livermore National Laboratory under contract W-7405-ENG-48.

laser. As discussed in Sec. 2, design goals of $\varepsilon_n \approx 10$ mm-mrad and $N_b \approx 5 \times 10^{10}$ are consistent with the performance of existing electron sources. Constraints on beam chacteristics imposed by cavity fill times, focusing, beam loading, wakefield control, and gas matching (in the case of plasma assisted wigglers) are presented in Secs. 3–6.

A major technological hurdle to be overcome in realizing high gradient rf accelerators is the gigawatt rf power source. Section 7 discusses the induction-linac-powered relativistic klystron and the possibility of deploying branched magnetics in order to provide macro-pulses as long as 1 μs duration should the community of linear light source users require such a characteristic. Finally, Sec. 8 presents the various elements of a cost estimation algorithm that can be used by the designer of a linear light source (or flavor factory) to access the financial resources needed to build a high gradient linac powered by relativistic klystrons.

1.1. Application-Dervied Constraints

Because of the diverse radiation output characteristics desired by the potential user communities, no specific facility or economic constraints have been formulated for linear light sources. It is useful, nonetheless, for the accelerator designer to adopt some rough guidelines as goals:

$$\text{Length/energy} > 5 \text{ m/GeV},$$
$$\text{Cost/energy/pulse duration} < 10 \text{ M\$ GeV}/\mu\text{s},$$
$$\text{Repetition rate} > 200 \text{ Hz} \quad .$$

Cast in this form the cost and size goals for the light source designer are found to be roughly equivalent to those of the designer of the next generation of linear colliders at TeV energies. Exploring design concepts consistent with these goals has been the object of active and steadily increasing efforts by the high energy physics research community throughout the world. Consequently, a substantial literature exists concerning the physics of accelerators relevant to compact linear light sources. Recent work by Wilson,[1,2] and Palmer,[3] and Amaldi[4] form the basis of the analysis of Secs. 2-6. A spread-sheet program including these analyses, those concerning cost, and those of FEL physics is now in active use at LLNL with a goal of system optimization.

1.2. Energy Constraints

In addition to setting size and cost constraints, the physics of free electron laser operation set several constraints on the characteristics of the beam such as the beam energy, E (or relativistic factor, γ), the normalized emittance, ε_n, and the energy spread, $\Delta E/E$ (or momentum spread, $\Delta p/p$). For an FEL with a planar wiggler the well-known design equations linking the beam characteristics with FEL performance are specified in terms of the wiggler field, B_0, the wiggler wavelength, λ_w, and the radiation wavelength, λ_s. The wiggler characteristics can be combined in terms of a dimensionless vector potential

$$K = e\ \lambda_w B_0/\sqrt{2}\ \pi\ m_e c^2 \approx 0.93\ B_0(T)\lambda_w\ (\text{cm})\ . \tag{1}$$

The wiggler and radiation wavelengths are connected to the beam energy by the resonance condition

$$\lambda_s = \left(\lambda_w/2\gamma^2\right)\left(1 + K^2/2\right)\ . \tag{2}$$

1.3. Mono-Chromaticity and Emittance Requirements

FEL physics place tight bounds on an important characteristic of the accelerator subsystem, namely, the allowable momentum spread, $\Delta p/p$, within the pulse or from bunch to bunch within a macro-pulse. The momentum spread has several sources, the largest of which originates from the duration of each pulse relative to that of the rf cycle. This spread, plus longitudinal wakefield effects, will make it difficult to produce high energy beams with $\Delta p/p < \pm 0.1\%$.

The performance of the FEL scales[5] with a single electron beam parameter, ρ:

$$\rho = \left(\frac{K\ \omega_p}{8\ \omega_w}\right)^{2/3}, \tag{3}$$

where ω_p ($\sim \gamma^{-3}\ N_b^{1/2}$) is the relativistic plasma frequency of the electron beam. After passing through a wiggler of length $L \sim \gamma_w/\rho$, the output from the FEL will saturate with the power in the radiation being ρ times the electron beam power. The constraints on energy spread in the beam and the normalized beam emittance are

$$\frac{\Delta E}{E} \leq \frac{\rho}{4} \; , \tag{4a}$$

and

$$\varepsilon_n \leq \lambda_s \gamma \; . \tag{4b}$$

If the XRFEL employs a plasma-assisted wiggler[6] in which the beam assumes its self-focused radius (or if the beam is to be focused to an extremely small radius as in a flavor factory), a magnet subsystem must match the beam size in the accelerator (~ 100 μm) to the self-focused size in the plasma (~ 1 μm). Relatively compact focal arrays with such capabilities have received considerable attention from the designers of linear colliders. Most notably, the Stanford Linear Collider (SLC) has such a final focus system in operation. For the SLC design, the size of the focal spot is related to the momentum spread by

$$a_{\text{focus}} \propto \left(\frac{\Delta p}{p}\right)^2 \; . \tag{5}$$

A focal system with quadrupoles with field strengths of less than 2 T and apertures of 1 cm will require that $\Delta p/p < \pm 0.5\%$.

2. CHARACTERISTICS OF EXISTING ELECTRON BEAM SOURCES

This section reviews the present state of the art concerning bright, high current sources of electron beams. These sources, which include electron injectors, synchrotron radiation sources, and linear colliders, form the data base for scaling accelerator design to the regime applicable to linear light sources and flavor factories.

2.1. High Brightness Sources

As the beam emittance plays a critical role in determining electron density and thereby the gain of a free electron laser, the examination of the scaling basis for accelerators for linear light sources begins with a discussion of the present status of high brightness electron sources. Over the past decade the accelerator community has expended considerable effort in the design of ever lower emittance sources of electron beams with large peak currents. This work has been motivated by three applications: high energy linear colliders, high average power FELs operating at visible or shorter wavelengths, and effective sources of incoherent synchrotron radiation. During this period the beam brightness (current density divided by angular divergence) has been increased by at least three orders of magnitude, with presently achieved performances still a considerable distance from fundamental limits.

In early 1987 Brookhaven National Laboratory hosted a workshop[7] concerning the production of low emittance beams for both collider and FEL applications. The most stressing goal, that for XRFELs as set for the participants by Pellegrini, was a factor of a few more demanding than that for linear colliders; namely,

Beam Energy, E   1 GeV,

Normalized horizontal emittance area, $\epsilon_{nx} = \gamma \epsilon_z$ 1×10^{-6} πm − rad,

Normalized vertical emittance area, $\epsilon_{ny} = \gamma \epsilon_y$ 1×10^{-8} πm − rad,

Longitudinal brilliance, $B_L = I_{\text{beam}}/\gamma$ 200 A.

Working groups studying beam injectors, damping rings, and the problems of emittance preservation during the acceleration process concluded that the injectors would have to be improved beyond the present state of the art to meet the stated goals for an FEL. They found no mechanism incompatible with this goal and expected that such a beam could be accelerated from a few tens of MeV to a few GeV without degradation.

The best performance in building high current, low emittance guns using photocathodes has been achieved by Fraser, Sheffield et al.[8,9] at Los Alamos. They have produced a beam at 1.1 MeV with a peak current of 130 A and a normalized rms emittance of $5 \times 10^{-6}\pi$ m-rad. This value is approximately twenty times smaller than that of the SLAC gun,[10] which produces a beam of ≈ 1 kA at an emittance of $2 \times 10^{-4}\pi$ m-rad. At SLAC this beam is accelerated to ≈ 1 GeV for insertion into a damping ring. Upon exiting the damping ring the beam emittance has been reduced to $1 \times 10^{-6}\pi$ m-rad.

Low emittance beams have also been produced with thermionic cathodes in rf guns, albeit at somewhat lower peak currents. For example, the Mark III used for FEL experiments at Stanford's High Energy Physics Laboratory delivers a peak current of 30 A at 44 MeV with normalized horizontal and vertical emittances of $\varepsilon_{nx} = 3.6 \times 10^{-6}\pi$ m-rad and $\varepsilon_{ny} = 1.8 \times 10^{-6}\pi$ m-rad, respectively.

2.2 Existing and Planned Synchrotron Beams

Existing and planned synchrotron sources[11] in the USA are (in alphabetical order) Argonne 6 GeV Ring, Berkeley 1-2 GeV Ring, Cornell High Energy Synchrotron Source (CHESS), National Synchrotron Light Source (NSLS) (Brookhaven), Stanford Synchrotron Radiation Laboratory (SSRL), Stanford Photon Research Laboratory (SPRL), Synchrotron Radiation Center (SRC) (Stoughton), and Synchrotron Ultraviolet Radiation Facility (SURF) (Gaithersburg). Their properties are detailed in Table 1.; note that the emittances given are unnormalized, i.e., $\varepsilon = \varepsilon_n/\gamma$.

Table 1. Existing and Planned Storage Rings

Status*	Name	Energy (GeV)	ε_{horiz} ($\pi \times 10^{-9}$m rad)	ε_{vert}
o	ALADDIN (Stoughton)	0.75	63	110
o	TANTULUS (Stoughton)	0.24	230	4000
p	ALS (LBL)	1.3	6.8	4.0
p	ANL (Argonne)	7.0	7.0	0.2
o	CESR (Cornell)	5.5	200	6.6
o	NSLS I(BNL)	0.75	130	230
o	NSLS II(BNL)	2.5	80	13
o	PEP (SLAC)	15	150	0.67
	" "	8.0	10	0.16
o	SPEAR (SLAC)	3.0	450	50
o	SURF II(Gaithersburg)	0.28	270	3400
c	SPRL (SLAC)	1.0	10	10

* o=operational, c=in construction, p=planned

2.3. The Stanford Linear Collider (SLC)

The most relevant data base for scaling the design of rf linac technology to the parameter regime suitable for linear light sources (and for linear collider flavor factories) is the 50 GeV SLC at the Stanford Linear Accelerator Center (SLAC). The SLC linac consists of a 3-km-long assembly of sections of S-band (2.87 GHz) disk-loaded waveguide (traveling wave). Each linac section, consisting of nearly 100 cavities with a length one-third the rf wavelength, is fed by a high peak power klystron. Although other types of rf structures (e.g., standing wave) may also be suitable, the disk-loaded structure is amenable to analytical calculation and has received considerable study. Consequently, scaling the SLC design is a convenient way to scope the characteristics of the accelerator. The following analysis requires the definition of several quantities listed in Table 2. along with their respective values for the SLC. Units are mks unless noted.

3. SCALING BASIS FOR ACCELERATORS FOR LINEAR LIGHT SOURCES

For use in a linear light source, a compact accelerator should both be highly efficient in transferring rf energy from the source to the high energy beam and highly conservative in use of physical space. As shown in the next section, these constraints push one toward high gradients, high frequencies, and short fill times.

3.1. High Voltage Breakdown

The physical phenomenon which forms the basis for scaling rf linacs from the SLC value of 17 MeV/m to the required > 200 MeV/m is the observation[12] that the peak electric field that can be sustained without breakdown rises with increasing rf frequency and with shortening duration of the rf power (see Fig. 1).

Not all of the potential for increased fields can be realized in practical high energy accelerators. At field values somewhat below breakdown values, small currents will flow near the cavity walls. These currents give rise to copious x radiation and, more significantly, to magnetic fields that exert random transverse forces on the beam, thereby exciting wakefield instabilities. To avoid this difficulty the peak

Table 2. SLC values of scaling variables

Parameter	Symbol	Value
Beam energy	E	50 GeV
RF frequency	f	2.87 GHz
RF wavelength	λ	10^5 mm
Iris aperture (radius)	a	11 mm
Cavity Q	Q	$1.39 \cdot 10^4$
RF Group velocity	v_g	$0.011\ c$
Fill time	t_f	921 ns
Length per feed	L_f	2.97 m
Attenuation time	T_o	1.4 μs
Fill length / attenuation	t	0.6
Average gradient	E_a	17 MeV/m
Peak field in cavity	E_{max}	41.3 MeV/m
Loading	h_b	3.12%
Peak RF power	P_{rf}	9.8 MW/m
Particles per bunch	N_b	$\leq 5 \times 10^{10}$
Bunch length (Gaussian σ)	s_z	1 mm (3.43°)
Momentum spread ($\Delta p/p$)	σ_p	0.39%
Beam size in linac	$\sigma_{x,y}$	77 μm

field should be reduced by a safety factor, F_s, with $0.5 \leq F_s \leq 0.75$. The interdependence model uses a value $F_s = 0.66$. Commonly, in the discussion of linear colliders, the scaling of E_{pk} with frequency is given as $E_{pk} \propto f^{7/8}$. This scaling takes advantage of the increase of sustainable peak field with decreasing rf pulse duration:

$$E_{pk} = 120 \text{ MV/m} \left(\frac{F_s}{0.66}\right)\left(\frac{f}{2.87 \text{ GHz}}\right)^{7/8}. \qquad (6)$$

Of course, this value can only be achieved if sufficient power is supplied to the structure; it is not in the case of SLC. Should the application require an output

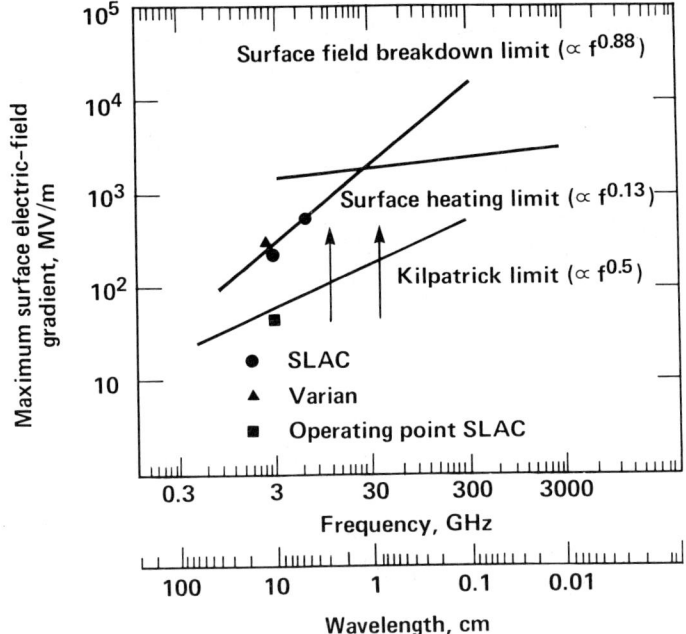

Figure 1. The peak electric field that can be sustained in rf structures without breakdown. Generally, only approximately half this value can actually be used.

period of ~ 1 μs indpendent of frequency, the exponent in Eq. 6 should be reduced to 1/2.

3.2. Cavity and rf Characteristics

The scaling characteristics of $2\pi/3$ disk-loaded structures as described in Refs. 1–3, 13, and 14 are based on computer calculations with Farkas's code TWAP. The disk-loaded structure has a phase velocity equal to c, the speed of light, and a group velocity that is a function of the rf wavelength and the aperture of the iris. Palmer's fits[3] to the TWAP calculations[13] as shown in Fig. 2(b) give

$$\beta_g = \frac{v_g}{c} \approx \exp\left[3.1 - 2.4\left(\frac{\lambda}{a}\right)^{1/2} - 0.9\left(\frac{a}{\lambda}\right)\right] \quad . \tag{7}$$

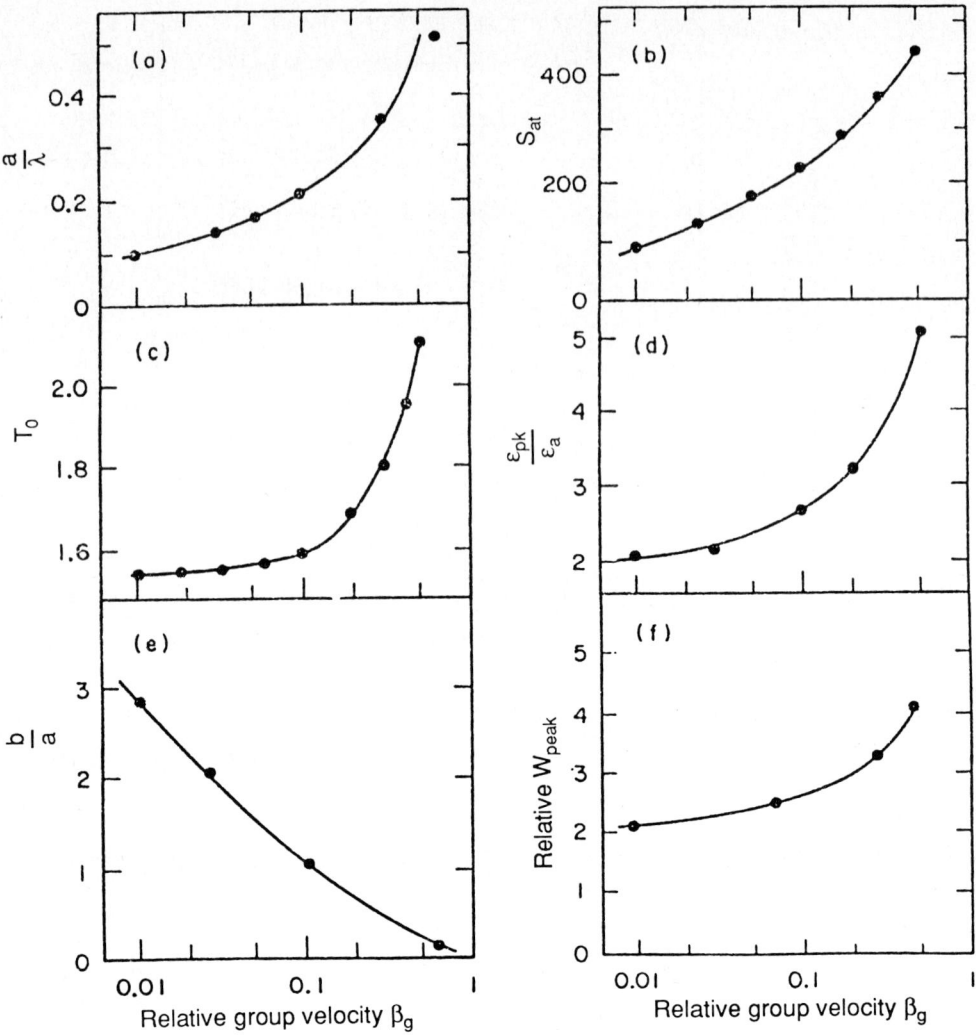

Figure 2. Palmer's[3] fits to Farkas's TWAP[13] calculations of SLAC-like accelerating structures. (a) iris radius a divided by wavelength; (b) normalized corrected elastance S_{at}; (c) attenuation time constant T_0 in μsec; (d) ratio of peak and average accelerating fields; (e) ratio of outer cavity and iris radii; (f) the relative peak rf power.

With the group velocity specified the remainder of the cavity characteristics can be calculated. Initially assume that no rf energy is lost as the wave travels through the structure. In that case the energy per unit length, w_a (J/m), required to produce a given acceleration of the beam is related to the average accelerating gradient by a monotonically increasing function of the group velocity, namely, the corrected elastance, s_t:

$$s_t = \frac{E_a^2}{w_a} = 5.7 \cdot 10^{14} \text{ V/mC} \left(\frac{10 \text{ mm}}{a}\right)^2 \left(\frac{v_g}{c}\right)^{0.4} . \qquad (8a)$$

Frequently accelerator designers prefer to quote a "normalized" elastance that is independent of the iris aperture, i.e.,

$$s_{at} = s_t a^2 . \qquad (8b)$$

Because both the beam and the rf move through the structure with a finite velocity, the length of the accelerating pulse can be less than that of the section. Therefore, the amount of energy, w_f, needed to fill the structure is greater than w_a by an amount

$$w_f = \frac{w_a}{(1-\beta_g)} . \qquad (9a)$$

Similarly, the uncorrected elastance is related to s_t by

$$s = (1-\beta_g) s_t . \qquad (9b)$$

In actual structures the rf wave is attenuated as it travels down the section. The characteristic attenuation time,

$$T_o = 1.45 \text{ }\mu s \left(1 + 1.29 \beta_g^{1.5}\right) \left(\frac{\lambda}{105 \text{ mm}}\right)^{1.5} , \qquad (10)$$

is related to the Q of the cavities by

$$Q = \pi f_{rf} T_o . \qquad (11)$$

The fill time of the section is

$$t_f = \frac{L_f}{v_g} . \qquad (12)$$

In terms of the ratio of the fill time to attenuation time,

$$t = \frac{t_f}{T_o} \;, \tag{13}$$

the rf energy needed for the same acceleration in a constant impedance structure is

$$w_{rf} = w_a \left[\frac{\tau^2}{(1 - e^{-\tau})^2} \right] = \frac{w_a}{h} \;, \tag{14}$$

where h is defined as the structure efficiency. Note that as the length of the section is decreased the fill time goes to zero, but $w_{rf} \to w_a$. Therefore, the required power increases without bound:

$$P_{rf} = w_{rf}/t_f \;. \tag{15}$$

Combining Eq. (7) with eq. (15), one can express the rf power per unit length as

$$P_{rf} = \frac{E_a^2 a^2}{h T_o \tau s_{at}} \;. \tag{16}$$

The maximum gradient is determined from the peak permissible value of the electric field:

$$E_a = \frac{E_{pk}}{2 + 6\beta_g} \;, \tag{17}$$

where E_{pk} is given by Eq. (6). Hence,

$$E_a = 1.2 \cdot 10^8 \text{ V/m} \; (2 + 6\beta_g)^{-1} \left(\frac{F_s}{0.66} \right) \left(\frac{f}{2.87 \text{ GHz}} \right)^{1/2} \;. \tag{18}$$

In designing a linear light source of minimum length for a given beam energy, one can choose only the frequency, iris aperture, and the attenuation parameter; all the other cavity characteristics are determined. Completing the cavity parameter descriptions is

$$\frac{b_c}{a} \approx 1.04 - 0.29 \ln \beta_g + 0.068 \ln^2 \beta_g \;, \tag{19}$$

where b_c is the inner radius of the accelerating cavity.

4. FOCUSING IN THE LINAC

The design of the focusing system for a high energy linac has two impacts. The first and least important of these is to increase the overall length of the accelerator. In a transport design scaled from the SLC FODO lattice this consideration can be quantified by a single number, F_q, the fraction of the linac occupied by the quadrupoles (typically <10%). The second implication is much more important. The strong focusing needed to control wake field instabilities can lead to severe tolerance constraints on the linac design. The beam transport in the linac is described by the beta function of the focusing lattice. The average strength of the focusing in a FODO structure consisting of quadrupoles with field strength B_q and aperture a_q is

$$\langle \beta_y \rangle = \left(\frac{\sin \mu}{\mu^2} \frac{E}{ec} \frac{2a_q}{B_q F_q} \right)^{1/2} \quad , \tag{20}$$

where 2μ is the phase advance per cell. Following the design of the SLC linac, one chooses

$$\mu = 45^\circ \quad , \tag{21a}$$

$$B_q = 1.5 \text{ T} \quad , \tag{21b}$$

$$a_q = 1.2 \, a \quad . \tag{21c}$$

With these choices one can rewrite Eq. (20) as

$$<\beta_y> = 2.49 \text{ m} \left(\frac{E}{10 \text{ GeV}} \frac{a}{10 \text{ mm}} \frac{1.5 \text{ T}}{B_q} \frac{0.1}{F_q} \right)^{1/2} \tag{22}$$

Of course, the local value of the focusing strength will vary throughout the accelerator as the beam gains energy. Using ζ to denote distance along the accelerator, one can compute the number of quadrupoles in the accelerator:

$$N_q = \int_0^L \frac{d\zeta}{\beta(\zeta)\mu} \quad , \tag{23}$$

where L is the length of the accelerator. The beam size produced by this lattice is

$$\sigma_y(\zeta) = \left(\frac{\beta_y(\zeta)\varepsilon_n}{\gamma(\zeta)}\right)^{1/2} \tag{24}$$

Palmer[3] shows that the alignment tolerances along the linac can be kept constant, if the strength of the quadrupoles is varied weakly ($\approx \gamma^{1/3}$) along the linac; whence,

$$\beta(\zeta) \propto \gamma^{1/3} \propto \zeta^{1/3} \,. \tag{25}$$

Then, Eq. (23) gives

$$N_q = 1.5 \left(\frac{L}{\beta(L)\mu}\right) \tag{26}$$

Calculations[15] of the ratio of beam displacement to beam radius at the end of the linac indicate that in order to preserve the beam emittance, the quadrupoles must be aligned with an rms accuracy of

$$<dy> = \frac{\sigma_y}{\mu\sigma_p}\left(\frac{2}{N_q}\right)^{1/2}, \tag{27}$$

where σ_p is the momentum spread, ($\Delta p/p$). Probably the largest source of momentum spread in the main body of multi-GeV linacs will be a head-to-tail energy variation which is deliberately introduced and maintained throughout the accelerator in order to damp transverse wakefield effects. This technique is often erroneously called Landau damping, which refers to an instantaneous spread in betatron frequencies. The head-to-tail spread with the $4k_\beta$ increasing at the tail is an idea of Balakin, Novokhatskii, and Smirnov and should be called BNS damping. The linear component of the head-to-tail spread can be removed in a small accelerating section at the end of the accelerator that operates at a phase advance of $90°$. This section has a length

$$\begin{aligned} l_{\text{corr}} &= \frac{\sigma_p p \lambda}{2\pi E_a \sigma_z} \tag{28}\\ &= 0.81\text{ m}\left(\frac{E}{10\text{ GeV}}\right)\left(\frac{\lambda}{100\text{ mm}}\right)\left(\frac{E_a}{200\text{ MeV/m}}\right)^{-1}\left(\frac{\sigma_z}{10\text{ mm}}\right)^{-1}, \end{aligned}$$

where σ_z is the rms bunch length.

Wilson[2] has showed that the tolerance on alignment of the focusing system in large linacs is generally more severe than the tolerance on alignment of the accelerator sections. A prudent approach, however, may be to adopt Eq. (27) as establishing the tolerance requirements for all beamline components. Similarly the position of the beam in the linac must be controlled to this level. For linacs of applicability to linear light sources and flavor factories, the alignment tolerances should not be worse than ≈ 100 μm. The difficulty may be compounded by the fact that vibrational tolerances in such linacs tend to be an order of magnitude tighter than the static alignment tolerances. Tolerance requirements may be relaxed if alternatives to BNS damping such as rf focusing with elliptical irises prove practical; however, such means of controlling wake fields in high gradient linacs must still be considered to be untested at this time.

5. WAKEFIELDS AND BEAM LOADING

In passing through the accelerating structure each beam pulse will both remove energy from the fundamental accelerating mode of the cavity and excite higher order modes in the cavity. Via the excitation of these higher order modes (wakefields), the head of a beam pulse can affect the energy of (longitudinal wakes) and can deflect (transverse modes) the tail of the pulse. The results will be an energy spread in the beam and a head-to-tail growth of emittance as the beam travels through the accelerator.

5.1. Transverse Wakefields

The dipole (deflecting) modes have an E_z component that varies linearly with distance from the cavity axis and with the cosine of the azimuthal angle, f. The nth traveling wave modes generated by a charge q traveling parallel to the cavity axis at a radius r_q have an electric field at a distance $\zeta = ct$ behind q of

$$E_{zn} = -2k_n q \left(\frac{r}{a}\right)^{m-1} \left(\frac{r_q}{a}\right)^m \cos m\phi \cos \omega_n \tau \quad , \qquad (29)$$

where k_n is Wilson's[1] loss parameter which relates the field intensity to the stored energy for the nth mode (E_n^2/W_n or one-fourth the elastance for the mode). The interaction of the beam with these modes can be characterized by a transverse shunt impedance $r_\perp$. For typical disk-loaded structures

$$\frac{r_\perp}{Q} \approx \frac{100}{\lambda} \text{ ohms} \qquad (30)$$

gives a rough estimate of the transverse shunt impedance. According to Wilson,[1] the lowest order (beam break-up) mode has a frequency $\approx 40\%$ higher than the fundamental accelerating mode. Therefore, one expects the Q of the dipole mode to be only slightly lower than that of the accelerating mode. Consequently, without special precautions the transverse wakefield of a pulse will persist for many tens of nanoseconds. Via these long-range wakes, both the transverse motion and the emittance can increase from one micro-pulse to another.

A theorem by Panofsky and Wenzel shows that the total deflecting kick on the particle at ζ can be expressed solely in terms of the E_{zn} (or the corresponding potentials). The difficulty however, in computing the transverse deflection of the particle is that one must include as many modes as possible. The values of ω_n and k_n in Eq. (29) are obtained by solving the boundary value problem for a charge-free cavity or structure with a computer code such as SUPERFISH.[16] Figure 3 shows the results of such calculations of the dipole wake per unit accelerating cell for the SLAC disk-loaded structure for three different time durations. In particular, Fig. 3(a) indicates the size of the contribution of the higher modes to the total dipole wake.

For the purposes of estimating the characteristics of high gradient linacs for linear colliders, Palmer[3] has fitted such computer calculations of the initial linear rise, $_1W_t(\zeta)$, and the maximum of the wakefield, $_2W_t$, to obtain the scaling laws (in mks units):

$$_1W_t(\zeta) = 6.64 \cdot 10^{10} \text{ V/(C}-\text{m}^2) \; \zeta \; a^{-7/2} \; \lambda^{-1/2} \qquad (31)$$

for $\zeta \ll a$ and

$$_2W_t(\zeta) = 3.28 \cdot 10^{10} \text{ V/(C}-\text{m}^2) \; a^{-2.2} \; \lambda^{-0.8} \quad , \qquad (32)$$

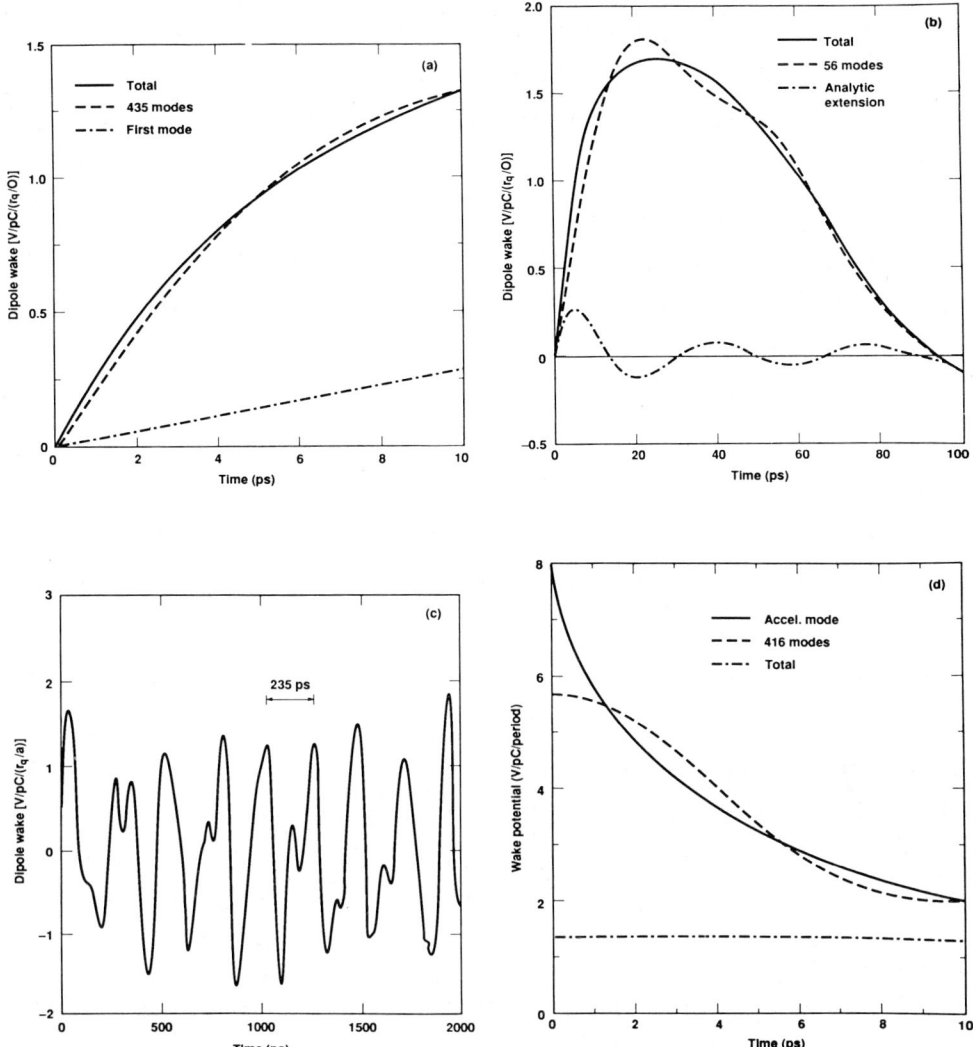

Figure 3. (a)-(c): Dipole wake per cell for the SLAC disk-loaded structures over various time intervals; (d) longitudinal wake per cell with a cell length of 3.5 cm and beam aperture radius of 1.163 cm.

which occurs at $\zeta \approx a$. For the SLAC geometry a reasonable fit to W_t at intermediate values of $\zeta < a$ is obtained from

$$W_t(\zeta) = \left[{}_1W_t^{-2}(\zeta) + {}_2W_t^{-2}(\zeta)\right]^{-1/2} \tag{33}$$

As one can see from Fig. 3(c), the dipole fields persist long after each beam pulse. Using an asymptotic approximation valid for long pulse trains, Yokoya[17] has computed the value of the the dipole wakefield for multi-bunch macro-pulses in long linacs. He finds that the multi-bunch wake field has the same form as the single bunch wake with the growth determined by the average current during the macro-pulse. For the case of no spread in betatron frequency and strong focusing ($k_\beta = 1$ m^{-1} $[E/2\text{ GeV}]^{-2/3}$), a fit to Yokoya's calculations of the wake of the jth pulse on the kth rf cycle yields

$$W(j) = 0.36 \; {}_2W_t \; j^{-0.8} \exp\left[\frac{-k}{Q^*} + 1.38 j^{1/2}\right] , \tag{34}$$

where ${}_2W_t$ is given by Eq. (32) and Q^* is the effective quality factor of the dipole modes. This expression is overly pessimistic for short (≈ 1 GeV) linacs designed to accelerate ~ 10 pulses in a macro-bunch.

As argued above, Q^* is approximately 70% of the value for the fundamental ($\approx 10^4$); hence, the decay term is of little help in preventing disastrous growth of the dipole mode. However, the frequency difference between the deflecting modes and the fundamental accelerating mode can, in principle, be used to out-couple the energy of these modes. Such a procedure is used in low-frequency cw linacs and has been cold-tested for disk-loaded structures by Palmer and his collaborators at Brookhaven and SLAC. They have been able to design structures with a Q of 20 for the dipole modes. In assessing system requirements for a XRFEL accelerator, one can characterize the requisite level of energy out-coupling by a dequeing parameter, D, where

$$Q^* = 0.7 \frac{Q}{D} . \tag{35}$$

This requirement is easily satisfied by the slotted structures cold-tested at BNL.

5.1.1. *Damping transverse motion.* The transverse wakes from a single pulse can be effectively controlled by BNS damping the transverse motion through the introduction of a spread in betatron frequency increasing from head-to-tail within the beam. This spread can be obtained by appropriately centering the pulse in phase with respect to the rf so as to maintain an energy spread between the front and back of the bunch. If the bunch is approximated as two point charges separated by $2\sigma_z$, the energy spread required is

$$\Delta E \approx 2E\sigma_{p,\text{Landau}} \approx \frac{eN_b}{4}(2\sigma_z)W_t\beta_y^2 \quad , \tag{36}$$

where ΔE and the betatron function are specified at the full energy of the linac. Solving for σ_p and substituting in Eqs. (27) and (28) yields the alignment tolerance requirements and the length of the correction section, respectively. If one uses Eq. (34) in Eq. (36), it is apparent that BNS damping will be insufficient to control multi-bunch transverse wakes in multi-GeV flavor factories. This technique may, however, be sufficient for several pulses in short (≈ 1–GeV) linacs suitable for early linear light sources.

5.2. Longitudinal Wakes and Beam Loading

As the beam particles move through the waveguide structure they will lose energy into the structure. For the disk-loaded structure this loss can be considered a diffraction loss by a plane wave with the same power spectrum and Poynting vector at the iris radius as the actual field due to the charge. The wake potential due to this process can be calculated using the same procedures as are used to compute transverse wakes; again as many modes as possible should be included. Fig. (3b)[1] shows the results of such a calculation for the SLAC structure. The importance of the higher order modes can be seen by the comparison of the total wake with that due to the excitation of the fundamental accelerating mode. (Note that a positive potential is decelerating.)

The beam loading potential is the convolution integral of the wake potential with the beam current. The value of the beam loading voltage, $E_b(t)$, within a

Gaussian bunch (centered at $t = 0$) is shown in Fig. 4 for bunches of three different lengths. The total energy gain of a particle at time t in the bunch is the sum of the accelerating and beam loading voltages:

$$E(t) = E_a \cos(\omega t - \theta) - E_b(t) \quad , \tag{37}$$

where θ is the phase angle by which the center of the bunch leads the crest of the rf accelerating wave. Both components of the energy gain give rise to an energy spread within a bunch of finite length. To some extent this spread can be reduced by choosing the q so that the negative-going beam loading balances the increasing accelerating voltage. This effect is illustrated schematically in Fig. 5(a) and quantified in Fig. 5(b). The penalty for reducing the energy spread in this manner is reducing the average value of the energy gain of the pulse, $<E>$, by a small amount.

As the acceptable energy spread of the beam is restricted (Sec. 1.) by FEL physics constraints (or luminosity constraints and energy resolution in flavor factories, one must be able to estimate how the wakefields affect ΔE. One begins with an estimate[3] of the longitudinal wake that depends on rf wavelength and iris aperture, a. For very short bunches, that is, for $\zeta \ll a$,

$$_1W_1(\zeta) = 1.78 \cdot 10^{10} \frac{\text{V}}{(\text{C}-\text{m})} a^{-2} \quad . \tag{38}$$

For $\zeta \approx a$,

$$_2W_1(\zeta) = 1.25 \cdot 10^{10} \frac{\text{V}}{(\text{C}-\text{m})} \zeta^{-1/2} a^{-1} \lambda^{-1/2} \quad . \tag{39}$$

For intermediate values of ζ,

$$W_1(\zeta) = \left[_1W_1^{-3}(\zeta) + {}_2W_1^{-3}(\zeta) \right]^{-1/3} \quad . \tag{40}$$

Next, following Palmer, one notes four distinct contributions to σ_p: (1) the average energy of the bunch is reduced (zeroth order); (2) the tail of the pulse has a greater loss than the head of the pulse (first order); (3) long pulses have a significant second order term with the rate of change of momentum falling at the

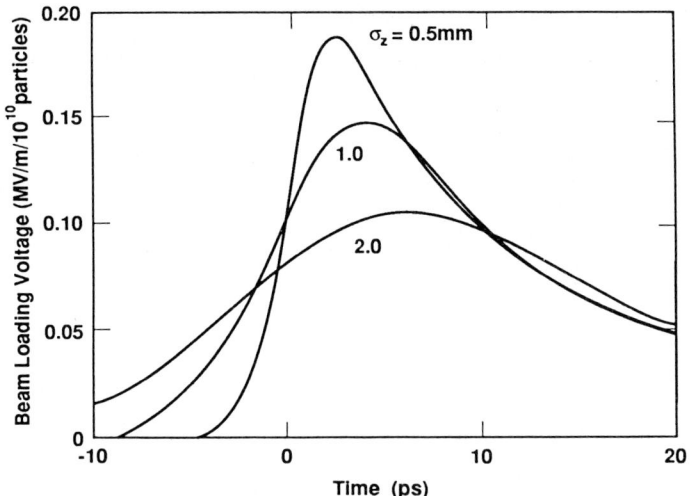

Figure 4. Beam loading voltage within a Gaussian bunch for the SLAC structure for three values of bunch length (adopted from Ref. 1).

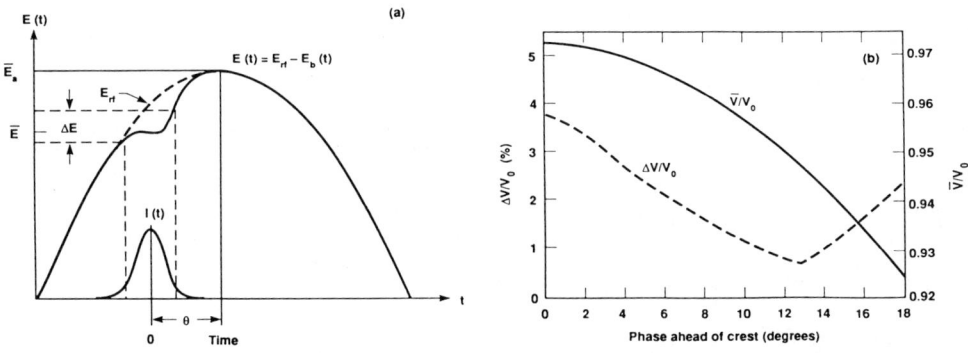

Figure 5. (a) Relationship between single-bunch beam loading $E_b(t)$, rf accelerating wave E_{rf}, and net gradient $E(t)$; (b) energy spread and average gain per particle for the SLAC structure with $E_a = 17$ MV/m, $N_b = 5 \times 10^{10}$, and $\sigma_z = 1.0$ mm (adopted from Ref. 1).

tail; (4) Gaussian bunches have a significant third order term. To account for each of these contributions in computing the beam loading, Palmer divides the bunch into four equal sub-bunches: two at $\sim 0.2\sigma_z$ and two at $\sim 1.4\sigma_z$. The energy losses of the four bunches are

$$V_1 = MW_l(0) \,, \tag{41a}$$

$$V_2 = M\left[W_l(0) + W_l(1.2\sigma_z)\right] \,, \tag{41b}$$

$$V_3 = M\left[W_l(0) + W_l(0.4\sigma_z) + W(1.6\sigma_z)\right] \,, \tag{41c}$$

$$V_4 = M\left[W_l(0) + W_l(1.2\sigma_z) + W(1.6\sigma_z) + W_l(2.8\sigma_z)\right] \,, \tag{41d}$$

where

$$M = \frac{eN_b}{4E_a} \,. \tag{42}$$

From these beam loading voltages, Palmer then computes the first, second, and third order momentum spreads in the four-bunch approximation:

$$_1\sigma_p = 0.05(V_3 - V_2) + 0.35(V_4 - V_1) \tag{43a}$$

$$_2\sigma_p = 0.25(V_3 + V_2) + 0.25(V_4 + V_1) \tag{43b}$$

$$_3\sigma_p = 0.22(V_3 - V_2) + 0.033(V_4 - V_1) \tag{43c}$$

A continuous pulse convolution integral gives more accurate results, but the four-bunch approximation should be sufficient for parameter surveys.

As noted above, the first order effect can be canceled by the rf with an appropriate choice of θ. Since some of this momentum spread is required for BNS damping of the transverse wake, one should choose

$$\tan\theta = \frac{1}{2\pi}\frac{\lambda}{\sigma_z}\left(_1\sigma_p - \sigma_{p,\text{BNS}}\right) \,, \tag{45}$$

where $\sigma_{p,\text{BNS}}$ is given by Eq. (36). Recall that $\sigma_{p,\text{BNS}}$ is removed at the end of the linac. Including the effects of the rf accelerating field, the total second order spread is

$$_2\sigma_{p,\text{tot}} = {_2\sigma_p} - 0.5\left(\frac{2\pi\sigma_z}{\lambda}\right)^2 \,. \tag{46}$$

For the purpose of designing the beam matching sections into the wiggler, one can use as the final momentum spread

$$\sigma_p = \left(2\sigma_{p,\text{tot}}^2 + 3\sigma_{p,\text{tot}}^2\right)^{1/2} . \tag{47}$$

5.2.1. *Beam Efficiency.* A quantity of considerable interest to the designer of linear colliders is the efficiency, η_b, with which the beam extracts energy from the rf structure:

$$\eta_b = esN_b \frac{<E>}{E_a^2} \approx \frac{esN_b}{E_a} \propto \frac{N_b}{\lambda^2} , \tag{48}$$

where the uncorrected elastance s is given by Eq. (2.9b). From Eq. (37) $<E>$ differs from E_a by the beam loading voltage, E_b. An approximation to Eq. (48) for structures with the ratio a/λ roughly the same as the SLAC structure is

$$\eta_b \approx 0.13\% \left(\frac{N_b}{10^{10}}\right) \left(\frac{100 \text{ mm}}{\lambda}\right)^2 \left(\frac{100 \text{ MeV/m}}{E_a}\right) . \tag{49}$$

The efficiency of transferring energy from the rf source to the beam is the product of the beam efficiency and the structure efficiency, i.e., $h_s\eta_b$.

A XREL or collider macro-pulse will contain B bunches of equal charge. In order that each bunch extract the same fraction of the the stored energy from the structure, the beam loading must be compensated for in one of several possible ways (Fig. 6). For example, the first pulse can be injected before the section is completely filled, and the bunch-to-bunch spacing and power flow can be adjusted so that the energy added between bunches just compensates for the pulse-to-pulse sag.

An alternative procedure suitable for use with pulse trains of duration much greater than T_{fill} is to recirculate the rf power that exits from a section by adding it back at the input of the same section. A steady-state condition can be achieved after $\sim 3-5\,T_{fill}$, at which point the beam is injected.

Although both schemes of compensating for beam loading will have the side effect of reducing the gradient of the linac, they allow one to raise the overall

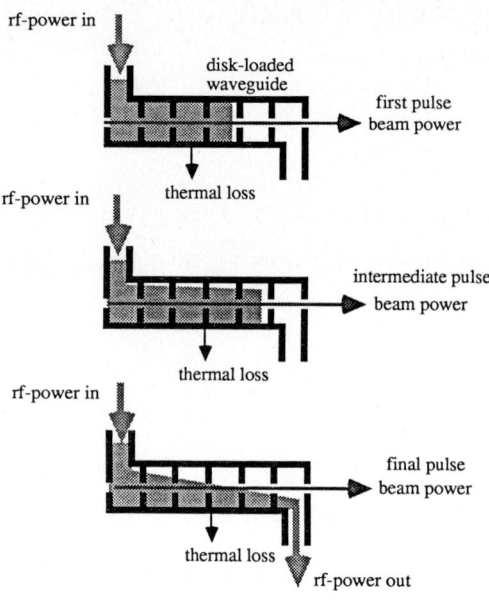

Figure 6. Efficient energy extraction from a travelling wave linac section for short pulse trains

electrical efficiency of the accelerator to

$$\eta_{\text{tot}} = h_s \eta_b \eta_{rf} \left(\frac{B T_{\text{fill}}}{T_{rf}} \right) ,\qquad (50)$$

where η_{rf} is the efficiency of converting electrical power to rf and T_{rf} is the total period during which rf is supplied to the high gradient structure. With both efficient rf power sources and beam-loading compensation it should be practical to obtain $\eta_{\text{tot}} \approx 10\%$.

From the analysis of the previous section one should note that increasing N_b to increase the beam efficiency is limited by wake field effects. Via longitudinal wakefields raising N_b will lead to the simultaneous introduction of an irreducible, nonlinear momentum spread in the beam. Raising the beam loading has the additional drawback of increasing the amount of energy spread needed to BNS damp the single bunch transverse wakefield. The length of the section designed to remove this energy spread will, therefore, increase with increased beam loading. Fortunately, however, in accelerators for linear light sources this increase in overall linac length is a 1–2% effect.

6. PULSE MANIPULATION

Linear light sources will probably have several output beamlines. In addition to directing the beam to the appropriate beamline, the final beam manipulation subsystem may be required to perform other functions:

1) Stretch the length of the bunch from its value in the high frequency, high gradient accelerator to a much longer pulse matched to user requirements.

2) Focus the beam from its ≈ 100 μm size in the accelerator to the ≈ 1 μm size appropriate to self-focused plasma wigglers or other plasma experiments.

3) Provide for a vacuum-to-gas transition in the case of plasma devices.

6.1. Pulse Stretching

The stretching of the pulse can be accomplished by taking advantage of the irreducible momentum spread acquired during the acceleration process. Although many forms of dispersive transport may be used as a debuncher, for scaling purposes it is convenient to choose the four dipole chicane arrangement (Fig. 7), typically used as a debuncher in conjunction with damping rings.

Dipoles 1 and 2 and dipoles 3 and 4 are separated by L_d meters; the distance between dipoles 2 and 3 is J_d meters. With four dipoles of equal strength and with the beam trajectory normal to the entrance and exit edges of the dipoles, the

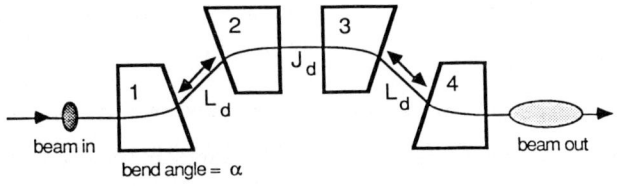

Figure 7. Four dipole pulse stretcher in a chicane arrangement

chicane can be made doubly achromatic to second order. The effect of the chicane is described[18] by a transfer matrix, M, with unit diagonal elements and two nonzero off-diagonal terms. The dispersion of the chicane is described by a debunching parameter (in units of cm / %$\Delta p/p$):

$$M_{34} = 2L \tan^2\alpha + 4\rho \tan\alpha + 4\rho \tan\alpha - 8\rho \sin\alpha \quad , \qquad (51)$$

where α is the angle of bend and ρ is the radius of curvature in the magnetic field. The only other nonzero matrix relates the entrance position to the exit angle; that is

$$M_{12} = 4\rho \tan\alpha + \frac{2L}{\cos^2\alpha} + J_d \quad . \qquad (52)$$

The magnetic field and the radius of curvature are related by

$$B\rho = 33 \text{ T}-\text{m} \left(\frac{E}{10 \text{ GeV}}\right) \quad . \qquad (53)$$

Precision, small aperture (4 to 5 cm) dipoles are practical with a field strength of 6.5 T. With such magnets, pulse stretching to ≈ 1 cm with $\Delta p/p = \pm 1\%$ is achievable in a chicane of overall length less than 7 m. Thus, compact linear light sources can tailor the duration of the pulse up to that available with circular synchrotron sources.

The drawback to stretching the pulse by more than a factor of several is that the peak current in the linac is raised to more than 1 kA. Although increasing the peak current does not lead to a critical constraint in the linac proper, it does make the task of designing the low emittance injector ever more difficult. In tradeoff studies it is probably unwise to drive peak current from the injector much above 1 kA.

6.2. Beam Matching

The matching section for a flavor factory or a plasma-focused wiggler XRFEL system is analogous to the final focus of a linear collider flavor factory and scales accordingly. In the final focus the beam radius must be reduced by a factor ≈ 100 in as short a distance as possible; i.e., the design must have the minimum achievable β^*. Let β_o^* be the minimum that can be achieved for a given momentum spread by a focal array with no chromatic correction. Chromatic correction elements can reduce this value by a factor S which scales inversely with σ_p. Thus,

$$\beta^* \geq \frac{\beta_o^*}{S} = \frac{\beta_o^* \sigma_p}{S_o} = 25 \, \beta_o^* \sigma_p \quad . \tag{54}$$

Unlike the case of the linear collider, the focal spot for a plasma focused XRFEL may not be highly elliptical; therefore, quadrupole doublet designs are ruled out. An appropriate design for scaling is a quadrupole triplet with a layout (Fig. 8) based on the SLC final focus, which brings the beam to a circular focus. One scales such a design by modifying all lengths by one factor and all transverse dimensions by another. The scaling factors are respectively f^*, the "ideal" focal length, and a^*, the aperture radius of the first quadrupole, which has pole tips of field strength B^*. The "ideal" focal length is

$$f^* = \left[\left(\frac{a^*}{B^*} \right) (B\rho) \right]^{1/2} \quad , \tag{55}$$

where $B\rho$ is given by Eq. (53). Palmer[3] expresses the performance of the focal system in terms of f^*, a^* and constants that depend on details of the magnet system: T, A_x, A_y and Λ.

In terms of the scaling parameters, the β^* of the focal array is

$$\beta_o^* = 2\sigma_p f^* T \quad . \tag{56}$$

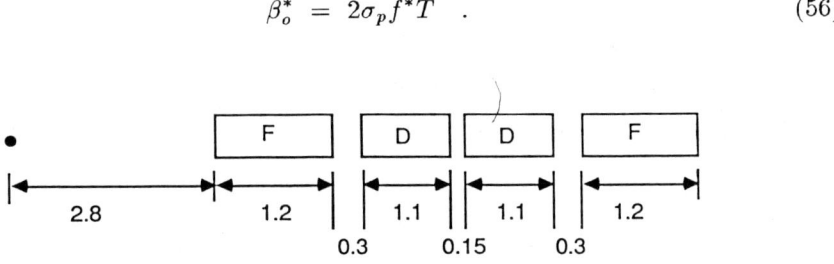

Figure 8. Quadrupole triplet array for beam matching based on the SLC design. All quadrupoles have a field gradient of 50 T/m. Distances are given in meters.

The maximum angular acceptance of the focal array and the free space before the first quadrupole are

$$\Theta_{x,y} = \frac{a^*}{A_{x,y} f^*} \qquad (57)$$

and

$$l_1 = \Lambda f^* \,, \qquad (58)$$

respectively. Combining Eqs. (54), (55) and (56), one obtains

$$\beta^* = \left(\frac{T}{S_o}\right) \beta \rho \left(\frac{a^*}{f^* B^*}\right) \sigma_p^2 \,. \qquad (59)$$

For this design, $T = 2.96$, $A_x = 4.3$, $A_y = 3.2$, $\Lambda = 1.36$, and $S_o = 0.04$.

In order to focus the beam to the very small radii, the pole tip fields must either be very strong or the aperture must be very small. Electro-mechanical engineering considerations are likely to limit the minimum aperture of superconducting quadrupoles to 2–3 cm. In that case the field strength will have to be raised to > 6 T. The alternative is to choose a conventional magnet design with iron pole pieces. For this case the pole tip fields will be limited to ≈ 1.5 T. More than compensating for this limitation is the fact that the field can be led into a very small region of space (apertures ≈ 1–2 mm) at the expense of magnet efficiency. Therefore, the system designer should opt to use conventional magnets, even though cryogenic systems will be almost mandatory in the design of a pulse stretcher (if that is desired).

6.3. Gas Matching for Self-Focused XRFELs

If the size of the focal spot is not equal to the self-focused radius of the beam in gas, the equilibrium beam size will differ from the size of the focal spot. This consideration is especially important in the case that the vacuum-air transition is a very thin foil rather than a differentially pumped transition. Foils may be useful in bringing the accelerator to early operational status at low repetition rates.

If the focal system produces a spot of radius a_f, then in the gas the beam will suffer radial oscillations, eventually settling to an equilibrium radius given by

$$a_{eq}^2 = a_f^2 \exp\left(\theta_{tot}^2 \frac{\gamma I_o}{I_b} - 1\right) \,, \qquad (60)$$

where I_o=17 kA. The mean betatron angle, θ_{tot}, is determined by a combination of the emittance and the foil induced scattering:

$$\theta_{tot}^2 = \left(\frac{\varepsilon_n}{\gamma a_f}\right)^2 + \theta_{scat}^2 \quad . \tag{61}$$

For very thin foils of thickness X, the Moliere-corrected rms scattering angle is

$$\theta_{scat} = \left(\frac{28}{\gamma}\right)\left(\frac{X}{X_{rad}}\right)^{1/2}\left[1 + 0.111\log_{10}\left(\frac{X}{X_{rad}}\right)\right] \quad , \tag{62}$$

where X_{rad} is the radiation length of the foil material or of the gas in the transition region in the case of a plasma assisted XRFEL.

7. CHOICE OF FREQUENCY AND RF POWER SOURCES

The generation of bright beams at multi-GeV energies requires the accelerator designer to consider new classes of machines that can operate with gradients exceeding 200 MeV/m in structures suitable for high-repetition rate-operation. The approach must be highly reliable and energy efficient, if the electrical power demands of accelerator operation are to remain within currently accepted bounds for most facilities. As discussed in the Sec. 3, the accelerating gradient can be increased and the energy required per meter of accelerator structure reduced by scaling the (effective) rf structure of the accelerator to high frequencies ($\geq$10 GHz.) The practical limits of this approach for conventional rf cavities are set by the electron induced breakdown limit and the surface heating limit, with little gain in gradient being achievable for frequencies exceeding 30 GHz. Considering that both the cost of fabricating the rf structure and the difficulty of maintaining the alignment of the structure is likely to increase once the cavity is miniaturized beyond a scale size of $\approx$ 2 cm, one can select 10–20 GHz as the best frequency range for a compact accelerator. If standing-wave linacs with geometries more complex than that of the disk-loaded waveguide are used to accelerate long pulse trains, ease of fabrication will tip the balance even further in the direction of choosing longer rf wavelengths.

The peak rf power needed to drive an accelerator based on a generic $2\pi/3$ disk-loaded waveguide can be determined by substituting Eqs. (6)–(13) into Eq. (16).

A useful, though less exact, estimate can be obtained by ignoring the dependence of the elastance on the group velocity. Scaling the SLC linac design at constant β_g–that is, constant a/λ– one can rewrite Eq. (16) as

$$P_{rf} = 74 \text{ MW/m} \left(\frac{E_a}{100 \text{ MeV/m}}\right)^2 (h\tau)^{-1} \left(\frac{\lambda}{105 \text{ mm}}\right)^{1/2}. \qquad (63)$$

For the SLC design $h\tau$, the product of the structure efficiency and attenuation parameter is ≈ 0.3; halving the fill time will reduce $h\tau$ to ≈ 0.2. Thus, a rough estimate of the total peak power needed to drive a accelerator is

$$P_{rf} \approx 60 \text{ GW} \left(\frac{E}{10 \text{ GeV}}\right) \left(\frac{\lambda}{105 \text{ mm}}\right)^{1/2}. \qquad (64)$$

One concludes that a new class of rf power sources will be needed to drive compact accelerators of beams for compact linear light sources and flavor factories.

7.1. Relativistic Klystrons

The high energy accelerator community has considered many approaches to providing the large amounts of rf power at X-band and higher frequencies needed to achieve ultra-high accelerating gradients. Recently a consensus[19] has emerged in the U.S., Western Europe, and the USSR that the range from 10 to 30 GHz holds the most promise for the next generation TeV linear collider. The same considerations apply to the accelerator for a compact linear light source. At the low end of this frequency range one can perhaps combine many conventional klystrons with an advanced rf pulse-compression network to power the accelerator. More novel approaches, however, seem to offer the required performance levels at substantially lower cost and complexity. One of these approaches, the relativistic klystron, offers particular promise, especially if high repetition rates are desired as in the case of flavor factories. The relativistic klystron[20] (Fig. 9) is a concept for modulating a multi-kA, multi-MeV beam produced with an linear induction accelerator (LIA) at the desired rf frequency (10 to 15 GHz) and using the modulated high current beam to excite an rf generating transfer structure. The high peak power rf is then fed via waveguides to the miniaturized rf cavities of the high gradient (200–400 MeV/m) accelerator of the high energy beam.

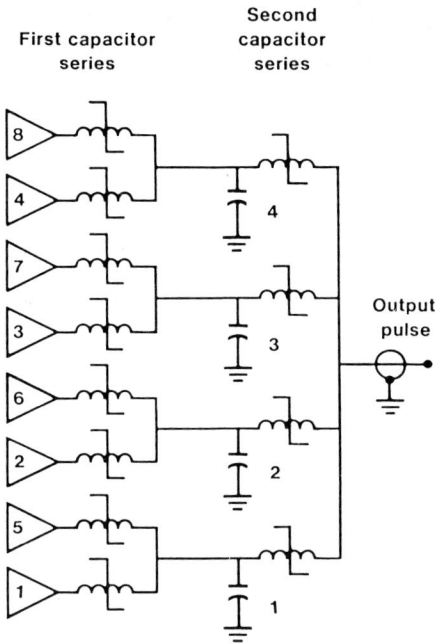

Figure 9. Conceptual design of the front end of a relativistic klystron powered by a multi-kA, multi-MeV beam from an induction linac.

Three basic variants of the relativistic klystron concept are now under active investigation by a SLAC/LLNL/LBL collaboration. These are:

1) Extrapolation of conventional (velocity modulation), high gain klystron design to 1–3 MV operation ($\approx$ 1 GW).

2) Adiabatic deceleration of an energy modulated, bunched, 5–10 MeV beam with multiple klystron output cavities ($\approx$ 10 GW).

3) Alternating multiple cavity deceleration with induction re-acceleration of the high current beam to form a single power "tube" ($\approx$ 100 GW).

Early experiments[21] at LLNL have already yielded record power levels ($\approx$ 80 MW) for X-band klystrons at high conversion efficiency ($\approx$ 55%). Over 200 MW have been produced at higher frequency (11.4 GHz) although at lower efficiency. This power was sufficient to yield an accelerating field of 130 MeV/m in a small section of high gradient structure. The most convincing demonstration of the practicality of the relativistic klystron approach would be the construction and operation at the earliest possible date of an ultra-compact (5-m), 1-GeV linac powered by the beam from the 7-MeV, 3-kA ETA-II at LLNL.

Before such an ambitious goal can be come a reality, the collaboration must address several fundamental physics issues relating to (1) rf generation and power handling and (2) stability of the drive beam in the rf generating structures. Still, for the purposes of scoping studies, the design space for the induction linac beam source can already be roughly delimited:

1) Both the acceleration and deceleration rates for the high current beam can exceed 1 MeV/m.

2) Using present design techniques, the maximum rf current that can be handled efficiently with standing wave klystron cavities is 3–4 kA.

3) The minimum voltage beam that can be efficiently decelerated in a multi-cavity transfer structure is 2–3 MeV.

4) Maximum beam-to-rf conversion efficiencies of 50–70% seem possible.

5) The efficiency of converting electrical energy into usable beam energy with induction linacs is $\approx 50\%$.

Thus, a flavor factory using the relativistic klystron approach would require $\approx$60 MeV of LIA cells operating with a 2-kA beam to provide sufficient power to drive two 6-GeV, compact rf linacs operating at $\approx$11 GHz.

The large power flow carried by LIA beams is not the sole reason for their interest to the high energy accelerator designer. Because the magnitude of the transverse wake varies as a^{-3}, one is unlikely to scale the disk-loaded structure to X-band or higher frequencies while maintaining $A_l \equiv a/\lambda$ constant. Rather one should increase A_l by a factor of 2 to 3 to control wakefields. Increasing A_l has several additional effects. First, by Eq. (7), raising A_l will increase the rf group velocity thereby shortening the fill time of the structure. The rf attenuation time of the structure also varies as $\lambda^{3/2}$. Therefore, scaling with constant attenuation parameter t implies shortening the fill time from the ≈ 1 μs at SLC to ≈ 50 ns for an X-band linac.

These short fill times match the pulse durations of modern induction linacs. The second effect of increasing A_l is to raise the maximum surface field that is associated with a given accelerating field. As long as sufficient rf power is available, one will probably choose to operate the linac at as close to the maximum surface field as is practical. Therefore, shortening the fill time by opening the iris aperture comes at the expense of reducing gradient.

With respect to the design of its rf power source, a linear light source could have one significant difference from the linear collider. The collider's power source must deliver rf for approximately one fill time. In contrast, some applications of compact linacs may require hundreds of bunches in a macro pulse. In that case, the rf power must be supplied for a time, T_{rf}, several times longer than the fill time; i.e., ≈ 300–1000 ns.

Modern induction linacs with ferrite cores produce pulses with a duration of several tens of nanoseconds. From the design principles described in Ref. 22, one sees that scaling ferrite loaded LIAs to produce continuous pulses much longer than ≈ 100 ns is probably impractical. By changing the ferromagnetic material to a dielectric coated metallic glass, the pulse duration may be extendible to ≈ 200–300 ns. However, the beam break-up effects of this change are not known. The first induction linacs such as the Astron with soft iron tape cores produced 300 ns pulses. Indeed, a 1-μs pulse core was built at NBS in the early 1970s. At 1 μs such cores are hardly small or inexpensive. Moreover, eddy current losses in the metallic cores will be high. The alternative is to recycle the induction cores at multi-MHz rates with branched magnetic drives.[23]

7.1.1. *Branched Magnetic Drives.* The magnetic pulse compressors that supply pulsed power in modern LIAs consist of a repetitive pulse source followed by several serial stages of pulse compression. These networks can drive the LIA at any repetition rate up to the maximum firing rate of the primary commutator. In the ETA-II, for example, this rate is set by the recovery time of the thyratron switches. If the application were to require a fixed number (3–10) of output pulses from the LIA in a burst (1–2 μs), a new scheme would become feasible. The technique, called branched magnetics, involves deploying several capacitive stores and switches in parallel that are fired in a sequence at any repetition rate permitted by the load. One such circuit was tested with an eight-pulse burst (Fig. 10). This proof-of-principle experiment demonstrated pulse generation into a resistive load at 16 MHz.

In the application of this approach to synthesize microsecond bursts of LIA beam power, the firing rate will be limited by the time required to reset the induction cores in the linac. Unless the reset times are at least twice as long as the original pulse time of the core, the amount of power consumed by the reset circuits will be comparable to that of the beam drive circuits. Hence, the highest practical duty factor of the LIA during the macro-pulse is $\approx 30\%$. Therefore, even with branched

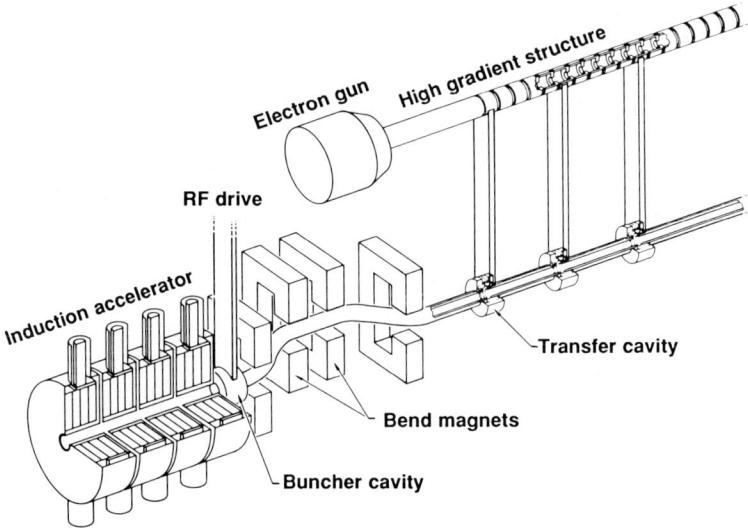

Figure 10. Simplified cicuit diagram of the burst mode generator designed to provide any desired repetition rate for a fixed number of pulses (in this case, eight). The numbers indicate the order in which the capacitors are discharged.

magnetics, supplying rf power for $\approx$ 1 μs will either require multiple LIAs or an innovation in rf power handling. One such idea is recycling the rf power that leaves one linac section into another section. Maintaining adequate rf phase control and stability at high power is a critical physics issue for this latter approach.

From the point of view of making the LIA beam source as simple and economical as possible, the accelerator designer should endevour to make T_{rf} as small as possible. In particular, the fill time must be kept below 100 ns without making L_f so short that both the peak power and number of rf feeds are raised excessively. Therefore, even if the size (gradient) constraints were eased considerably, the accelerator designer would be strongly motivated to build the linac at X-band or higher frequency.

8. COST AND SIZE ESTIMATES

The primary capital cost element of the accelerator will be the LIA driven, relativistic klystron power sources for the rf linac. The considerations that govern the installed capital cost of such rf sources is discussed at length in Ref. 22 for the case of single branch magnetic drives. It remains only to indicate how to modify those scaling relations to the case of a an induction drive accelerator with branched magnetics.

One begins by determining the total active length of high gradient accelerator, including the section to correct energy spread:

$$L_{\text{act}} = 50 \text{ m} \left(\frac{E}{10 \text{ GeV}}\right)\left(\frac{E_a}{200 \text{ MeV/m}}\right) + L_{\text{corr}} \quad . \tag{65}$$

The overall length of the linac is then

$$L_{\text{acc}} = (1 + F_q) L_{\text{act}} \quad . \tag{66}$$

Using Eq. (16), one then calculates the total voltage to be supplied by the induction cells as

$$V_{\text{ind}} = \frac{\Pi_{rf}}{I_{\text{ind}} \eta_{\text{rf}}} = 10^{-3} P_{\text{rf}} L_{\text{acc}} \left(\frac{2 \text{ kA}}{I_{\text{ind}}}\right)\left(\frac{0.5}{\eta_{\text{rf}}}\right) \quad . \tag{67}$$

where current in the LIA is I_{ind}. The LIA pulse length, T_{ind}, the number of LIAs, N_{ind}, and the total number of magnetic drive branches in all LIAs, N_{br}, depends on the maximum pulse length, T_{imax}, assumed for the induction cores and on the induction reset duty factor during the macro-pulse, Δ_{ind}. In terms of the standard number theoretic functions,

$$N_{ind} = \text{Min} \left[\left(\text{Int} \left(\frac{T_{rf}}{T_{imax}} \right) + 1 \right), \left(\text{Int} \left(\Delta_{ind}^{-1} \right) + 1 \right) \right] \quad ; \quad (68)$$

$$N_{br} = \text{Int} \left(\frac{T_{rf}}{T_{imax}} \right) + 1 \quad ; \quad (69)$$

$$T_{ind} = \frac{T_{rf}}{N_{br}} \quad . \quad (70)$$

Once these quantities are determined the cost scaling algorithm can be applied with the modifications indicated in the Appendix. For quick estimates of the cost of powering the one can use the graph of rf cost vs total "tube" power given in Fig. 11.

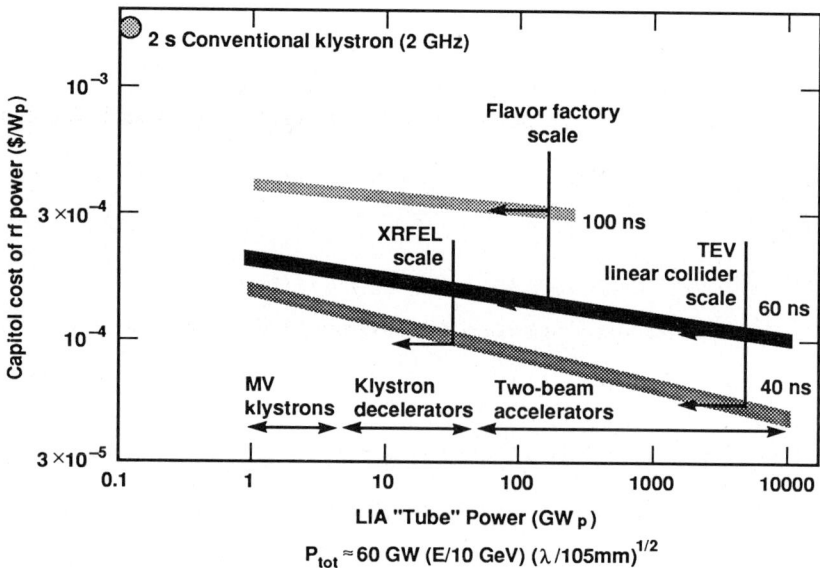

Figure 11. Rf cost vs. "tube" power for various LIA pulse lengths

With respect to scaling the cost per active meter of the disk-loaded waveguide, actual data at many frequencies are scarce. The following heuristic equation seems, however, to summarize the experience of the accelerator community:

$$C_{\text{wave}} = 45 \text{ k\$/m} \left(\frac{\lambda}{500 \text{ mm}}\right)^{1/3} + 2.5 \text{ k\$/m} \left(\frac{50 \text{ mm}}{\lambda}\right)^{5/2} + 5 \text{ k\$/m} \quad . \quad (71)$$

The first term, representing a cost of materials and machining, scales as the waveguide volume. The second term is an alignment cost, which scales as the transverse wake. The final term is a fixed handling cost.

Klystron transfer cavities and power feeds to the high gradient structure are estimated to cost

$$C_{\text{cav}} = 20 \text{ k\$/m} \quad . \quad (72)$$

The final beam manipulation beamline is estimated to cost

$$C_{\text{bmln}} = 200 \text{ k\$/m} \quad . \quad (73)$$

As is done for the cost of the rf power, the costs of Eqs. (71), (72), and (73) should be multiplied by ≈ 1.35 to account for the associated manpower needed for installment, procurement, and project management.

ACKNOWLEDGMENT

The physics reported here has been borrowed liberally from many sources. With respect to the design of high gradient rf accelerators, the work of Robert Palmer (SLAC and BNL) and Perry Wilson (SLAC) in analyzing linear colliders has provided indispensible guidance. Andrew Sessler (LBL) has been an effusive source of ideas concerning the use induction technology to power high gradient accelerators. Dan Birx (Science Research Laboratory) has provided valued assistance in extrapolating induction linacs into the realm of branched magnetics and hitherto unexplored cavity geometries. Others from LLNL who have contributed importantly include William Fawley, Kelvin Neil, and Art Paul.

Appendix A. COST SCALING ALGORITHM FOR BRANCHED DRIVES

Variables used in the cost scaling Eqs. (22) for the induction driver with branched magnetics include repetition frequency f in Hz, the total accelerating voltage V in MV, the total volt-seconds of ferrimagnetic core $W = VT_{\text{ind}}$, the peak gap stress E_g in kV/cm, the single pulse energy $E = VI_{\text{beam}}T_{\text{ind}}$ in joules, the average power $P = fE$ in MW, the effective gradient G in MV/m, and the inner radius of the induction cells, r_i, in cm. In the scaling equations italicized quantities refer to injector voltage, power, pulse energy, etc. Costs are specified in constant FY87 k\$. The branched magnetic drives are composed of N_{br} branches and N_{ind} induction linacs.

A.1. Injector Subsystem

The scaling equation for the injector is divided into five separate components:

$$C_{\text{inj}} = \left\{ 825\left(\frac{W}{0.17}\right)_{\text{cells}} + 475\left(\frac{E}{450}\right)_{\text{mag}} + 450\left(\frac{P}{2.25}\right)_{\text{isps}} \right.$$
$$\left. + 165\left(\frac{V}{3}\right)^3_{\text{vac}} + (60)_{\text{fixtures}} + 90\left(\frac{I}{3000}\right)^{1/2}_{\text{cathode}} \right\}. \quad (A1)$$

As the branched drives may comprise as many as three separate induction linacs, the injector cost has been multiplied by N_{ind} in the estimate of total hardware cost.

A.2 Beam Transport Subsystems

For solenoidal transport, the cost has three separate components:

$$C_{\text{sol}} = \left\{ \left[\left(57 + 81\left(\frac{r_i}{7.5}\right)\left(\frac{I}{3000}\right)\right)\left(\frac{V}{1.5}\right)\left(\frac{I}{3000}\right)\left(\frac{r_i}{7.5}\right)\left(\frac{0.75}{G}\right) \right]_{\text{focus}} \right.$$
$$\left. + \left[42\left(\frac{V}{1.5}\right)\left(\frac{r_i}{7.5}\right)\left(\frac{0.75}{G}\right) \right]_{\text{steer}} + [78]_{\text{match}} \right\}. \quad (A2)$$

For the alternative laser guiding scheme the scaling equation is

$$C_{\text{laser}} = \left(\frac{7.5V}{G}\right) N_{\text{ind}}, \quad (A3)$$

which includes laser, gas handling, and matching magnet costs for $f \leq 250$ Hz.

A.3. Accelerator Cell Subsystems

The cell block cost is

$$C_{\text{block}} = \left[234 \left(\frac{V}{1.5}\right) \left(\frac{r_o}{20}\right)^2 \left(\frac{0.75}{G}\right) \left(\frac{175}{E_g}\right)\right] N_{\text{ind}} , \quad (A4)$$

where r_o is related to r_i by

$$r_o = r_i + \left(\frac{GT_{\text{ind}}}{p}\right) \Delta B . \quad (A5)$$

Here p is the ferrite packing fraction ~ 0.8 and ΔB is the total flux swing ~ 0.6 Wb/m². The ferrite cost is given by

$$C_{\text{ferrite}} = \left[140 \left(\frac{W}{0.225}\right) \left(\frac{r_o + r_i}{27.5}\right)\right] N_{\text{ind}} , \quad (A6)$$

where we have used the relation between the ferrite area A and volume $\pi A(r_o + r_i)$. The cost of intermediate stores and power supplies scales as

$$C_{\text{isps}} = \left[678 \left(\frac{P}{3.2}\right) + 5 \left(\frac{E}{630}\right)\right] N_{\text{br}} . \quad (A7)$$

If the repetition rate of the system is less than 600 Hz, the power in Eqs. (A1) and (A7) should be replaced by a value corresponding to 600 Hz operation. The scaling for the magnetic pulse compressors is

$$C_{\text{mag}} = \left[241 \left(\frac{E}{630}\right)\right] N_{\text{ind}} . \quad (A8)$$

For power sytems delivering pulses at repetition rates $\approx$ 100 Hz the magnetic modulators and intermediate stores can be re-engineered to reduce costs by nearly factor of two. The cost of the strongback alignment structure is

$$C_{\text{strong}} = \left[450 \left(\frac{V}{12.5}\right) \left(\frac{r_o}{20}\right)^2 \left(\frac{0.75}{G}\right) K\right] N_{\text{ind}} . \quad (A9)$$

where the constant K depends on the focusing scheme; namely,

$$K = 1.0 \text{ for quadrupoles,}$$
$$= 0.6 \text{ for solenoids,}$$
$$= 0.5 \text{ for laser guiding.}$$

A.4. Ancillary Subsystems

The cost for low and extremely low conductivity water will scale as

$$C_{lcw} = 90 \left(\frac{P}{13.1}\right) N_{br} \quad , \tag{A10}$$

for $f > 100$ Hz. For $f < 100$ Hz one should use the value of C_{lcw} at $f = 100$ Hz. The vacuum system is scaled as if it were pumping speed limited (valid for $r_i < 2.5$ cm):

$$C_{vac} = 660 \left(\frac{V}{12.5}\right) \left(\frac{r_i}{7.5}\right)^2 \left(\frac{0.75}{G}\right) K N_{ind} \quad , \tag{A11}$$

where K is the same as in Eq. (A9). The cost of electrical fluids is proportional to average beam power;

$$C_{fluid} = 542 \left(\frac{P}{13.1}\right) N_{br} \quad . \tag{A12}$$

The scaling of dump costs is similar to that for reactors; *i.e.*, \$1 per watt of time average beam power into the dump, P_d. For the relativistic klystron assume that the average beam voltage at the dump, V_d, is 5 MeV; hence, $P_d = V_d T_{ind} I_{beam} f$. The cost of the dump is, therefore,

$$C_{dump} = 1000 P_d N_{ind} \quad . \tag{A13}$$

The cost of fixtures is proportional to the length of the induction linac:

$$C_{fixture} = 20 \left(\frac{V}{12.5}\right) \left(\frac{0.75}{G}\right) N_{ind} \quad . \tag{A14}$$

The cost of the instruments and controls scales as a fixed (buy-in) value plus a percentage of the cost of the hardware to be monitored and controlled:

$$C_{i\&c} = 1500 + 0.04 (C_{inj} + C_{cell} + C_{focus}) \quad . \tag{A15}$$

Summing the cost equations yields the total hardware cost for the induction linac driver.

A.5. Installation and Engineering Support

The installation costs for the base design were estimated for each component; for estimating purposes the installation costs can be taken as a fixed percentage of the total hardware costs adjusted to the fully loaded labor rate, R, in k$/man-month.

$$C_{install} = 0.09\, C_{hardware} \left(\frac{R}{10}\right) . \qquad (A16)$$

Similarly, a cost for engineering management and support is estimated as a percentage of the hardware costs; i.e.,

$$C_{engin} = 0.125\, C_{hardware} . \qquad (A17)$$

Supplies and equipment used for engineering and installation increase the total cost by 10%;

$$C_{s\&e} = 0.1\, C_{hardware} . \qquad (A18)$$

Adding these installation costs to the hardware cost yields a total cost which includes $\sim 10\%$ contingency distributed (unevenly) among the various cost centers.

REFERENCES

[1.] P. B. Wilson, "High Energy Electron Linacs," SLAC-PUB-4295, (1982); also contained in the Proceedings of the 1981 High Energy Particle Accelerator Summer School published by the American Institute of Physics.

[2.] P.B. Wilson, "Linear Accelerators for TeV Colliders", SLAC-PUB-3674 (Rev.) (1985)

[3.] R. Palmer, "The Interdependence of Parameter for TeV Linear Colliders", SLAC-PUB-4295 (1987).

[4.] U. Amaldi, "Introduction to the Next Generation of Linear Colliders", CERN Report EP/87- 28 August 1987.

[5.] R. Bonifacio, C. Pellegrini, and N. Narducci, Free Electron Generation of Extreme Ultraviolet Coherent Radiation, ed. J. Madey and C. Pellegrini (Am. Inst. Phys., New York, 1984) p. 236.

[6.] W. A. Barletta and A.M. Sessler, Radiation from Fine, Intense Self-Focussed Beams at High Energy, UCRL-98767 (1988).

[7.] Proceedings of the ICFA Workshop on Low Emittance e^-e^+ Beams, Brookhaven National Laboratory, March 1987, BNL-52090 (1987).

[8.] R. Sheffield, et al., Ibid. p.141

[9.] J. S. Fraser, et al., Proceedings of the 1987 IEEE Particle Accelerator Conference, p. 1705.

[10.] R. Steining, The Status of the Stanford Linear Collider, Proceedings of the 1987 Particle Accelerator Conference, p. 1.

[11.] H. Winick, Nuc. Instr. and Meth., A261, 9 (1987).

[12.] J.W. Wang, V. Nguyen-Tuong, and G. A. Loew, "RF Breakdown Studies in a SLAC Disk-loaded Structure", 1986 Linear Accelerator Conference Proceedings, SLAC-RPT-303, 461 (1986)

[13.] Z. D. Farkas, P. B. Wilson, "Comparison of High Group Velocity Structures", SLAC-PUB-4088 (1987)

[14.] Z. D. Farkas, "The Roles of Group Velocity, Frequency and Aperture in Traveling Wave Linear Accelerator Design", SLAC internal report AAS Note 33 (1987)

[15.] R. Ruth, "Emittance Preservation", SLAC-PUB-4436 (1987); The technique analyzed by Ruth was first described by V. E. Balakin, A. V. Noppvokhatskii, V. P. Smirnov, Transverse Beam Dynamics, Proceedings of 6th National Conference on Accelerators, Dubna, 1978.

[16.] K. Halbach and R.Holsinger, Particle Accel. 7, 213 (1976)

[17.] K. Yokoya, SLAC internal report AAS-Note 26 (February, 1987).

[18.] A.C. Paul, private communication (1987).

[19.] see, for example, Proceedings of Workshop on New Developments in Particle, ed. S. Turner, CERN 87-11 (1987).

[20.] A.M. Sessler and S.S. Yu, "Relativistic Klystron Two-Beam Accelerator", Phys. Rev. Lett. 58, 2439 (1986).

[21.] M. Allen et al., SLAC-PUB-4650 and UCRL-98843 (June 1988) submitted to Proceedings of European Particle Accelerator Conference (1988).

[22] W. A. Barletta, "Cost Optimization of Induction Linac Drivers of Linear Colliders", LLNL Report UCRL-95909 (1986). Proceedings of the 3rd International Symposium on Ultra High Energy Accelerators, Orsay (1987).

[23] D. L. Birx, L.L. Reginato, J. A. Schmidt, "Investigation into the Repetition Rate Limitations of Magnetic Switches", LLNL report UCRL-87278 (1982).

LINEAR COLLIDER REGIMES

Ugo Amaldi

CERN, Geneva, Switzerland

1. INTRODUCTION

By now it is well understood that the scaling laws of linear colliders are *not* very favourable: when in a 'Gedanken' experiment we increase the energy and pass from center-of-mass energies around W = 100 GeV (as today achieved, as far as energy is concerned, at the SLAC Linear Collider = SLC) to one TeV and, even more, to many TeV's, one is obliged to change the bunch-bunch *'regime'*. At low energies the electrons and positrons radiate classically, when bent by the megagauss magnetic fields produced by the opposite bunch. At many TeV's, to obtain the needed large luminosities (L > 10^{33} cm^{-2} s^{-1}) the dimensions and densities of the bunches have to be modified and the average critical energy of the radiated photons $\bar{E}_c$ becomes larger than the beam energy E_0; this is the so-called *quantum regime*, which entails bunches which must be few microns long and have transverse dimensions of some tens of Ångstroms. As for storage rings, synchrotron radiation is the cause of the poor scaling laws of linear electron-positron colliders which I want to discuss. More detailed arguments along the same lines can be found in the presentation which I gave at the CERN-USA Accelerator School[1] in 1986.

2. DISRUPTION

We call *disruption* the focusing (or defocusing) produced on the particles of one bunch by the electric and magnetic fields of the opposite bunch. Its intensity is described by the *'disruption parameter'* D.

An electron (positron) deflected by the opposite bunch – which is supposed to have a 'round' cross-section of r.m.s. radius σ_x and length σ_z and to contain N particles – radiates electromagnetic energy. A typical particle at a distance σ_x from the axis of the bunch sees a magnetic field proportional to the incoming current ($\propto$ N/σ_z) and inversely proportional to σ_x: B $\propto$ N/($\sigma_x \sigma_z$). The radius of curvature in this field, supposed to be uniform, is proportional to γ/B and thus

$$\rho = \gamma \sigma_x \sigma_z/(N r_e) , \qquad (1)$$

where r_e = 2.82 10^{-13} cm is the classical electron radius. The deflection angle is $\theta = \sigma_z/\rho$ and, for small deflections, the focal distance is F = $\sigma_x/\theta = \sigma_x\rho/\sigma_z$. The disruption parameter is defined as D = σ_z/F, so that it is small when the focal distance is large with respect to the length of the bunch. By combining D = $\sigma_z^2/\sigma_x\rho$ with Eq. (1) we obtain, for round bunches of r.m.s. radius σ_x and r.m.s length σ_z,

$$D = \frac{r_e N \sigma_z}{\gamma \sigma_x^2},\qquad(2)$$

where $\gamma = E_0/mc^2$. When $D \leq 0.5$ the focusing is very little and one can indeed consider the bunch as a lens of focal length F; for $D \geq 0.5$ there is a 'pinch' effect and the effective transverse dimensions of the bunches during the collisions are *smaller* than what one would compute from the properties of the final focus system.

The *natural* transverse area of the bunch is such that

$$\sigma_x^2 = \varepsilon_n \beta^*/\gamma,\qquad(3)$$

where ε_n is the *'invariant transverse emittance'* of the bunches (which does *not* change in a linear accelerator without wake field and space charge effects) and β^* the β-value of the final focus system at the interaction point. It can be shown[2] that, for a fixed magnetic field at the poletip of the final focusing quadrupoles, the value of β^* increases with the energy of the focused particles as $\gamma^{1/3}$.

The pinch effect reduces the transverse dimensions with respect to σ_x, so that the effective radius becomes $\sigma_x/\sqrt{H_D}$. The luminosity increases and can be written in the form

$$L = \frac{f_r N^2}{4\pi \sigma_x^2} H_D,\qquad(4)$$

where f_r is the repetition rate of the bunch-bunch collisions and H_D is the *'enhancement factor'* due to the pinch effect.

For round beams Chen and Yokoya — by using simulation methods[3] — have obtained simple formulae which express H_D as a function both of D and of the ratio $A = \sigma_z/\beta^*$:

$$\begin{array}{ll} 1 + 2D/3\sqrt{\pi}, & 0 \leq D \leq 0.5, \\ 1 + 2D/3\sqrt{\pi} + 0.43\,[\ln(D/A)]^2, & 0.5 \leq D \leq 2, \\ 1.6 + 0.43\,[\ln(D/A)]^2, & 2 \leq D \leq 100. \end{array}\qquad(5)$$

Theses formulae are valid for round beams: $\sigma_x = \sigma_y$. Usually $D \simeq 30$ is considered to be a maximum value for the disruption parameter. For $D \geq 2$ the focusing is so strong that some particles oscillate. The average number of oscillations performed in $n = \sqrt{D/2\pi}$ so that $n = 1$ for $D \simeq 12$. Chen and Yokoya found that the particles, coming from a narrow ring in the transverse plane, are focused on the beam axis within the oncoming bunch and give rise to the second term in the enhancement factor of Eqs. (5b) and (5c). For $D = 5$ and $A = 1/4$ Eq. (5c) gives $H_D = 5.5$, a sizeable effect indeed.

Recently Balakin and Solyak[4] contested Eq. (5c) by claiming that above $D \simeq 10$ the numerically calculated enhancement factor *decreases* with D, as initially found by Hollebeek[5]. At $D = 20$ the two results differ by a factor 2. In the following we shall not consider disruption parameters larger than 4, and the discrepancy does not influence our arguments.

By combining Eqs. (2) and (4) with the definition of *beam power*, i.e. the power of one beam,

$$P = N E_o f_r \qquad (6)$$

one gets the *first fundamental relation* of linear colliders

$$L/(10^{33} \text{ cm}^{-2} \text{ s}^{-1}) \simeq (D H_D/29) \text{ (mm}/\sigma_z) \text{ (P/MW)} \qquad (7)$$

which express quantitatively the fact that the luminosity is proportional to beam power.

Eq. (7) implies P > 1 MW with $\sigma_z \simeq 0.1$ mm and $DH_D \simeq 1.5 \times 3 \simeq 5$, to obtain $L > 10^{33}$ cm^{-2}s^{-1} as needed for *physics reason* at W > 1 TeV. This shows why powerful beams made of short bunches are a *must* in TeV linear colliders.

3. NATURAL SCALING

Since most of us have by now seen many times the parameter list of SLC, I reproduce it in the first column of Table 1. To scale these project parameters to (0.5 + 0.5) TeV it seems at first sight reasonable to keep the same type of accelerator, i.e. the same repetition rate f_r and the same bunch population N.

Table 1
SLC parameters and scaling to a (0.5 + 0.5) TeV Collider

Parameter	SLC (Project values)	Scaling factor (from Table 2)	Scaled NLC
Beam energy, E_o (GeV)	50	10	500
c.m. energy, W (GeV)	100	10	1000
Particles/bunch, N	7.2×10^{10}	1	7.2×10^{10}
Bunch radius, $\sigma_x = \sigma_y$ (μm)	1.65	0.1	0.165
Bunch length, σ_z (mm)	1.0	.1	0.1
Repetition rate, f_r (kHz)	0.18	0.18	0.18
Power/beam, P (MW)	0.1	10	1.0
β-value, β^* (mm)	5.0	2.1	10.5
Emittance, ε_n (m)	4×10^{-5}	0.046	$1.9\ 10^{-6}$
Disruption parameter, D	0.75	1	0.75
Enhancement factor, H_D	2.2	1	2.2
Luminosity, L (cm^{-2}s^{-1})	6×10^{30}	100	6×10^{32}
Beamstrahlung parameter, δ	9×10^{-4}	10^4	9

Luminosity is *proportional* to beam power; and beam power, for N and f_r constant, increases only *linearly* with γ. Since the *point-like* e^+e^- cross-section is proportional to γ^{-2}, to have a constant event rate while increasing γ, the luminosity has to be proportional to γ^2. The needed extra factor γ can be obtained by making $\sigma_z \propto \gamma^{-1}$, so that it seems natural to scale the various quantities as summarised in Table 2.

Table 2
'Natural' scaling laws from SLC to the Next Linear Collider (NLC)

Quantity	γ-dependence	Quantity	γ-dependence
N	1	L	γ^2
f_r	1	P	γ
$\sigma_x = \sigma_y$	γ^{-1}	$\varepsilon_{nx} = \varepsilon_{ny}$	$\gamma^{-4/3}$
σ_z	γ^{-1}	$\beta_x^* = \beta_y^*$	$\gamma^{1/3}$
D	1	δ (defined in Eq. 8)	γ^4

The results of the 'natural' scaling laws are collected in the last column of Table 1 which shows that for W = 1 TeV one would get L = 6 10^{32} cm^{-2} s^{-1}, close to our goal. The last line of the Table shows that the *new* parameter δ (to be defined in the next Section) has a frightening γ-dependence and passes from $\simeq 10^{-3}$ at SLC to $\simeq 10$ at the 1 TeV Next Linear Collider (NLC) we are considering. This is a very serious problem, as we want now to discuss.

4. BEAMSTRAHLUNG FOR ROUND BUNCHES

As anticipated, the problem is synchrotron radiation in the field of the opposite bunch. According to classical electrodynamics, the energy radiated by a particle per unit length is $P_\ell = 2\, r_e\, mc^2\, \gamma^4/(3\rho^2)$. Using for the radius of curvature the expression of Eq. (1), the average *fractional energy loss* in a length σ_z for a particle which is only *slightly* deflected is $\delta = P_\ell\, \sigma_z$ and can be written in the form

$$\delta = 2/9\, (r_e^3\, \gamma/\sigma_z)\, (N^2/\sigma_x^2)\,. \tag{8}$$

(The numerical factor, which equals 0.222, has a simple form useful for later simplifications. The exact form computed for gaussian bunches both in the transverse and the longitudinal directions is $8\,\pi^{1/2}/21 \simeq 0.215$.) The quantity δ has been dubbed *'beamstrahlung parameter'*.

Equation (8) shows that, as anticipated in Table 2, δ increases proportionally to γ^4, if one decides to scale up a collider with N = const, f_r = const, P $\propto \gamma$, $\sigma_x = \sigma_y \propto \gamma^{-1}$ and $\sigma_z \propto \gamma^{-1}$, so to have L $\propto \gamma^2$. At SLC, $\delta \simeq 10^{-3}$ (Table 1), so that at W = 1 TeV with the 'natural' scaling of Table 2 one would have $\delta \simeq 10$, clearly unphysical since a particle cannot loose 10 times the energy it possesses. This is the reason for which, starting from SLC parameters of Table 1, the scaling laws of Table 2 *cannot* be followed above $E_0 \simeq 300$ GeV.

In 1985 Himel and Siegrist showed[6] that at very large energies linear colliders have more favourable scaling laws than the ones implied by Eq. (8). The issue has been discussed and clarified in many recent papers[7] and I limit myself to the presentation of the main results.

In a uniform magnetic field the critical energy of the spectrum of the radiated photons has the form $E_c = (3 \hbar c \gamma^3) / (2\rho)$ so that, using Eq. (1), the *average fractional critical energy* $\overline{\Upsilon} = \bar{E}_c/E$ has the form

$$\overline{\Upsilon} \simeq 5/(12\alpha) \; (r_e^2 \, \gamma/\sigma_z) \; (N/\sigma_x) \, , \qquad (9)$$

where $\alpha = 1/137$ and the numerical factor comes from averaging on gaussian bunches[7]. Himel and Siegrist showed that for large values of $\overline{\Upsilon}$ the fractional energy loss is much less than the classical estimate given by Eq. (8), but before introducing their result we have to consider flat bunches.

5. DISRUPTION AND BEAMSTRAHLUNG FOR FLAT AND PINCHING BUNCHES

'Flat' bunches have $\sigma_y < \sigma_x$. By introducing the aspect ratio

$$R = \sigma_x / \sigma_y \geq 1 \qquad (10)$$

two disruption parameters determine now the focusing in the two planes.

$$D_x = (r_e N \sigma_z) \, (\gamma \sigma_x \sigma_y)^{-1} \, [2 / (1+R)] \, , \qquad (11)$$
$$D_y = (r_e N \sigma_z) \, (\gamma \sigma_x \sigma_y)^{-1} \, [2R / (1+R)] = D = RD_x > D_x \, . \qquad (12)$$

In the following the largest one, D_y, will be still called *'disruption parameter'* and indicated by the symbol D. For flat beams the enhancement factor is reduced with respect to the one given as a function of D in Eq. (5):

$$H(R) \simeq H_D \, (1+R)/2R \, , \qquad (13)$$

and the luminosity is

$$L = f_r N^2 H(R) / (4\pi \sigma_x \sigma_y). \qquad (14)$$

The *first fundamental relation* (7) becomes

$$L/(10^{33} \text{ cm}^{-2}\text{s}^{-1}) \simeq [DH(R)/29] \, [(1+R)/2R] \, (\text{mm}/\sigma_z) \, (P/\text{MW}) \, . \quad (15)$$

The fractional average critical energy $\overline{\Upsilon}$ gets a non-trivial modification:

$$\overline{\Upsilon} = 5/(12\alpha) \, (r_e^2 \gamma/\sigma_z) \, (N/\sqrt{\sigma_x \sigma_y}) \, H_b^{1/2}(D,R) \, , \qquad (16)$$

where, following Chen[8], I have introduced a new factor, which depends on D and R

$$H_b(D,R) \simeq 4R \, H_D \, [1 + R \, H_D^{(R-1)/2R}]^{-2} \, , \qquad (17)$$

and which intervenes also in the fractional energy loss when $\Upsilon << 1$ and there is pinch effect:

$$\delta = 2/9 \, (r_e^3 \gamma / \sigma_z) \, (N^2/\sigma_x \sigma_y) \, H_b \, (D,R) \, . \tag{18}$$

For round bunches ($R=1$) with pinch effect $H_b \simeq H_D$ and the parameters $\overline{\Upsilon}$ and δ are proportional to $H_D^{1/2}$ and H_D respectively. For very flat bunches ($R >> 1$) Eq. (17) gives $H_b \simeq 4R^{-1}$ and (with $\sigma_x \sigma_y$ = const) $\overline{\Upsilon} \propto R^{-1/2}$ while $<\delta> \propto R^{-1}$. *Flat bunches can thus be used to reduce $\overline{\Upsilon}$ and δ by large factors*, while Eq. (15) is modified *only* by a factor 2.

6. COLLIDER REGIMES

If $\overline{\Upsilon} << 1$ the collider runs in the so-called *'classical regime'*. By increasing the fractional critical energy $\overline{\Upsilon}$ one enters in the *'quantum regime'*, first studied by Himel and Siegrist. In Ref. 1 I introduced an handy approximation, valid to ~ 5%, which bridges the equations valid in the two regimes by writing:

$$<\varepsilon> = \delta \, F \, (\overline{\Upsilon}) \, , \tag{19}$$

with

$$F \, (\overline{\Upsilon}) \simeq (1 + \overline{\Upsilon}^{1/2} + 3 \, \overline{\Upsilon}/2)^{-4/3} \, . \tag{20}$$

The quantity $<\varepsilon>$ is the *average fractional energy loss* which is equal to δ only when $\overline{\Upsilon} << 1$ and can be much smaller than δ for Eq. (20) when $\overline{\Upsilon} \geq 1$. It is generally assumed that NLC can accept relatively large values of the fractional loss, as $<\varepsilon> \simeq 0.2$-0.3. This is due to the fact that such a value corresponds typically to cen-

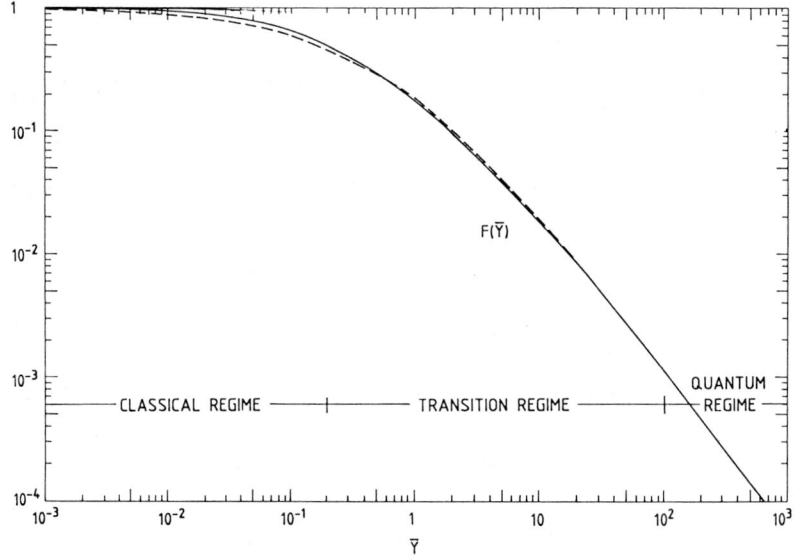

Fig. 1 The factor $F(\overline{\Upsilon})$, which multiplies the beamstrahlung parameter δ and gives the fractional energy loss $<\varepsilon>$. The continuous line is the result of Noble numerical calculations[7]; the dashed curves is the simple analytic expression of Eq. (20).

ter-of-mass spreads $\Delta W/W \simeq 10\%$, which is not too large when running in the "continuum".

The function $F(\bar{\Upsilon})$, called sometimes H_Υ, is plotted as dashed line in Fig. 1.

For the scaled-up version of SLC appearing in Table 1, $R = 1$ and $H_b = H_D = 2.2$, so that from Eq. (16) $\bar{\Upsilon} = 2.9$. Eq. (20) gives $F(\bar{\Upsilon}) = 0.093$ and the beamstrahlung parameter $\delta = 9$ of Table 1 implies a fractional energy loss $<\varepsilon> \simeq 0.85$. Such a loss is still *too* large with respect to what is generally considered acceptable ($<\varepsilon> \simeq 0.3$): as anticipated, the scaling laws of Table 2 are acceptable only up to $E_0 \simeq 300$ GeV.

The fractional energy loss $<\varepsilon>$ of Eq. (19) can be written in a convenient form by expressing δ of Eq. (18) as a function of $(L/f_r)^{1/2}$. One obtains the *second fundamental relation* of linear colliders[9]:

$$<\varepsilon> = (16 \pi^{1/2} \alpha/15) (H_b/H)^{1/2} (r_e^2 L/f_r)^{1/2} \bar{\Upsilon} F(\bar{\Upsilon}). \quad (21)$$

Since $<\varepsilon>$ is in practice fixed, the quantity $\bar{\Upsilon} F(\bar{\Upsilon})$ is very important because, together with R and D, it fixes L/f_r.

Noble's numerical results[7] for $\bar{\Upsilon} F(\bar{\Upsilon})$ are plotted as a continuous line in Fig. 2, while the dashed line represents the approximation given by the simple formula (20). Three regimes can be clearly distinguished:

$$\begin{aligned}
&\textit{classical regime:} && \bar{\Upsilon} < 0.2 \\
&\textit{transition regime:} && 0.2 < \bar{\Upsilon} < 10^2 \\
&\textit{quantum regime:} && 10^2 < \bar{\Upsilon}.
\end{aligned} \quad (22)$$

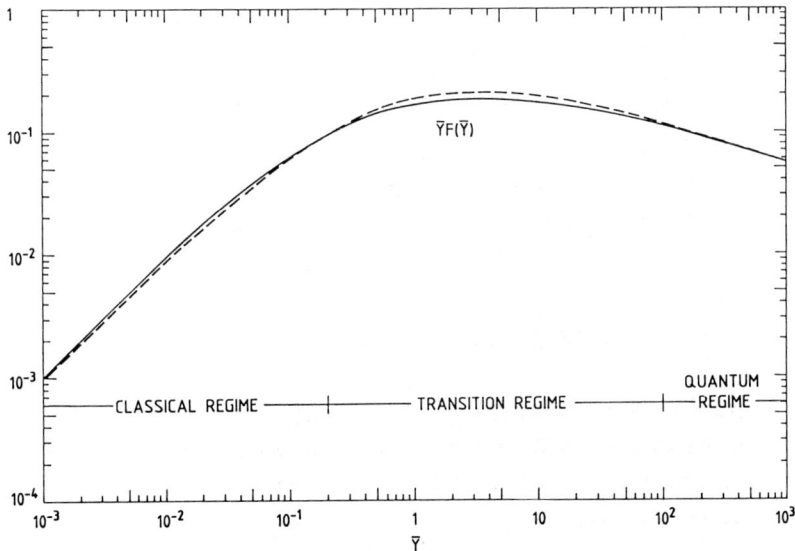

Fig. 2 *The function $\bar{\Upsilon} F(\bar{\Upsilon})$, which enters in the expression of $<\varepsilon>$ [Eq. (21)], is plotted versus $\bar{\Upsilon}$. The continuous curve is due to Noble and the dashed curve represents the simple approximation of Eq. (20): $\bar{\Upsilon}(1 + \bar{\Upsilon}^{1/2} + 3\bar{\Upsilon}/2)^{-4/3}$.*

In the first regime $<\varepsilon> \propto \bar{\Upsilon}$, in the second $<\varepsilon> \simeq$ constant within a factor 2, and in the third $<\varepsilon> \propto \bar{\Upsilon}^{-1/3}$. We shall conclude that the Next Linear Collider will run in the intermediate regime, where $\bar{\Upsilon}F(\bar{\Upsilon})$ is about constant. (Note that, since $\Upsilon = 2.9$, also the scaled version of SLC would have this property.)

7. RELATIONS CONNECTING COLLISION PARAMETERS

We have introduced a large number of quantities to describe bunch-bunch collisions: E_0 (or γ), L, P, R, D (and $H(R/H);H_b$)), σ_z, f_r, N, σ_x, $\bar{\Upsilon}$, $<\varepsilon>$. These *eleven* parameters are linked by the *five* equations (6), (12), (14), (16), (21), so that *six* parameters are enough to define any linear collider. To help the reader, the relevant formulae are collected in Table 3.

Table 3
Summary of collider formulae
(the symbol $\simeq$ signifies 5 − 10% accuracy)

Quantity	Formula	Eq.
Energy/beam	$E_0 = \gamma mc^2$	
r.m.s. radii	$\sigma_x = (\varepsilon_{nx}\beta_x^*/\gamma)^{1/2}$ $\sigma_y = (\varepsilon_{ny}\beta_y^*/\gamma)^{1/2}$ $[\sigma_x = R\sigma_y]$	(2)
Power/beam	$P = Nf_r E_0$	(6)
Disruption par.	$D = (r_e N \sigma_z)(\gamma \sigma_x \sigma_y)^{-1} [2R/(1+R)]$	(12)
Pinch factor (R=1)	$H_D = $ Eqs. (5)	
Pinch factor (R≠1)	$H(R) \simeq H_D^{(1+R)/2R}$	(13)
Luminosity	$L = f_r N^2 H(R)/(4\pi \sigma_x \sigma_y)$	(14)
Fract. critical energy	$\bar{\Upsilon} \simeq (5/12\alpha)(r_e^2\gamma/\sigma_z)(N/\sqrt{\sigma_x \sigma_y}) H_b^{1/2}(D,R)$	(16)
Beamstrahlung factor	$H_b(D,R) \simeq 4R\, H_D [1+RH_D^{(R-1)/2R}]^{-2} \to 4/R$	(17)
Beamstrahlung par.	$\delta = 2/9\,(r_e^3\gamma/\sigma_z)(N^2/\sigma_x\sigma_y) H_b(D,R)$	(18)
Fract. energy loss	$<\varepsilon> = \delta F(\bar{\Upsilon})$	(21)
Quantum factor	$F(\bar{\Upsilon}) \simeq (1 + \bar{\Upsilon}^{1/2} + 3\bar{\Upsilon}/2)^{-4/3}$	(22)

How do we orientate ourselves in such a complicated multidimensional space? In this Section we first discuss some relations which are simple consequences of the equations of Table 4. But to make full use of them, in the next Section we have to review the scaling laws of linar accelerators.

I address, on the basis of the equations of Table 3, the questions: (i) Will NLC run with sizeable pinch effect? (ii) In which beamstrahlung regime NLC will operate? The answers will be: (i) NLC will utilize the pinch effect to increase luminosity; (ii) NLC will either run in the transition regime or very close to it. The first point is clear when looking at Eq. (15): for a limited beam power the luminosity is proportional to HD!

Table 4
Relations derived from the equations of Table 3 (*)
(practical units are introduced for convenience of use in numerical calculations)

Relation	Eq.
$\sigma_z/\text{mm} = [DH/29] \, [(1+R)/(2R)] \, (10^{33}/L) \, (P/MW)$	(15)
$\langle \varepsilon \rangle \simeq 3.9 \, (H_b/H)^{1/2} \, (L/10^{33})^{1/2} \, (kHz/f_r)^{1/2} \, \Upsilon \, F(\Upsilon)$	(21)
$\Upsilon \simeq 0.32 \, (H_b/H)^{1/2} \, (E_0/\text{TeV}) \, (L/10^{33})^{1/2} \, (kHz/f_r)^{1/2} \, (\text{mm}/\sigma_z)$	(23)
$\Upsilon \simeq 9.1 (DH)^{-1} (H_b/H)^{1/2} [2R/(1+R)] (E_0/\text{TeV}) \, (L/10^{33})^{3/2} (MW/P)(kHz/f_r)^{1/2}$	(24)

(*) *In this Table $10^{33} = 10^{33}$ cm^{-2}s^{-1} and $H = H(R)$.*

To discuss the second question we have to write down some new equations. We have already met the two fundamental relations (15) and (21), which are rewritten in practical units in Table 4. The third relation in the Table is obtained by expressing Υ of Eq. (16) in terms of (L/f_r). The fourth relation derives from Eqs. (15) and (23) of Table 4. It relates the parameter Υ to the *six* most important physical parameters (E_0, L, P, D, R, f_r) which describe fully bunch-bunch collisions. (Note that D and R appear also implicitly in the ratio H_b/H.)

When designing a collider, E_0 and L are given by physics, and the present wisdom requires[10]

$$L \simeq (2 \, E_0/\text{TeV})^2 \, 10^{33} \, \text{cm}^{-2} \, \text{s}^{-1} . \qquad (25)$$

P is fixed by the acceleration technology and the money available, since given a certain efficiency η_{tot} of the energy transfer from wall-plug to the accelerated beam, the power (per linac) is P/η_{tot}. (Today $\eta_{tot} \leq 1\%$ and in future accelerators one can expect $\eta_{tot} \simeq 5\%$.)

The value of D defines the choice between *pinch effect* (D > 0.5) or *no pinch effect* (D < 0.5). The aspect ratio R can be varied by using both the transverse emittances ε_{ny} and ε_{nx} and/or the ratio β_x^*/β_y^*.

Eq. (24) of Table 5 shows that, if one chooses to scale the luminosity according to Eq. (25), $\Upsilon \propto E_0^4$ and at large enough energy the collider **has** to run in the quantum regime ($\Upsilon > 10^2$).

For a 'standard' NLC (W = 1 TeV and L = 10^{33} cm^{-2}s^{-1}) with P = 1 MW (so that the *total* plug power is $\sim$ 100 MW with $\eta_{tot} \simeq 2\%$) it follows from Eq. (24) that $\Upsilon = (C/DH) \, (kHz/f_r)^{1/2}$. It can be easily checked that the numerical factor is $1 < C < 10$ for all values of H and R. A standard NLC, which profits from the pinch effect

(D ≥ 0.5) *and* runs at the (small) SLC repetition rate ($f_r = 0.2$ kHz), will thus have $0.2 \leq \bar{Y} \leq 10$ for all reasonable values of R. However, higher rates and beam powers tend to reduce $\bar{Y}$, and one can get $Y < 10^{-1}$ if the aspect ratio R is large. At higher energy (for instance at $W = 2$ Tev, as for the CERN linear collider CLIC), the choice is again pushed to the transition regime because of the dependence $\bar{Y} \propto E_0^4$ even if f_r is few kHz.

In conclusion, a low repetition rate NLC profiting from the pinch effect will have to run in the transition regime. At higher repetition rates and for flat bunches one can choose to sit at the higher end of the classical regime. Anyway, by increasing the energy one is pushed towards the quantum regime, where the scaling laws change. I do not discuss it here, because this regime is not interesting for NLC; still it is worth mentioning that, in this extreme case, one would probably choose D small and no pinch effect because otherwise the required emittances would be unattainable small.

8. SCALING LAWS OF COPPER LINACS

It is by now generally accepted that the acceleration techniques of the next generation of linear colliders will be 'reasonable' extrapolations of the ones in use today. The most promising candidates are copper cavities running at high frequency ($10 \leq f \leq 30$ GHz). Superconducting cavities would be an ideal solution[11], but at present the achievable gradients are too small and imply long and expensive linacs to reach the TeV energy range[12].

Let us consider a normal conducting linac made of travelling wave sections of length L at frequency f. Given the *stored energy* W' per unit length, the *power dissipation per unit length* P_d' determines the decay time of the energy: W'/P_d'. The structure (unloaded) quality factor Q is defined as the ratio between this decay time and the characteristic time of the RF oscillation: $(2\pi f)^{-1} = 1/\omega$:

$$Q = \omega W' / P_d' . \qquad (26)$$

The quantity τ, proportional to the decay time, fixes the time it takes to build up the field:

$$\tau = W'/(2P_d') = Q/2\omega . \qquad (27)$$

(The factor 1/2 corresponds to the usual choice made for high energy colliders. In general $\tau = Q \alpha/\omega$ were α is the 'attenuation constant' of the structure.)

Each one of the sections of length L is excited in a resonant mode with an electric field component in the direction of the particle motion; the excitation is caused by a square power pulse of frequency f, duration τ and peak power $\hat{P}_L$. The group velocity v_g, which is the velocity of the energy flow, determines the filling time so that $\tau = L/v_g$. (In the structures to be considered typically one has $v_g/c \simeq 0.05$-0.1.)

The stored energy *per meter* W' (W/m) is clearly proportional to the square of the accelerating field E (V/m) and to the structure cross-section, i.e. to the square of the wavelength λ. One can thus write:

$$W' = E^2 \lambda^2/(2\pi cZ) . \qquad (28)$$

where Z is an impedance independent of λ for any given geometry of the structure. A typical value is $Z \simeq 300\ \Omega$, so that at $\lambda = 10$ cm and $E = 17$ MV/m (as for SLC) one needs $W' \simeq 5$ J/m.

The energy W' is pumped into the structure with a repetition rate f_r in pulses of duration τ. Typically only $\eta_t = 75\ \%$ of the pumped energy is still there at the moment in which a bunch of Ne particles crosses the structure, is accelerated by the field E and extracts energy with efficiency

$$\eta = \text{Ne E}/W' = 2\pi\ cZ\ \text{Ne}\ /\ (E\lambda^2)\ . \tag{29}$$

Of course some power is spent in accelerating the electron and positron beams. When a bunch of charge Ne interacts with the structure it induces a field $E_i \propto$ Ne which cancels part of the accelerating field. The average particle of the bunch will thus see the field $(E - E_i/2)$ and this causes a momentum spread. If η is the *fraction* of the stored energy extracted by a bunch, when no particular attention is payed to the problem, the momentum spread is of the order of $\eta/2$. One can do better by choosing the phase of the bunch with respect to the RF wave such that, without beam loading, the particles in the tail of the bunch would see a larger accelerating field.

If b bunches of N particles each are accelerated during a single RF pulse, and extra power is poored in the structure in between bunches to compensate for the energy extracted, the efficiency of the system can be increased without augmenting the momentum spread. In this case the RF power repetition rate f_{rf} is smaller than the collider rate f_r

$$f_r = b\ f_{rf}\ , \tag{30}$$

and the fraction of stored energy which is extracted is roughly

$$\eta \simeq b\ \text{Ne E}/W' = 2\pi\ cZ\ b\ \text{Ne}/(E\lambda^2)\ . \tag{31}$$

For a single bunch $\eta \leq 10\%$, because of the momentum spread, and one can hope to have $\eta \simeq 30\%$ in the multibunch compensated scheme described above.

The total energy transfer efficiency from rf-power to beam is $\eta_t\eta$ and the RF power needed for *two* linacs to obtain two high energy beams of power P *each* is

$$P_{rf} = 2\ P/(\eta_t\eta) \simeq (2E_0 f_r\ E\lambda^2)/(2\pi\ cZ\eta_t b)\ , \tag{32}$$

where Eq. (6) has been used.

9. CONCLUSION

We can now make contact with the discussion of Section 7, and in particular with the equations of Table 4. By using the quoted value $Z = 300\ \Omega$, in practical units Eq. (32) becomes:

$$P_{rf}/MW \simeq (425/b)\ (E_0/TeV)\ (f_r/kHz)\ (E/MV\ m^{-1})\ (GHz/f)^2,\quad (33)$$

which clearly displays the gain in RF power implied by high RF frequencies (i.e. small wavelengths) and by the use of a multibunch scheme with $b > 1$. Eq. (33) could be called the *third fundamental relation* of copper cavity linear colliders. Applied to the NLC of Table 1 with $f = 3$ GHz, $E = 17$ MV/m and $f_r = 0.18$ kHz (as for SLC) it gives $P_{rf} \simeq 72$ MW, a not unreasonable number. The length of the collider would instead be much too large: 2×30 km. If the gradient was increased by a factor 5, so that the length becomes 2×6 km, the power would become $P_{rf} \simeq 350$ MW. This is unreasonable because the plug power P_{ac} has typically to be twice as large.

By combining the second fundamental relation (21) with Eq. (33) one finally gets:

$$b^{1/2}f/GHz = 80(H_b/H)^{1/2}(E_0/TeV)^{1/2}(L/10^{33})\ (MW/P_{rf})^{1/2}\ (E/MeVm^{-1})^{1/2}\ \overline{\Upsilon}F(\overline{\Upsilon})/<\varepsilon> \quad (34)$$

which does *not* contain the repetition rate. For a 'standard' NLC ($W = 2E_0 = 1$ TeV, $L = 10^{33}$ cm^{-2}s^{-1}, $P_{rf} = 50$ MW, $D = 2$, $H_D \simeq 3.5$, $<\varepsilon> = 0.3$, $b = 1$, $E = 100$ MV m^{-1}) which runs in the *transition regime*, (so that $\overline{\Upsilon} f(\overline{\Upsilon}) = 0.15 \pm .05$), Eq. (34) implies

$$f \simeq (40 \pm 13)\ (H_b/H)^{1/2}\ GHz$$

which, for $R = 1$, is ten times larger than the SLC frequency. The factor $(H_b/H)^{1/2}$, which is equal to 1 for $R = 1$, decreases as $2R^{-1/2}\ H_D^{-1/4} \simeq 1.3\ R^{-1/2}$ for $R \geq 3$. Flat bunches are thus useful to decrease either the RF frequency, or the plug power. For $R = 10$ and $H_D = 3.5$ from Eqs. (13) and (17) one computes $H_b/H \simeq 0.20$, so that the optimal frequency is $f \simeq 18$ GHz.

The main parameters are all fixed, but for the repetition rate f_r, which can be obtained from the second fundamental relation with the choice $<\varepsilon> = 0.30$ and the first order value $\overline{\Upsilon} F(\overline{\Upsilon}) = 0.15$: $f_r \simeq 0.75$ kHz.

In summary, we have seen that the Next Linear Collider will have to run with some pinch effect, to increase the luminosity, and in the transition regime. Then the other parameters are practically fixed by the requirement of not having a too large power consumption. In particular the frequency of the structure cannot be very different from 20 GHz and the repetion rate has to be of the order of 1 kHz.

REFERENCES

1) U.Amaldi, Introduction to the next generation of linear colliders, CERN-EP 87-169, in: Frontiers of particle beams, eds. M.Month and S.Turner, (Springer-Verlag, Berlin, 1988) p. 341.

2) P.B.Wilson, in: Proc. UCLA Workshop on linear collider $B\bar{B}$ factory conceptual design, ed. D.H.Stork (World Scientific, Singapore) p. 373.

3) P.Chen and K.Yokoya, SLAC-PUB-4339 (1987).

4) For a summary see: N.A.Solyak, Collision effects in compensated bunches of linear colliders, Inst. Nucl. Physics, Novosibirsk, 1988, Preprint 88-44.

5) R.Hollebeek, Nucl. Instr. and Meth. 184 (1985) 333.

6) T.Himel and J.Siegrist, SLAC-PUB 3572 (1985), in: Laser acceleration of particles, eds. C.Joshi and T.Katsouleas (AIP Conf. Proc., 1985) No. 130.

7) K.Yokoya, Nucl. Instr. and Meth. A251 (1986) 1.
R.J.Noble, Nucl. Instr. and Meth. A256 (1987) 427.
R.Blankenbecher and S.D.Drell, Phys. Rev. D 36 (1987) 277.
M.Jacob and T.T.Wu, Phys. Lett. B197 (1987) 253 and CERN TH 4907/87.
M.Bell and J.S.Bell, Part. Accel. 22 (1988) 301.

8) P.Chen, in: Frontiers of particle beams, eds. M.Month and S.Turner, (Springer-Verlag, Berlin, 1988) p. 495.

9) P.B.Wilson, SLAC-PUB 4310 and Proc. of the Part. Acc. Conference, Washington D.C., March 16-19, 1987.

10) U.Amaldi, Proc. of the Workshop on physics at future accelerators, La Thuile (Italy) and Geneva (Switzerland), 7-13 Jan., 1987, ed. J.Mulvey, (CERN, Geneva, 1987) Vol. 1, p. 323.

11) K.Johnsen et al., Report of the Advisory Panel on the prospects for e^+e^- linear colliders in the TeV range, CERN 87-12, May 1987.

12) U.Amaldi, H.Lengeler and H.Piel, CERN EF 86-8 and CLIC Note 15 (1986).

THE CLIC PROJECT AND THE DESIGN FOR AN e^+e^- COLLIDER

Simon van der MEER

CERN, European Organization for Nuclear Research
1211 Geneva-23, Switzerland

1. INTRODUCTION

The name CLIC (CERN Linear Collider) originally referred to an advisory subpanel of CERN's long range planning committee in the years 1985-87. CLIC (presided by K. Johnsen) dealt with linear electron-positron colliders in the TeV range, trying to survey the field and arrive at some understanding of future possibilities. The results of this work were published in a formal report[1].

Although this panel was dissolved, the studies (which had mobilised many individuals at CERN on a part-time basis) were continued under the leadership of W. Schnell, a small budget was obtained and the name CLIC now covers these activities, carried out by a varying number of people, not all from CERN, and most of whom also have other responsibilities.

Whereas the original CLIC panel considered many different ways that might lead to a practical collider, most of the work by far is nowadays concentrated on a two-beam scheme proposed in 1986 by W. Schnell[2].

Other directions are also pursued to a certain extent (e.g. the switched power linac) but I will only discuss the two-beam approach here. Before doing this, I must first explain the most important problems for collider designers.

2. LUMINOSITY, DISRUPTION AND BEAM RADIATION

The only existing example of a linear collider is the SLC at Stanford, now in the commissioning stage. Compared to this machine, we have set ourselves the goal of a 20 times higher energy (1 TeV + 1 TeV). Since the interaction cross-sections scale roughly as $1/E^2$, this means that we need a luminosity of the order of 400 times higher. It is this that causes the main problems, rather than the high energy by itself.

Lecture given at the International School on Electromagnetic Radiation and Particle Beams: Physics and Applications, Varenna, Italy, June 1988.

The luminosity of a linear collider is equal to

$$L = \frac{fN^2}{4\pi\sigma^2}$$

where f is the repetition frequency, N the number of particles per bunch and σ the r.m.s. size of the beam spot (assumed to be round and gaussian) at the interaction point. Since increasing f or N will increase the average beam power, and this will be uncomfortably high in any case, the required L can only be made with a very small beam size at the collision point.

This is limited by two effects: disruption and beam radiation, sometimes called "beamstrahlung". The two colliding bunches will be pinched by each other's focusing fields (space charge). This effect, called disruption, will be beneficial up to a point, increasing the luminosity. However, the beams may destroy each other before they have had a good chance to interact, especially if they are not perfectly aligned[3]. Also, the strong divergence of the disrupted beams after the collision will make it very difficult to avoid hitting the last focusing elements of each beam and causing excessive heating[4].

The pinching will also result in radiation of photons, which may cause a considerable energy loss and energy spread before the electrons and positrons interact. This beam radiation and the disruption depend on the various beam parameters in different ways and both effects together will constrain the design in important ways.

Two extreme regimes of beam radiation exist: the classical and the quantum regime. In the first, the beam parameters are such that the typical photon energy, derived in a classical way, is less than the electron energy. If this isnot the case, a quantum-mechanical approach is needed. The radiated power then depends on the beam parameters in a different way; and the more we get into this regime, the smaller the effect will be. Unfortunately it turns out that extremely short bunches, beyond present possibilities, are then needed. Most present-day designs and especially the CLIC one are working in the wide transition region between the classical and quantum regimes.

The beam radiation may be reduced by using flat beams at the interaction point. With equal area, the luminosity is the same, but space-charge fields are smaller.

3. MAIN CLIC PARAMETERS

It would be beyond the scope of this talk to discuss the detailed relationships between the various parameters following from the desired luminosity and the limited allowable beam power, disruption, beam radiation and many other effects[5,6,7]. The main CLIC parameters (very tentative) are at present as follows[8]:

Table I

Main Linac Parameters

Energy	1.0 TeV
Luminosity	1.1×10^{33} cm^{-2} s^{-1}
Accelerating gradient	80 MV/m
Final focus aspect ratio	5
Final focus beam height	12 nm
Fractional energy loss by beam radiation	0.27
Fractional average critical energy	0.71
Pinch enhancement	2.37
Repetition rate	1.69 kHz
Number of bunches per pulse	1
Bunch population	0.50×10^{10}
Beam power (per beam)	1.35 MW
Energy extraction	5.0%
Iris aperture over wavelength	0.2
Shunt impedance over Q per unit length	28 kΩ/m
Fill efficiency	0.78
RF frequency	29 GHz
RF power (average per beam)	35 MW
Bunch length	200 μm
Disruption	3.3
Vertical Emittance (normalised)	10^{-6} rad m
Emittance ratio	3 (say)
Vertical amplitude function	282 μm
Ratio of amplitude functions	8.3
Length, per Linac	12.5 km

Many of these parameters are close to practical limits. The luminosity should not be much lower (in fact, we would prefer 10^{34} cm^{-2} s^{-1}). The focus aspect ratio cannot be much increased without reducing even further the extremely small beam height required. The fractional energy loss by beam radiation should certainly not be any higher and increasing the pinch-enhancement might be very difficult. The RF frequency is uncomfortably high. However, relaxing any of these parameters would lead to even higher beam power and RF power.

It would be very attractive to accelerate several bunches per pulse. This could be done by injecting the first bunch when the accelerating structure is not yet entirely filled. Successive bunches would see a more completely filled structure; this would balance the gradient decrease caused by beam loading from the preceeding bunches. A power extraction efficiency of 30% might be reached[2] but multi-bunch wakefield effects might not allow the scheme to work.

4. SUPPLYING THE RF POWER - THE CLIC TWO-BEAM APPROACH

The peak power required during each pulse is about 3.6 TW per beam. To provide this with klystrons as in conventional linacs would be out of the question financially, even if high-power 29 GHz klystrons would exist.

The solution proposed is to combine all klystrons into a single high-intensity electron beam (drive linac) running in parallel with the main linac (Figure 1).

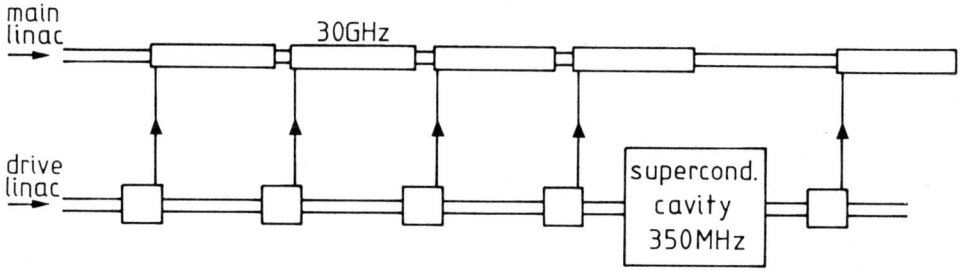

FIGURE 1
Two-beam accelerator

This beam is accelerated by superconducting cavities filling only a fraction of its length and the beam energy is transmitted to so-called transfer structures that feed the main linac at 29 GHz. The superconducting cavities work at a much lower frequency (350 MHz) and with a much lower average field strength (6 MeV/m over the cavity length) than the main linac (80 MeV/m).

To make the various transfers of energy possible, each pulse of the drive beam must contain a number of bunches at 350 MHz and each of these must be subdivided into bunchlets at 29 GHz. The drive beam must have an energy of a few GeV to avoid phase slippage with respect to the main linac. Of course, the total charge per pulse must be much higher in the drive linac than in the main one, since the gradient is lower and energy is conserved.

The transfer structures have a filling (or rather emptying) time of one 350 MHz period, so that the first bunchlet of each 350 MHz bunch will see an empty structure. Each successive bunchlet will see a higher decelerating gradient as the transfer structures charge up. However, the acceleration by the superconducting cavities will also increase, since the bunches will pass these on the rising slope of the sine wave (Figure 2). These two slopes can be made to balance.

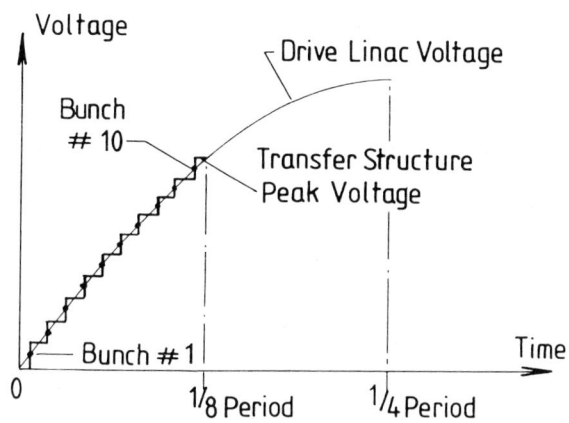

FIGURE 2
Voltage on transfer structure and accelerating voltage in drive linac. Only 10 of the 40 bunches are shown.

The main linac's structures will have a longer filling time, corresponding to the total duration of the drive linac's bunch train.

It turns out that the ratio of the accelerating gradients of main linac and drive linac scales with the frequency ratio. (Some other factors, such as the fraction of the drive linac occupied by superconducting cavities, also play a role). This makes the 350 MHz drive frequency near-optimum. As it happens, CERN has developed such cavities for LEP. Their cost will have to be reduced for use in CLIC, but otherwise they are just right.

To obtain the relatively low deceleration of the drive beam together with a large energy transfer, the transfer structures must have a low impedance (a few hundred times less than the main linac's structures). They will thus have a large aperture which helps to avoid problems with the intense drive beam such as transverse wakefield effects. The losses in these structures also turn out to be low (90% efficiency).

It would, in principle, be possible not to dissipate the power at the end of the main linac's sections in terminating resistors, but to feed it back to the next transfer structure of the driving linac, which could then accelerate additional bunchlets injected at the correct phase. These might in turn give back their gained energy to the superconducting cavities. In this way, the efficiency might be increased by a factor 2.

Typical parameters for the drive linac are given below.

Table II

Drive Linac Parameters

Fraction of main linac active length occupied by drive linac	0.2
Drive linac active length	2.5 km
Drive linac frequency	350 MHz
Number of bunches per pulse	4
Number of bunchlets per bunch	10
Bunchlet population	4×10^{11}
Bunch length	1 mm
Transfer efficiency assumed	0.9
Drive linac energy extraction	0.1
Drive linac R over Q parameter	270 Ω/m
Drive linac accelerating field	6.0 MV/m
Drive linac total voltage gain	15.0 GV
Drive linac quality factor	5×10^9
Cryogenic efficiency assumed	2×10^{-3}
Total cryogenic input power	33.5 MW

The main problem with the drive linac will be to generate the short, intense bunchlets. The emittance need not be small, and in principle we could imagine to combine a number of SLC-like beams in transverse phase space. However, a more elegant solution may well be found. Some preliminary work on a possible test facility has been done by Y. Baconnier and colleagues[9], but this is still at a very early stage.

In the following sections, I shall describe some of the detailed work that has been done lately on various aspects of this scheme.

5. WAKEFIELD EFFECTS AND RF FOCUSING

A serious problem in high-energy linacs is the so-called beam break-up caused by the transverse wakefield generated by a bunch displaced from the central axis. Such a bunch will execute transverse oscillations because of the focusing (quadrupoles) around the beam and the wakefield will oscillate in the same way. The wake caused by the front of the bunch may now excite the transverse movements of the tail in resonance if front and tail have the same oscillation frequency. The effect scales inversely with the 3rd power of the structure's aperture and is therefore especially dangerous at high accelerating frequency.

This effect would be catastrophic in the main linac if nothing were done to suppress it. One way to do this is to use very strong transverse focusing. This, however, is expensive and makes the beam very sensitive to small misalignments and jitter (the tolerance could be a fraction of a μm), which is dangerous because it might prevent the two opposing beams to meet precisely in the very small focal spot.

Another way to reduce the effect is to make the focusing field for the trailing particles stronger than for the leading ones. The phase of the wakefield excitations will then be such as to damp the effect rather than to excite it. This is very important because without this damping any misalignments would be rapidly transformed into emittance increases because of the filamentation caused by the spread in transverse wavelength.

One way to achieve this is to allow a strong beam loading so that the trailing particles will have less energy than the leading ones and will thus be more strongly focused. The amount of energy spread needed to suppress the effect turns out to be uncomfortably large for the CLIC parameters[10,11]. It would make the final focus design extremely difficult (chromatic aberrations).

A better solution appears to be the use of radio-frequency focusing obtained by using slits instead of round apertures in the irises. This method, first proposed by R. Palmer and analysed in detail by Schnell[12] and Henke[13] is interesting because the focusing effect scales both with frequency and with accelerating gradient and thus becomes appreciable in the present application.

The particles have to pass on the rising slope of the waveform fo compensate the energy spread arising from the longitudinal wakefield. Trailing particles then are also focused stronger (the focusing is zero at the crest of the wave and increases away from it), which is just what is wanted. The calculations suggest that the problem is solved: the centre of the bunch is strongly damped and alignment tolerances are relaxed to 10-20 µm. Also the energy spread of the beam may now be reduced as much as is possible, without causing problems from beam break-up. Finally, RF focusing will reduce the cost (fewer or no quadrupoles).

Numerical studies of slotted-iris structures[14] have shown that earlier rough estimates of their performance were correct within 15%.

6. WAKEFIELDS IN THE DRIVE LINAC

Study of the wakefield effects has just started[15]. It has appeared that the longitudinal effects of cross-section changes along the line may be important and that such changes will have to be minimized. In the super-conducting cavities the longitudinal wakefield may also be harmful, as it will change the linear voltage increase for successive bunchlets.

In the transfer structures the resistive wall impedance of the smooth sidewalls may cause beam break-up; to suppress this, either an energy spread of 5% over the entire bunch train will be needed, or the aperture of the structure will have to be increased.

7. TRANSFER STRUCTURES

Model tests have been made on scaled-up (2 GHz) models of the low-impedance transfer structure. Properties of a 4-cell and later a 12-cell model (Figure 3) were measured[16], such as dispersion curves and coupling impedance.

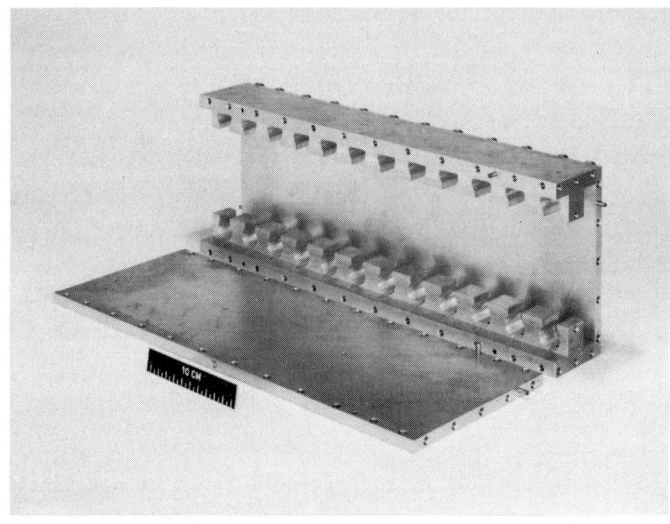

FIGURE 3
Test model (2 GHz) for a transfer structure with its cover taken off

The structure is a rectangular waveguide with teeth far away from the beam to obtain the required low impedance. The teeth do not extend across the full width of the guide so that adjacent cells couple both electrically and magnetically.

The model measurements are encouraging: the impedance, the group velocity and the efficiency of energy transfer to the fundamental mode ($\pi/2$) seem about right. Further optimisation is now being done by computer modelling[17], using the MAFIA codes developed by Weiland at DESY.

8. MAIN LINAC STRUCTURES[18]

Tests on different geometries for the main linac structures[19] and calculation[20] have shown that it is difficult to find a better structure than the classical disc-loaded waveguide. This has therefore been adopted for further study. The centre hole has to be relatively large compared to existing linacs to minimize wakefield effects and to obtain a high group velocity; this is necessary to keep the section length reasonable. Despite the short wavelength, the tolerances (mainly because of the high group velocity) turn out to be reasonable ($\pm$ 3μm).

A complication is the provision of longitudinal slots in the structure's sidewalls, foreseen (following a suggestion by R. Palmer) to increase the dissipation of undesired transverse modes. This might make the use of multiple bunches possible. The slots would be somewhat longer than the inter-iris distance and would therefore have to cut through the irises.

Many different fabrication methods for the 25 cm long, 9 mm outer diameter sections have been considered, but it is far too early to make a choice among them. The large number (~ 100'000) of structures to be manufactured means that a very careful consideration of all details is essential.

The structures proper will have to be embedded into a larger diameter cylinder provided with longitudinal and transverse holes for vacuum and water cooling, and for feeding in and out the power. Beam position monitors will have to be included in the structures. The study of all this has only just started.

Calculations of the dissipation show that the maximum average temperature increase on the copper surface will be only 1.9 °C. Slightly more worrying is the instantaneous increase of 5 °C during each 11 ns pulse and it still has to to be shown that this is acceptable from the fatigue point of view.

9. DAMPING RINGS

Three groups have worked on damping rings in connection with the CLIC studies. All groups propose rings with wigglers in zero-dispersion straight sections.

Bassetti, Guiducci and Palumbo[21] studied various types of lattices for a proposed beauty factory. These would require a higher repetition rate (12 kHz) and a larger bunch population (2×10^{10}) than CLIC and are therefore not directly applicable, but many results of this study seem useful for CLIC also. Transverse normalised emittances of 2×10^{-6} m and a transverse damping time of 1.5 ms are obtained with a ring of 646 m circumference.

A similar approach by Delahaye, Krejcik and Potier[22] at CERN has so far resulted in a ring of half the size, damping times of 4.5 ms and similar final emittances, now adapted to the CLIC parameters.

Both approaches would need further study on collective effects, dynamic aperture, etc.

Evans and Schmidt[23] have investigated the possibility of using the CERN SPS tunnel for a damping ring. Because of the large circumference a longer damping time (30-60 ms) is now acceptable. A lattice similar to the SPS one (although far from optimum) is assumed for a start.

However, to obtain a sufficiently low longitudinal emittance, the RF frequencies and voltage would have to be increased, and this would indirectly lead to higher transverse tunes. With the parameters adopted, the longitudinal impedance would have to be less than 0.2 ohm, which might be difficult. In any case, if a CLIC machine would eventually be built, it is not sure that it would be at CERN or that the SPS tunnel would be available.

10. FINAL FOCUS

The very small beam height at the focus given in Table I was postulated in order to have an acceptable beam power, at a stage where no corresponding design for the final focus existed.

This still does not quite exist, and it may therefore be said that we have no consistent set of parameters. Nevertheless much work is being done in this area and it is hoped that a satisfactory solution may be found.

The most important problem is chromatic aberration. Using telescope optics (phase shift π) for demagnification, designs may be found that minimise the chromatic effects[24,25] but the remaining aberration is still too high; it would require a reduction of the normalised emittances to less than 10^{-7} m, which might be very difficult.

Better results have been found by using bending magnets and sextupoles, as was done at SLC. This allows correction of the first-order chromatic terms; but - at least for present designs - the higher-order aberrations (both chromatic and geometric) remain somewhat too high (about a factor 2 in each plane). In addition, the quantum fluctuations in radiation loss, caused by the bending magnets will increase the emittance and cause an increase of final beam height that is still unacceptable in present designs. It may well be that this may be improved by reducing the bending field and perhaps using combined function magnets having both dipole and sextupole components.

A final energy spread of $\pm$ 0.25% was assumed for these studies. To obtain this with an energy extraction of 5% from the main linac will require a reduction of the "natural" energy spread caused by beam loading by an order of magnitude. This may be possible, but it has not been proven conclusively.

11. CONCLUSION

In spite of the low level of funding (1 MSF for 1988) and the fact that nearly all work is done on a part-time basis, some interesting progress has been made during the past two years. The main point has been the better understanding of the beam break-up effect and its suppression by RF focusing. Some hardware tests and calculations on models of the transfer structure have increased our confidence in this unusual component. Studies on the main structures and their fabrication are just starting. Some damping ring designs now exist that nearly promise the required performance and it is thought that some further improvement may be possible. Lastly, the final focus design is improving, although the nominal performance specification has not yet been met. It may be possible that multi-bunch operation will be one way of solving some of the outstanding problems. It is hoped that during the coming years a better-understood, consistent design for a 1 TeV + 1 TeV collider will emerge.

REFERENCES

1) K. Johnsen et al., CERN 87-12 (May 1987).
2) W. Schnell, CERN-LEP-RF 86-06 (CLIC Note 13).
3) R. Hollebeek, Nucl. Instrum. Methods **184**, 333 (1981).
4) W. Schnell, CLIC Note 27 (Dec. 1986).
5) U. Amaldi, CERN-EP 87-169 (CLIC Note 51).
6) R. Palmer, SLAC-PUB-4295 (Apr. 1987).
7) W. Schnell, SLAC/AP-61 (Nov. 1987).
8) W. Schnell, CLIC Note 56 (Dec. 1987).
9) Y. Baconnier et al., CLIC Note 65 (June 1988).
10) H. Henke, W. Schnell, CERN-LEP-RF 86-18 (CLIC Note 22).
11) H. Henke, CERN-LEP-RF 87-36 (CLIC Note 40).
12) W. Schnell, CLIC Note 34 (March 1987).
13) H. Henke, CLIC Note 48 (Aug. 1987).

14) I. Wilson, H. Henke, CLIC Note 62 (May 1988).
15) H. Henke, private communication.
16) T. Garvey, G. Geschenke, W. Schnell, I. Wilson, CERN-LEP-RF 87-46 (CLIC Note 50).
17) T. Garvey, CLIC Note 57 (Feb. 1988).
18) I. Wilson, private communication.
19) J.-P. Boiteux et al., CERN-LEP-RF 87-25 (CLIC Note 36).
20) I. Wilson, CLIC Note 46 (Aug. 1987).
21) M. Bassetti, S. Guiducci, L. Palumbo, CLIC Note 60 (Apr. 1988).
22) J.-P. Delahaye, P. Krejcik, J.-P. Potier, private communication.
23) L. Evans, R. Schmidt, SPS/DI Note 88-1 (CLIC Note 58).
24) B. Zotter, CLIC Note 64 (June 1988).
25) J.E. Spencer, B. Zotter, Proc. European Part. Acc. Conf., Rome 1988 (to be published).

PLASMA ASSISTED INVERSE FREE ELECTRON LASER

J.L. Bobin

Laboratoire de Physique et Optique Corpusculaires
Université Pierre et Marie Curie
T 12, E 5, 4 place Jussieu 75252 Paris (France)

Abstract

It is shown that using beating between a laser wave and an undulator in presence of a relativistic electron beam, inside a background plasma, can generate high amplitude electron waves suited to subsequently accelerate electrons up to high energies. Acceleration gradients of order 1 G.e.V./m are obtained with moderate laser intensities.

1. Introduction.

Laser undulator beats are commonly used in Free Electron Lasers (F.E.L.) in the Compton [1] or the Raman [2] regime. In such devices, a relativistic electron beam is passed through a transverse spatially periodic magnetic field (the undulator) together with an electromagnetic (laser) wave. In the reference frame of the electrons, beating between the laser wave and the pseudo-electromagnetic wave equivalent to the undulator, results, via the Lorentz force, in a longitudinal electric field. It has been proposed [3] to use this field to accelerate charged elementary particles up to very high energies over distances much shorter than in the conventionnal R.F. linacs: that is the essence of the Inverse Free Electron Laser.

Other contenders for a future generation of particle accelerators, include the beatwave scheme [4] in which the accelerating longitudinal field is provided by the resonant beating of two laser waves acting on a plasma. This state of matter is made of charged microscopic objects, with opposite signs: ions and electrons. At equilibrium, it is electrically neutral. Very high electric fields accompanied with charge density perturbations, can propagate through the medium as longitudinal waves (electron plasma modes) whose frequency is is the characteristic plasma frequency

$$\omega_p = \sqrt{\frac{n_0 e^2}{\varepsilon_0 m_0}} \qquad (1\text{-}1)$$

where n_0 is the background electron density which ensures neutrality at equilibrium; $-e$ and m_0 are the electron charge and rest mass respectively. An electron wave may have any phase velocity in a cold plasma. In the beatwave accelerator, the laser frequencies are chosen with their difference equal to the plasma frequency in order to drive resonantly the electron plasma wave [5][6]. The phase velocity of the created longitudinal mode is equal to the group velocity of the electromagnetic waves. It is close to c.

It turns out that some improvements of the I.F.E.L. can be obtained in presence of a background plasma [7]. In order to discuss this problem, the present paper is organized as follows: F.E.L. and I.F.E.L. properties will first be presented from a plasma physicist's viewpoint. Then, results are to be given about electron acceleration in travelling longitudinal electric field oscillations. The next topic deals with the plasma physics and the dynamics of the laser undulator beat wave in a plasma. Finally, numbers for a proof of principle experiment will be evaluated.

2. The plasma physicist's F.E.L. and I.F.E.L.

The starting point is the dispersion relation for the system made of a cold background plasma with plasma frequency ω_p and a relativistic electron beam with plasma frequency ω_{pb}, and velocity u_b. Denoting by $\varepsilon(k,\omega)$ the high frequency dielectric constant,

$$\varepsilon(k,\omega) = 1 - \frac{\omega_p^2}{\omega^2} - \frac{\omega_{pb}^2}{\gamma_R^2 (ku_b - \omega)^2} = 0 \quad \text{with} \quad \gamma_R^2 = \frac{1}{1 - \frac{u_b^2}{c^2}} \quad (2\text{-}1).$$

In the F.E.L. or the I.F.E.L. case, there is no background plasma and the dispersion relation reduces to

$$\varepsilon(k,\omega) = 1 - \frac{\omega_{pb}^2}{\gamma_R^2 (ku_b - \omega)^2} = 0 \quad \text{or} \quad \omega = ku_b \pm \frac{\omega_{pb}}{\gamma_R} \quad (2\text{-}2).$$

$\gamma_R (ku_b - \omega)$ is the Doppler shifted frequency of the electron wave ω. In the Brillouin diagram (ω,k) the dispersion relation (2-2) is represented by two parallel lines with slope u_b. The upper branch has a positive energy ($\partial\varepsilon/\partial\omega>0$ at $\varepsilon=0$) whereas the lower one has a negative energy ($\partial\varepsilon/\partial\omega<0$ at $\varepsilon=0$)[8]. Now, one may look at the possible couplings of these plasma modes in the beam, to an electromagnetic wave (the laser) and a zero frequency oscillation (the undulator). In figure 1, the relevant (ω,k) diagrams are shown for both the laboratory frame and the reference frame in which the beam is at rest: moving frame.

When the plasma mode belongs to the upper branch of the dispersion relation, a single mechanism takes place: the **stimulated Raman scattering** in which an incident electromagnetic wave interacts with a backscattered pseudo-mode (undulator) and an electron plasma wave. The selection rules are best expressed quantum-mechanically as

photon (ω_1, k_1) ↔ pseudo-photon $(0, k_2)$ + plasmon $(\omega_1, k_D = k_1 - k_2)$

in the laboratory frame. Since the phase velocity of the plasma wave is larger than the beam velocity this is an **inverse Cerenkov** process. The corresponding combinations of vectors in the (ω, k) diagrams are displayed on figure 2. The mechanism goes either way. All three modes have a positive energy. Consequently, the amplitude of the laser wave decreases whereas the amplitudes of the two others increases and conversely.

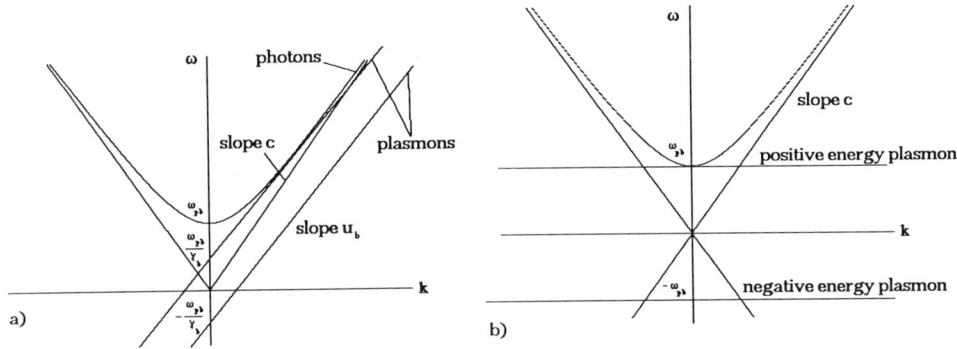

Figure 1.
Dispersion relations for electromagnetic waves and plasma waves in a relativistic beam. a) laboratory frame; b) moving frame.

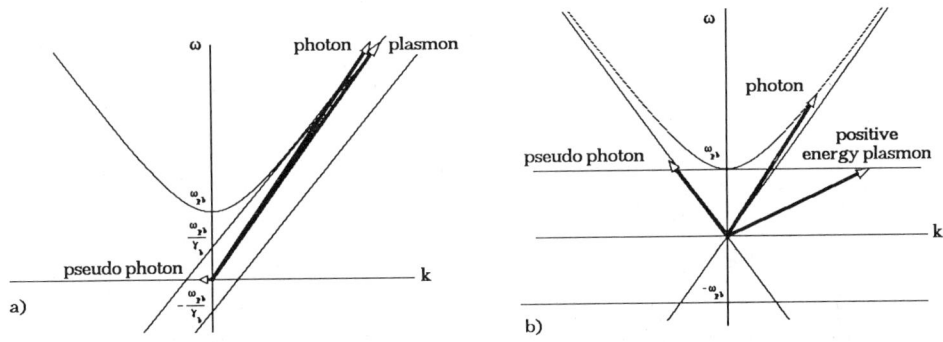

Figure 2.
**Raman backscattering in the inverse Cerenkov regime.
a) laboratory frame; b) moving frame.**

When the plasma mode belongs to the lower branch of the dispersion relation (slow wave), the phase velocity of the wave is smaller than the beam velocity: **Cerenkov** processes. Stimulated Raman scattering with the same selection rules is still possible (figure 3). However, since the plasma mode has a negative energy the variations of its amplitude are the same as those of the laser amplitude. Both can be simultaneously amplified. Another coupling is allowed in this case with a pseudo-mode in the same direction as the laser (figure 4). The selection rule now is

plasmon $(\omega_1, \mathbf{k}_D)$ ↔ photon $(\omega_1, \mathbf{k}_1)$ + pseudo photon $(0, \mathbf{k}_2 = \mathbf{k}_D - \mathbf{k}_1)$

The three modes can grow simultaneously. This is known in plasma physics as an **explosive instability**. A random noise can start the process which bears similarities with the so called Self Amplification of Spontaneous Emission (S.A.S.E.)[9].

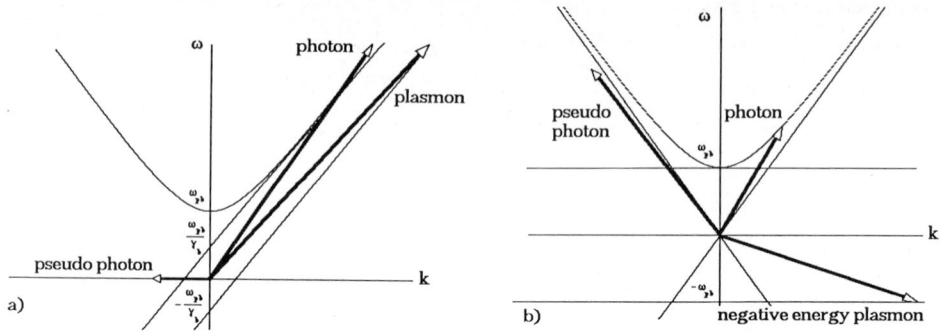

Figure 3.
**Raman scattering in the Cerenkov regime.
a) laboratory frame; b) moving frame**.

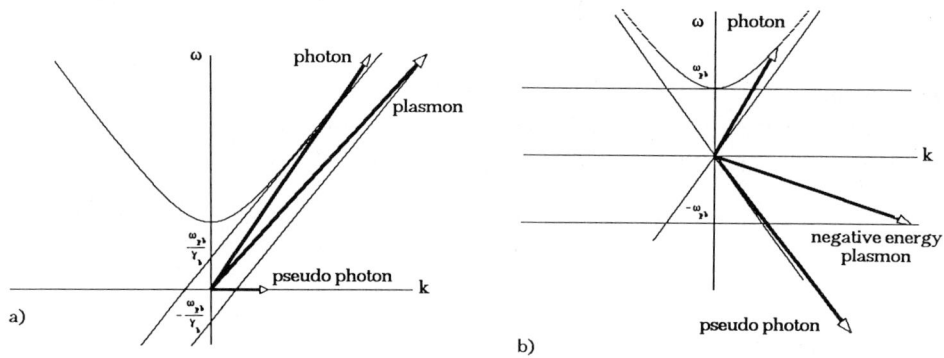

Figure 4.
Explosive instability. a) laboratory frame; b) moving frame.

Now, the Free Electron Laser can work in the Compton regime in which interaction deals with individual electrons rather than with a collective oscillation. In plasma physics, it is considered that individual effects happen over a length smaller than the Debye length

$$\lambda_{De} = \frac{<v_e>}{\omega_p} \quad (2\text{-}3)$$

where $<v_e>$ is an average (thermal) electron velocity. On the contrary, collective effects act on a scalelength larger than λ_{De}. In terms of wavenumbers and as an order of magnitude:

$|k| \gg |k_{De}| \approx \omega_p/\langle v_e \rangle$ holds for individual (Compton) interactions,
$|k| \ll |k_{De}|$ for collective (Raman) ones.

In a beam in which electrons are shaken by waves, the resulting energy spread can be used to define the thermal velocity [10]

$$\langle v_e \rangle = c\Delta\gamma_e/\gamma_\varepsilon \qquad (2\text{-}4)$$

in which γ_e is the Lorentz factor of individual electrons. Now, in a reference frame moving with the phase velocity of the plasma wave, the incident and backscattered electromagnetic modes have the same frequency

$$\omega'_1 = ck'_D/2 \qquad (2\text{-}5).$$

After a Lorentz transformation, one has in the laboratory frame

$$\omega_1 = (1 + \beta_e)\gamma_e\omega'_1 \approx 2\gamma_e\omega'_1 = c\gamma_e k'_D, \quad \text{hence} \quad k'_D = \omega_1/c\gamma_e \qquad (2\text{-}6).$$

The condition for a collective behaviour (Raman scattering) thus implies

$$\omega_1 < \omega_c = \frac{\gamma_e^2 \omega_p}{\Delta\gamma_e} \qquad (2\text{-}7).$$

It should be noted that, even in the Compton regime, electrons are periodically bunched as a consequence of a free electron laser process. The period is that of the laser-undulator beats, indicating a self organization of the electron beam in the form of a plasma wave.

A final remark is noteworthy: since the plasma wave and the laser have the same frequency in the laboratory frame, **the quantum efficiency is equal to 1**.

3. Electron acceleration in a longitudinal electrostatic oscillation.

Assume a longitudinal electric field propagates with a well defined amplitude and phase in the z direction. Let ω_D and k_D be the frequency and the wavenumber respectively. The phase velocity $u_R = \omega_D/k_D$ is close to c. Denoting by ϕ_0 an initial phase, the relativistic equation of motion for an electron is

$$\frac{dv_z}{dt} = -\frac{e}{m_0}\left(1 - \frac{v_z^2}{c^2}\right)^{\frac{3}{2}} E_0 \sin(k_D z - \omega_D t + \phi_0) \qquad (3\text{-}1).$$

The equation is better rewritten in terms of the Lorentz factor γ and with the similarity variable

$\xi = k_D z - \omega_D t$. One then gets the differential system

$$\frac{d\gamma}{dt} = -\frac{e}{m_0 c^2} \sqrt{1 - \frac{1}{\gamma^2}} E_0 \sin(\xi - \xi_0)$$

$$\frac{d\xi}{dt} = k_D v_z - \omega_D = ck_D \sqrt{1 - \frac{1}{\gamma^2}} - \omega_D$$

(3-2).

The phase portrait in the (γ,ξ) plane displays the usual trapped and passing trajectories. Now, the first order ordinary differential equation in $d\gamma/d\xi$ resulting from (3-2) is readily integrated to give

$$ck_D\gamma - \omega_D\sqrt{\gamma^2 - 1} - ck_D\gamma_0 + \omega_D\sqrt{\gamma_0^2 - 1} = \frac{eE_0}{m_0 c}(\cos\xi - \cos\xi_0)$$

(3-3).

The maximum value of the bracket in the right hand side is 2. It corresponds to the largest $\Delta\gamma$ along a passing trajectory which assuming that both γ_0 and γ are much greater than 1, is

$$\Delta\gamma = \gamma - \gamma_0 \approx -\frac{2eE_0}{m_0 c \omega_D} \frac{\beta_R}{1 - \beta_R} \quad \text{with} \quad \beta_R = \sqrt{1 - \frac{1}{\gamma_R^2}} = \frac{\omega_D}{ck_D}$$

(3-4).

The maximum energy that an electron may acquire in the process is then

$$W_A = m_0 c^2 \Delta\gamma \approx \frac{4ecE_0}{\omega_D} \gamma_R^2$$

(3-5).

Now, there are two situations of interest. Consider first an electron bunch extending over several wavelengthes of the electric field. Initially the particles have the same γ_0, high enough so that they are all to follow passing trajectories. After a while, some are accelerated, but more are decelerated (figure 5a). Energy has been lost by the electrons and gained by the field. This is the **Free Electron Laser regime**. On the contrary, when the extension of the bunch along the z axis is less than a small fraction (e.g. 1/10th) of the wavelength $2\pi/k_D$, all electrons are accelerated if they have the proper initial phase (figure 5b). This is the **Inverse Free Electron Laser regime** in which energy is gained by the particles with a maximum given by (3-5).

In a reference frame moving with the phase velocity ω_D/k_D, electrons are accelerated over at most half a wavelength. Since there is no phase locking, electrons in the wave are to be decelerated whenever they propagate beyond that distance. The corresponding acceleration length l_A in the laboratory frame is given by

$$W_A = 2eE_0 l_A \quad \text{hence} \quad l_A = 2\gamma_R^2 \frac{c}{\omega_D}$$

(3-6).

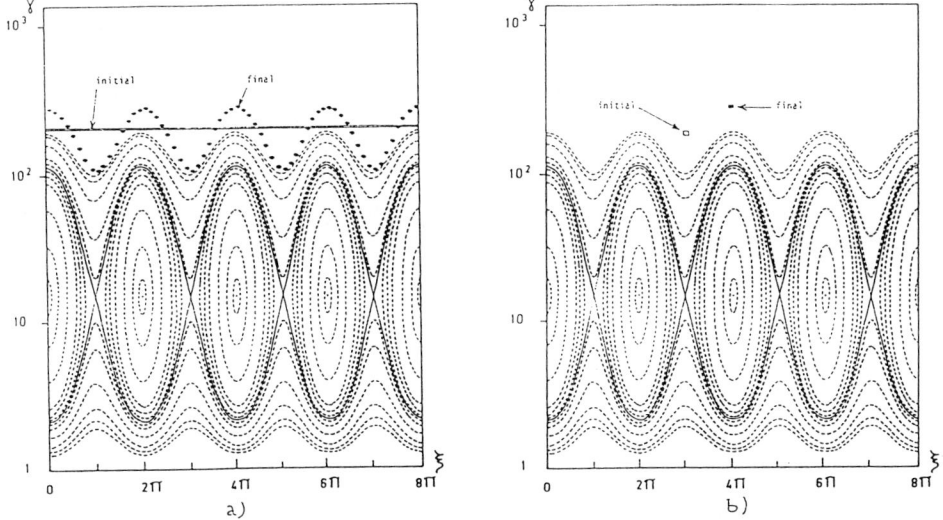

Figure 5.
Phase space behaviour of an electron bunch. a) long bunch, Free Electron Laser regime; b) short bunch, Inverse Free Electron Laser.

4. Plasma physics: cut-off and wave dynamics.

The presence of a background plasma does not changes the three wave couplings involved in the F.E.L. or the I.F.E.L. In the case of Raman backscattering, the phase velocity of the laser-undulator beats is

$$u_R = \frac{\omega_1}{k_1 + k_2} \quad (4\text{-}1).$$

From (2-1) a relationship between u_R and the beam velocity u_b is derived viz.

$$u_b = u_R \pm \frac{\omega_{pb}}{\gamma_b k_D \sqrt{1 - \frac{\omega_p^2}{\omega_1^2}}} \quad (4\text{-}2).$$

The upper and lower sign refer to the Cerenkov and inverse Cerenkov regimes respectively. In both cases the Lorentz factor of the beat is

$$\gamma_R^2 = \frac{1}{1 - \frac{u_R^2}{c^2}} = \frac{(k_1 + k_2)^2}{\left[(2k_1 + k_2)k_2 - \frac{\omega_p^2}{c^2}\right]} \quad (4\text{-}3)$$

which, since

$$\omega_1^2 = \omega_p^2 + k_1^2 c^2 \quad (4\text{-}4)$$

implies a divergence for

$$\omega_p^2 - 2c\omega_1 k_2 + c^2 k_2^2 = 0 \quad (4\text{-}5).$$

In other words, there exists a critical density for the background plasma or a cut-off value for the laser frequency, with the double inequality:

$$n_0 \leq n_{0c} = \frac{\varepsilon_0 m_0 (2\omega_1 - ck_2)ck_2}{e^2} \quad \text{or} \quad \omega_1 \geq \omega_{1\text{off}} = \frac{(\omega_p^2 + c^2 k_2^2)}{2ck_2} \quad (4\text{-}6).$$

The Lorentz factor γ_R depends upon the plasma frequency (i.e. the density) and upon the laser frequency ω_1 as depicted on figure 6.

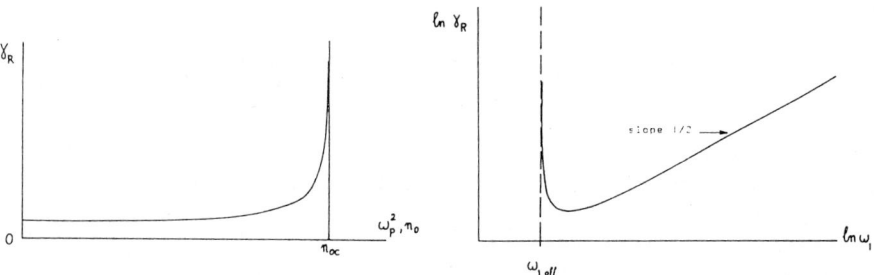

Figure 6.
Lorentz factor for the B.R. frame versus:
a) the plasma density; b) the laser frequency.

The dynamics of the plasma wave generation by resonant beating was thoroughly investigated in the case of two electromagnetic (laser) waves. This problem was first considered by Montgomery [5] and dealt with in more details by Rosenbluth and Liu [6] whose methods can be extended to the case of laser-undulator beating. Denoting by $\mathbf{A}_1$ and $\mathbf{A}_2$ the vector potentials (purely transverse), electrons in the plasma satisfy the equations of motion

$$\frac{dv_z}{dt} = \frac{e}{m_0 \gamma^3} E - \frac{e^2}{m_0^2 \gamma^2}\left(\frac{\partial |A|^2}{\partial z} + \frac{v_z}{c^2}\frac{\partial |A|^2}{\partial t}\right)$$

$$\frac{d\gamma}{dt} = \frac{e}{m_0 c^2} E + \frac{e^2}{m_0^2 c^2 \gamma}\frac{\partial |A|^2}{\partial t} \quad (4\text{-}7).$$

with $\quad |A|^2 = A_1^2 + A_2^2 - 2A_1 A_2 \cos \xi$

In a reference frame moving with the velocity u_R, i.e. the Bambini Renieri (B.R.) frame [11], the vector potential resulting from the laser (electric field E_1; $A_1=E_1/\omega_1$) and the undulator (magnetic field B_2; $A_2=B_2/k_2$), has no time dependance. The longitudinal electric field E obeys

$$\frac{\partial^2 E}{\partial t^2} = -\frac{1}{\varepsilon_0}\frac{\partial j}{\partial t} \qquad (4\text{-}8)$$

in which the longitudinal current j has to be calculated after the density perturbations and velocities both in the background plasma and in the beam. Linearized fluid dynamical equations are used for the velocity and the electron density perturbations in the background plasma and in the beam as well. The result is a second order partial differential equation for the electric field in the B.R. reference frame (no time dependance, non relativistic motion):

$$\frac{d^2 E}{dz'^2} = -k'^2_D E + 2\frac{m_0 c \omega_1}{e}\Lambda\left[\left(\frac{\omega_p}{\omega_1}\right)^2 + \frac{1}{\gamma_R}\right]\cos(k'_D z') \qquad (4\text{-}9)$$

where

$$\Lambda = \left(\frac{e}{m_0 c}\right)^2 A_1 A_2 \qquad (4\text{-}10)$$

is the **interaction parameter**. In the right hand side of (4-9), the first term in the [ ] bracket is the non resonant contribution of the background plasma and the second one is the resonant contribution of the beam.

Now, there are two ways of dealing with (4-9). First, in the case of a constant Λ, this equation represents a forced oscillator. The solution is sought as a cosine function with slowly varying amplitude and phase. At resonance, the amplitude diverges with phase locking at $\pi/2$. Alternatively, the variations of A_1 and A_2 are considered in compliance with the selection rule. A_1, A_2 and E obey propagative second order equations which, assuming slowly variable amplitudes, can be reduced to a first order system of 3 coupled differential equations. The undulator is equivalent to a very high intensity electromagnetic wave. Λ is made large with a comparatively moderate laser intensity. Since the quantum efficiency is unity, the plasma wave amplitude is obviously bounded by the laser amplitude.

5. Saturation of the plasma wave, electron acceleration.

The divergenge pointed out in the preceding section is unphysical. An unavoidable saturation results from the expected contributions of many mechanisms: pump depletion, cascading, collisions, relativistic effects... Since the static magnetic field of the undulator is set up over a limited number of periods, the finite size might also influence the growth of the plasma wave. This can be accounted for by introducing a phenomenological damping term in the left hand side of equation (4-9) in which the reciprocal of the undulator length L' (measured in the moving frame) appears as an effective collision frequency. This approximation was proved to work for parametric instabilities in bounded plasmas [12]. Now, a large amplitude plasma wave drives the electrons up to relativistic oscillatory velocities. Their mass increases accordingly inducing a detuning from the resonance conditions. In the weakly relativistic case, this detuning is represented by a quadratic ($\propto |E|^2$) correction to the eigen frequency of the forced oscillator, thus introducing a cubic non linearity in the equations. Consequently, (4-9) is replaced by

$$\frac{d^2E}{dz'^2} + \frac{1}{L'}\frac{dE}{dz'} = -(1-\beta|E|^2)\,k'^2_D\,E + 2\,\frac{m_0 c \omega_1}{e}\,\Lambda\left[\left(\frac{\omega_p}{\omega_1}\right)^2 + \frac{1}{\gamma_R}\right]\cos(k'_D z') \qquad (5\text{-}1)$$

where β is a small density dependant coefficient. With constant Λ, (5-1) is a Duffing's equation. The effective damping and the relativistic detuning can also be included in the first order system which accounts for pump depletion.

When the effective damping and pump depletion are small, the plasma wave amplitude is limited by the relativistic detuning. A maximum electric field is calculated after the Duffing's equation in the case of growth from a very low level

$$E_{max} = \frac{c m_0 \omega_1}{e}\left[\left(\frac{\omega_p}{\omega_1}\right)^2 + \frac{1}{\gamma_R}\right]\left(\frac{16}{3}\Lambda\right)^{1/3} \qquad (5\text{-}2)$$

The corresponding acceleration energy and length for an electron in the wave are then, in the laboratory frame,

$$W_A = 2\gamma_R m_0 c^2\left[\left(\frac{\omega_p}{\omega_1}\right)^2 \gamma_R + 1\right]\left(\frac{16}{3}\Lambda\right)^{1/3} \quad \text{and} \quad l_A = 2\gamma_R\frac{c}{\omega_1} \qquad (5\text{-}3).$$

The variation of W_A/l_A and l_A with respect to the laser frequency ω_1 are shown on figure 7.

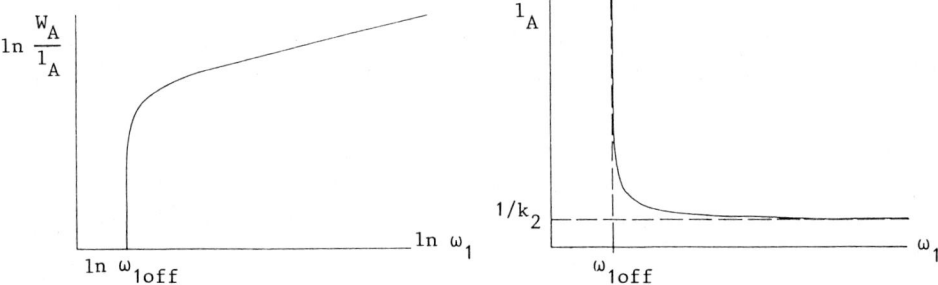

Figure 7.
Acceleration energy a), and length b), versus laser frequency.

The accelerating field is the smaller of two: the electric field given in (5-2) and the laser electric field. Furthermore, it has to be larger than the driving longitudinal field, otherwise there would be no plasma enhancement of the I.F.E.L. (5-2) holds provided

$$\frac{cm_0\omega_1}{e}\frac{\Lambda}{\gamma_R} < \frac{cm_0\omega_1}{e}\left[\left(\frac{\omega_p}{\omega_1}\right)^2 + \frac{1}{\gamma_R}\right]\left(\frac{16}{3}\Lambda\right)^{1/3} < E_1 = \frac{cm_0\omega_1}{e}\frac{cm_0}{eA_2}\Lambda \quad (5\text{-}4)$$

which implies

$$\gamma_R > \gamma_{Rcrit} = \frac{eA_2}{m_0c} = \frac{eB_2}{m_0ck_2} \quad (5\text{-}5).$$

When the Lorentz factor of the B.R. frame surpasses the critical value, one has the situation depicted in figure 8 which shows plasma enhancement of the I.F.E.L. when the interaction parameter Λ is smaller than about 0.6.

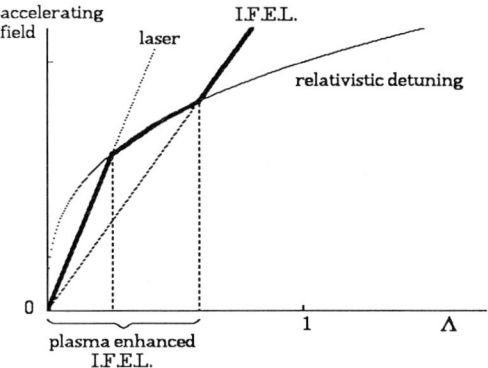

Figure 8.
Accelerating field versus interaction parameter.

6. A proof of principle experiment.

As far as the hardware is concerned, the only difference with the I.F.E.L. is the use of a plasma with the right density. It is then convenient to start from an actual project such as the one under study in the U.S. at Brookhaven National Laboratory [13]. There, the Center for Accelerator Physics is building the Accelerator Test Facility (A.T.F.), which includes a 50 M.e.V. high brightness linear electron accelerator whose characteristics are given on table 1, and a high power (100 GW) CO_2 laser. The facility is aimed at testing new ideas in accelerator physics.

Table 1.
A.T.F. Linac Parameters

	Low Intensity	High Intensity
Linac energy (M.e.V.)	50	50
Numer of Electrons per Bunch	$2\ 10^5$	$6\ 10^9$
Pulse Length (picoseconds)	6	6
Relative Energy Spread	$\pm 0.1\%$	$\pm 0.2\%$
Beam Radius (mm)	≤ 0.3 at wiggler entrance	

This electron beam and the associated CO_2 laser are well suited for a demonstration stage I.F.E.L. accelerator using a 60 cm long tapered wiggler. The parameters and expected performances are shown on table 2.

Table 2.
I.F.E.L. Accelerator Project.

Wiggler:
Length (cm)	60
Period Length (cm)	2.27 to 4.34
Gap (cm)	1
Maximum B Field (T)	1.5 (N.B. constant field amplitude)

Laser:
Power (watts)	$2\ 10^{11}$
Electric Field Amplitude (V/m)	10^{10}
Wavelength (μm)	10
ω_1 (s^{-1}), k_1 (m^{-1})	$1.78\ 10^{14}$, $1.93\ 10^5$
Rayleigh Length (cm)	30
Photon Beam Radius (mm)	.69 at wiggler entrance

Electrons:
Maximum Added Energy (M.e.V.)	103.5
Maximum Accelerator field (MV/m)	89

With these figures in mind and using the formulas presented in this paper, a plasma enhanced I.F.E.L. can be designed. A 60 cm long untapered wiggler with the shortest period length of the tapered one, i.e. 2.67 cm, $k_2=2.35\ 10^2$ m^{-1}, is convenient for this purpose. The corresponding interaction parameter is

$$\Lambda = \left(\frac{e}{m_0 c}\right)^2 \frac{E_1}{\omega_1} \frac{B_2}{k_2} = 0.126 \qquad (6\text{-}1).$$

This value falls in the interval for which the growth of the plasma wave is limited by the relativistic detuning. Now, the accelerator length l_A can be chosen in such a way that electron acceleration takes place over the full length of the undulator. The plasma density is then close to the critical value. Alternatively, since the plasma wave is propagative, it can be launched through a plasma longer than the undulator. The extra length can act as an accelerating section. In the asymptotic case $l_A=1/k_2$ (see figure 7) the gradient is large. However, the maximum electron energy is comparatively small. Evaluations were made for both possibilities. The results are shown on table 3.

Table 3.
Plasma Enhanced I.F.E.L.

	inside the undulator	outside the undulator
Acceleration Length l_A (cm)	60	.37
γ_R	422	34
Background Plasma Density (cm^{-3})	2.8 10^{15}	1.7 10^{15}
Acceleration Energy (M.e.V.)	414	34
Gradient (G.e.V./m)	0.69	9.1

Evidently the parameters of the I.F.E.L. accelerator at Brookhaven are convenient for a proof of principle experiment on plasma enhancement. The electron density in the background plasma has to be a few 10^{15}cm^{-3}. The best method to generate long cold homogeneous hydrogen plasmas with this density seems to be direct multiphotonic ionisation as tested in Great Britain at Rutherford Appleton Laboratory [14]. In order to perform a 2 laser beatwave experiment, the beam of a high intensity Nd glass laser was focused into a gas, producing a 1cm long, clean homogeneous plasma column with a density about 10^{17}cm^{-3}. The technique migth be made more efficient thanks to a tunable laser. This would allow resonant multiphoton ionization.

A final remark is noteworthy. This paper deals primarily with stimulated Raman scattering and its consequences on electron acceleration. Now, through the explosive instability, a background plasma can also be used as a way to enhance the S.A.S.E. mechanism [15]. This possibility could fit into any F.E.L./I.F.E.L. project.

References.

1) J.M.J. Madey, *J. Appl. Phys.* **42** (1971) 1906.
2) V. Granastein et al. *Appl. Phys. Lett.* **30** (1977) 384.
3) C. Pellegrini, Proc. ECFA-RAL *The Challenge of Ultrahigh Energies.* J. Mulvey ed. (1982).
4) T. Tajima and J.M. Dawson, *Phys. Rev. Lett.* **43** (1979) 267.
5) D.C. Montgomery, *Physica* **31** (1965) 693.
6) M.N. Rosenbluth and C.S. Liu, *Phys. Rev. Lett.* **29** (1972) 701.
7) J.L. Bobin, *Optics Comm.* **55** (1985) 413.
8) see e.g. L.D. Landau and E.M. Lifshitz, *Electrodynamique des Milieux Continus* Mir Moscou (1969).
9) R. Bonifacio and F. Casagrande, *Optics Comm.* **50** (1984) 251,
   J.B. Murphy and C. Pellegrini, *Nucl. Instrum. Methods* **A257** (1985) 159.
10) see e.g. J.L. Bobin, *Le Laser à electrons Libres, une Introduction Théorique* Rapport C.E.A. n° **R-5141** (1981) and references therein.
11) A. Bambini and A. Renieri, *Lett. Nuov. Cim.* **21** (1978) 239.
12) R. White et al. *Nuclear Fusion* **14** (1974) 45.
13) C. Pellegrini, private communication (1987).
14) A.E. Dangor et al. *IEEE Transactions Plasma Science* **PS-15** (1987) 161.
15) J.L. Bobin, *J. de Physique* Colloque C6, **47** (1986) C6-159.

RADIATION FROM FINE, INTENSE, SELF-FOCUSSED BEAMS AT HIGH ENERGY

William A. BARLETTA

University of California, Lawrence Livermore National Laboratory,
Livermore, California 94550

Andrew M. SESSLER

University of California, Lawrence Berkeley Laboratory,
Berkeley, California 94720

A theoretical analysis is presented of the radiation emitted when an intense, relativistic, low emittance electron beam propagates through a channel of pre-ionized gas. The beam will self-focus to a small radius and radiate, in the self-induced magnetic field, and it will emit an intense burst of gamma rays. If the beam is subject to a conventional wiggler field, the result will be a "wiggled channel," and a subsequent pulse of high energy electrons will radiate, more effectively than in the absence of ions, a substantial fraction ($\approx 0.1\%$) of its power coherently as an X-ray laser beam.

I. Introduction

Fine beams (normalized emittance $e_n \approx 10^{-5}$ m-rad) of high intensity ($I_{peak} = 1 - 3$ kA) and ultra-relativistic energy ($E \approx 1 - 3$ GeV) will radiate strongly when passing through an ionized channel. The radiation is incoherent as we describe in the analysis of Sec. II

Coherent radiation can be generated by sending the beam through a wiggler. The resulting free electron laser action, which grows from the incoherent radiation from the beam, can be enhanced by focussing the beam more strongly than the "natural" focussing of the wiggler through the introduction of a tenuous ion channel through the wiggler. Consideration of this configuration is given in Sec. III.

II. Incoherent radiation

Synchrotron radiation ("beamstrahlung" or magnetic bremsstrahlung), which is emitted by electron and positron beams as they pass through and focus each other, has become familiar to the designers of $e^+ - e^-$ colliders. The same physical process can take place if the positron beam is replaced by the strong focusing field provided by ions in a pre-ionized gas.[1,2] E. Lee has analyzed in detail the motion of of high-current beams propagating through dense gases.[3] In particular, he has computed the energy loss rate due to synchrotron radiation in the self-field of the beam. Because individual electrons emit radiation parallel to their trajectories, the radiation is concentrated in a narrow cone whose width is equal to the mean betatron angle. The e-folding distance for energy loss via beamstrahlung can be made very short; propagation paths ($\approx$ 20 m) can produce an extremely bright ($\approx 10^{18}$ W/St) burst of hard ($E_c \approx$ 1 GeV) gamma rays.

The linear beamstrahlung source, schematically illustrated in Fig. 1, is divided into two regions. Differential pumping in Region 1 matches the beam into a converter region of length L (Region 2). The converter is filled with a gas at a pressure that should be just high enough for the space charge of the beam to be completely neutralized.

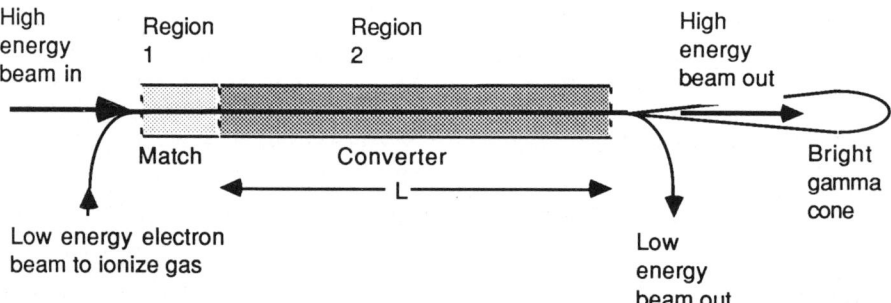

Figure 1 Schematic of an incoherent radiation source with effective megagauss fields.

For illustrative purposes we make the length of the converter, L, equal to the energy e-folding length due to beamstrahlung, L_D. Rewriting the scaling law of Ref. 3, one finds

$$L_D = 25 \text{ meters } (a / 1 \text{ } \mu m)^2 (1 \text{ kA} / f_c I)^2 (10 \text{ GeV} / E), \qquad (1)$$

where a is the radius of the beam, f_c is the charge neutralization fraction, I is the beam current, and E is the beam energy. During the radiation process the beam will remain in quasi-static equilibrium if L>>λ_β, where λ_β is the betatron wavelength. Unless the beamstrahlung damping length is much smaller than the scattering length in the gas, beam-gas scattering will increase the emittance, enlarge the beam, and decrease the radiation rate.

The gas identity and pressure are chosen to minimize the beam-gas scattering, the scale length for which is the Nordseick length [3]:

$$L_N = 0.135 \text{ meters } (E/1 \text{ GeV}) (I/1 \text{ kA}) X_R (760 \text{ Torr}/P), \quad (2)$$

where E is the beam energy, I is the beam current, P is the gas pressure, and X_R is the radiation length. The gas with the longest radiation length at STP is helium, for which X_R = 750 m.

The relationships among beam radius, energy, current, normalized emittance ε_n, and betatron wavelength and mean betatron angle, $\langle\Theta^2\rangle^{1/2}$, for self-focused beams are well known [4]:

$$a = (\varepsilon_n / \gamma) (17\gamma / f_c I)^{1/2}, \quad (3)$$

$$\langle\Theta^2\rangle^{1/2} = (17\gamma / f_c I)^{-1/2}, \quad (4)$$

$$\lambda_\beta = 2\pi a / \langle\Theta^2\rangle^{1/2}. \quad (5)$$

If the beam density exceeds the gas density, Eq. (3) should be replaced by

$$a = 35 \text{ μm } (\varepsilon_n / 5 \times 10^{-5} \text{ m-rad})^{1/2} (100/\gamma)^{1/4} (10^{15} \text{ cm}^{-3} / Z^* n_i)^{1/4}, \quad (6)$$

where Z^* is the average ionic charge state, and n_i is the ion density. The strength of the azimuthal focusing field at the beam's edge is B = (0.2 kG) (I / 1 kA) (1 cm / a).

Recalling that ε_n / γ is invariant in the cooling process, one sees that the beam radius decreases as the square root of energy loss in the converter cell. Combining this fact with Eq. (1), one sees that the e-folding length for radiation emission remains invariant as the electrons lose energy. This feature of the converter has been proposed as a means of producing a linear damping section for TeV-class linear colliders.[2]

Equations (3) - (6) describe the beam equilibrium state after the gas has been ionized and after the detached plasma electrons have been ejected from the beam vicinity. One must, therefore, specify how the gas is to be ionized and must determine whether equilibrium occurs in a short time compared to the pulse length of the high energy beam. A beam passing through helium of pressure P will achieve charge neutralization in a time T, where P(Torr) T(ns) = 4. The beam pulse lengths of interest in linear colliders, or in linear XUV light sources, are less than 10 ps. Therefore, the high energy beam cannot by itself ionize the low density gas in the converter unless P is significantly greater than atmospheric pressure.

Another possibility is to ionize the gas with a laser pulse and to follow the laser pulse with a high energy beam.[5] The space charge of the beam ejects the detached electrons with a time scale characterized by the beam radius, which is much smaller than the pulse length. Unfortunately, the ionization potential of helium is so high that no practical lasers are available for this purpose. However, a variant of this ionization scheme, where one replaces the pulsed laser with a low energy ($\approx$ 5 - 100 MeV) electron beam ($\approx$100 A), is suitable for use in the beamstrahlung converter. This low energy beam can be transported through the converter by low field strength magnets. Beams with such characteristics can easily be generated at repetition frequencies of 100 to 1000 Hz.

The properties of the electron beam produced at the Stanford Linear Collider at SLAC may be employed to analyze the characteristics of the beamstrahlung converter. At the output of the damping ring the beam has a normalized emittance of 3×10^{-5} m-rad, consists of $\approx 5 \times 10^{10}$ electrons, and has a peak current of ≈ 1 kA. If this beam is accelerated to 10 GeV before it is injected into the gas cell, then the self-pinched size of the fully charged neutralized beam will be 1 µm. The betatron wavelength in the damping region will be 4 mm with a focusing field of 2 MG. One finds the length of the damping region to be 25 m. From Eq. (6) one sees that the beams size varies slowly with the background ion density; full charge neutrality is not needed to produce a short damping region. Helium at $\approx$ 40 Torr will provide sufficient background ions to focus the beam, while producing only 10^{-3} radiation lengths of scatterer (the Nordseick length of the cell is $\approx 10^{4}$ m). Thus multiple scattering in the beamstrahlung converter leads to a scattering angle of 20 µrad, a value much smaller than the betatron angle.

The matching section [Region 1 of Fig. 1] must provide a slow transition ($\sim 5 \lambda_\beta$) between the equilibrium radius of the beam in the accelerator transport and the radius in the conversion section. Final focus telescopes of conventional design can provide such strong focusing. An alternative is to use a differentially pumped gas cell of length $\approx$ 5 m to allow smooth matching without increasing beam emittance through phase space filamentation.

The radiation converter is a remarkably efficient (> 65%) means of converting the energy of the electron beam into an extremely bright source of hard gamma radiation. In contrast with the output from bremsstrahlung converters, the most startling feature of the beamstrahlung converter is the hardness of the gamma radiation. The critical frequency, ω_c, is given by $\omega_c = 3 (\omega_0 \gamma^3)$, where ω_0 is the oscillation frequency of the electrons' motion in the laboratory frame. For the SLAC example, the oscillation frequency is

≈7 x 10^{10} Hz, which implies a critical frequency of 1.6 x 10^{24} Hz, or a critical energy of ≈ 8 GeV. Although this value is nearly equal to the incident beam energy, the quantum mechanical correction to the classical synchrotron distribution is not substantial because nearly all frequencies are kinematically allowed. The frequency distribution peaks at ≈ 1.5 GeV with substantial radiation down to 50 MeV. The electron beam pulse initially contains 30 J, of which 20 J are transformed into incoherent synchrotron radiation. The radiation is emitted at an angle (≈ 1 mrad) slightly smaller than the betatron angle, which will increase by a factor of $E^{1/2}$ as the beam passes through the converter. The peak brightness of the gamma pulse is, therefore, ≈ 10^{18} W/St.

III. Coherent Radiation

Electron beams can produced coherent optical radiation via the free electron laser (FEL) mechanism. Ion focussing channels may be employed to enhance the performance of free electron lasers as we shall now show. That the performance of an FEL may be enhanced beyond that occurring "naturally" has been realized by Pellegrini [6]. The extent to which the additional focussing can be supplied by ions is explored in this section.

The usual free electron laser resonance condition is

$$\lambda = (\lambda_w / 2\gamma^2)(1+a_w^2), \qquad (7)$$

where λ_w is the wavelength of the wiggler, λ is the wavelength of the radiation, and a_w is the dimensionless vector potential of the wiggler with peak magnetic field strength, B;

$$a_w = \frac{e B \lambda_w}{2\sqrt{2}\,\pi\, m_e c^2}. \qquad (8)$$

The wiggler provides natural focussing in the plane transverse to the wiggle plane. By appropriately shaping the pole face [7] this natural focussing can be extended to the wiggle plane. Then the betatron wavelength, λ_β, will be given by

$$\lambda_\beta = \frac{\sqrt{2}\,\gamma\,\lambda_w}{a_w} H, \qquad (9)$$

where the factor H = 1. The effect of ions is to decrease the betatron wavelength; that is to make H < 1.

It is useful to introduce the relativistic plasma frequency, ω_p, where

$$\omega_p^2 = \frac{4\pi n_e r_o c^2}{\gamma^3},$$

(10)

where r_o is the classical radius of the electron. The beam density, n_e, is given in terms of the beam current, I, and the beam radius, r_b, by the relation

$$I = \pi c e n_e r_b^2.$$

(11)

The radius of the electron beam is related to the normalized emittance, ε_n, of the beam by

$$\varepsilon_n = \frac{2\pi \gamma r_b^2}{\lambda_\beta}.$$

(12)

According to the one dimensional theory in the cold beam limit [8], the performance of the FEL is given in terms of the Pierce parameter, ρ, which is defined by

$$\rho = \left(\frac{a_w \omega_p \lambda_w}{8\pi c}\right)^{2/3}.$$

(13)

The power in the FEL grows exponentially with a gain length of

$$L_G = \frac{\lambda_w}{4\pi \rho}.$$

(14)

The amplifier will saturate in a length $L_u \approx \lambda_w/\rho$ at which point the power in the radiation field will be $P_{FEL} = \rho P_{beam}$.

To observe the conditions for proper FEL operation one must augment the one dimensional theory with conditions that account for two dimensional effects. The first condition [9] is

$$\varepsilon_n = \frac{\lambda \gamma}{2\pi} f_1,$$

(15)

where $f_1 \leq 1$. A condition for coherence over a gain length is

$$\frac{1}{2}\frac{\varepsilon_n^2}{r_b^2(1+a_w^2)} = \frac{\rho}{4}f_2, \qquad (16)$$

where $f_2 \leq 1$. A third design condition comes from requiring that diffraction does not remove radiation from the vicinity of the electron beam in a distance shorter than a gain length.[10] The Rayleigh length, Z_R, should be comparable to the gain length; that is,

$$L_G = Z_R f_3, \qquad (17)$$

where

$$Z_R = \frac{\pi r_b^2}{\lambda} \qquad (18)$$

and $f_3 \leq 1$.

The resonance condition and constraints (15), (16), and (17) are not all independent. One can easily show that

$$f_2 = f_1^2 f_3. \qquad (19)$$

Studies with two dimensional particle simulation codes indicate that the f_i can exceed unity by factors of 2 or 3 without appreciable degradation of the FEL performance. We will not be explore this topic further in this report, but rather will keep the f_i near unity. Note, however, that increasing the focussing increases the values of the f_i.

Equations (7) and (9) – (18) are eleven equations relating 17 quantities. We can, therefore, choose six independent quantities: the beam properties γ, I, ε_n, the wavelength of the radiation, λ, the wiggler potential, a_w, and H, the amount of ion focussing. It would be interesting to systematically explore the range of input parameters for a fixed λ and for $f_i > 1$ to compare the predictions of the scaling model with simulations. In that manner one might augment the one-dimensional theory with a semi-heuristic function, $F(f_i)$, multiplying L_G that describes decrease in performance with increasing violation of conditions (15), (16), and (17). As we have not yet done such a survey, we confine ourselves in this paper to an example of the potential benefits of ion focussing of an x-ray FEL. Table I compares the performance of a 5nm FEL without and with ions present. Note that the normalized emittance is rather small in both cases, but such values are not considered unrealistic by researchers developing high brightness guns with photocathodes. Furthermore, these low emittances seem required to achieve efficient lasing at 5 nm.

Table I An FEL for Generating X-rays at 50Å

I =	1 kA
E =	2.1 GeV
ε_n =	$6 \times 10^{-6} \pi$ m-rad
λ =	5 nm
a_w =	2.1

	No ions	With ions
H =	1	0.1
r_b =	144 µm	46 µm
λ_β =	91.2 m	9.1 m
λ_w =	3.2 cm	3.2 cm
B =	1.0 T	1.0 T
n =	3.2×10^{14} cm^{-3}	3.2×10^{15} cm^{-3}
ω_p =	3.7×10^6 sec^{-1}	1.2×10^7 sec^{-1}
ρ =	1.1×10^{-3}	2.4×10^{-3}
L_{Ray} =	13.1 m	1.3 m
L_G =	2.31 m	1.1 m
P_{sat} =	2.3 GW	5.04 GW
f_1	1.80	1.80
f_2 =	0.57	2.68
f_3 =	0.18	0.82

We find that the beneficial effects of ions can be exploited in several different ways: lowering the peak current, relaxing the requisite emittance, etc. In this example, we have kept the beam parameters unaltered; the benefit of the ion focussing is in reducing the L_G. In that sense, the FEL "works better."

Note that a factor of two reduction in the gain length is a non-trivial improvement. To attain that improvement by increasing beam current would require increasing I from 1 kA to 8 kA. Similarly, were it possible to reduce the emittance from 6×10^{-6} m-rad to 8×10^{-7} m-rad (!), L_G would decrease from 2.37 m to 1.25 m. Alternatively, the same performance could be achieved by improving the wiggler (making it much more difficult to construct) by increasing B to 2.6 T and decreasing λ_w to 1.75 cm.

Because ρ is reasonably large, the FEL can operate as an amplifier. In ths case the initial signal is incoherent synchrotron radiation in the first few bends of the wiggler. If the wiggler is long enough ($\approx 10\ L_G$), the X-ray power from the amplifier will saturate at $\approx \rho P_{beam}$; i.e., in the example of Table I at 5 GW.

In Table II we give a parameter set for an experimental test of the use use of ions in an FEL amplifier. This test requires only a 100 MeV electron beam such as may be available at the Accelerator Test Facility at Brookhaven.

Ultra high-gradient linacs could be based on the relativistic klystron scheme of Sessler and Yu.[11] These linacs can be fed with a very high brightness photo-cathode electron gun and, when combined with the gas cells discussed in this report, could provide a compact source of either incoherent or tunable coherent X-ray radiation with pulse lengths ≈ 1 ps or less.

Table II   An FEL Experiment to Test Ion Focussing in Wigglers

I =	300 A
E =	100 MeV
ε_n =	$1 \times 10^{-5}\ \pi$ m-rad
λ =	300 nm
a_w =	0.6

	No ions	With ions
H =	1	0.1
r_b =	257 μm	81.2 μm
λ_β =	8.3 m	0.83 m
λ_w =	1.76 cm	1.76 cm
B =	0.52 T	0.52 T
n =	3.0×10^{13} cm^{-3}	3.0×10^{14} cm^{-3}
ω_p =	1.1×10^8 sec^{-1}	3.5×10^8 sec^{-1}
ρ =	3.0×10^{-3}	6.5×10^{-3}
L_{Ray} =	69 cm	6.9 cm
L_G =	46 cm	21.6 cm
P_{sat} =	0.9 MW	1.95 MW
f_1	1.04	1.04
f_2 =	0.73	3.4
f_3 =	0.67	3.1

VI. Acknowledgements

We are grateful to William Fawley, Donald Prosnitz, and David Whittum for valuable conversations. This work was performed under the auspices of the U.S. Department of Energy by the Lawrence Livermore National Laboratory under contract No. W-7405-ENG-48 and by the Lawrence Berkeley Laboratory under contract No. DE-AC03-76SF00098.

REFERENCES

1) G.I. Budker,"Relativistic Stabilized Electron Beams," in Proceedings of the CERN Symposium on High Energy Accelerators (CERN, Geneva, Switzerland, 1956), p. 68.

2) W. A. Barletta, Linear Emittance Damper with Megagauss Fields (Lawrence Livermore National Laboratory, Livermore, CA, 1987), UCRL- 96947.

3) E. P. Lee, Radiation Damping of Betatron Oscillations (Lawrence Livermore National Laboratory, Livermore, CA, 1982), UCID-19381.

4) E. P. Lee and R. K. Cooper, Particle Accel. 7, 83 (1976)

5) W. E. Martin et al, Phys. Rev. Lett. 54, 685 (1985).

6) C. Pellegrini, "Progress towards a Soft X-ray FEL", Brookhaven national Laboratory report, BNL-40985, Submitted to Proceedings of the 9th International Conf. on Free Electron Lasers, 1987

7) E.T. Scharlemann, "Wiggle plane focussing in linear wiggler", LLNL Report UCRL-92429, (1985)

8) R. Bonifacio, C. Pellegrini, and L. M. Narducci, Opt. Commun. 50, 373 (1984).

9) C. Pellegrini, Nucl. Instrum. Methods 177, 227 (1980).

10) E.T. Scharlemann, A.M. Sessler, and J.S. Wurtele, Nucl. Inst. and Meth. A239, 29 (1985); G.T. Moore, ibid,p.19.

11) A. M. Sessler and S. S. Yu, Phys. Rev. Lett. 58, 2439 (1987).

The ELFA project: Guidelines for a high-gain FEL with short electron bunches

R. Bonifacio, I. Boscolo, F. Casagrande, G. Cerchioni, R. Corsini, L. De Salvo Souza,
D. Fadini, M. Ferrario, C. Maroli, P. Pierini, N. Piovella

Dipartimento di Fisica dell'Università
and INFN, Sezione di Milano
Via Celoria, 16, 20133 Milano, Italy

ELFA (Electron Laser Facility for Acceleration) has both a fundamental and a technological novel goal: i) the fundamental goal is to test with short bunches the existence of three different high-gain regimes at the heart of FEL physics: the already observed <u>steady-state</u> regime and the two novel regimes of cooperative synchrotron radiation, i.e., <u>weak</u> and <u>strong superradiance</u>. ii) the technological goal consists in exploring the possibilities of matching the advanced technologies of high-gain FEL's and of superconductive acceleration. The applications of ELFA should range from high-gradient particle acceleration to plasma heating and condensed matter physics.

1. INTRODUCTION AND GENERAL OUTLINE.

ELFA is a project which intends to exploit the capabilities and the flexibility of the FEL as a source of tunable, coherent, high peak-power radiation in the 30-300 GHz range, focussed on the development of new high-gradient accelerating structures and as a valuable tool for plasma heating in fusion research. In particular, since about 20 GHz is the upper limit on frequency accessible to conventional radiation sources for RF linacs (klystrons), the use of FEL radiation in the frequency range above this limit should allow very high acceleration gradients (200-300 MeV/m) with an increased efficency and a reduction in the length of future 1TeVx1TeV linear colliders. We believe that FEL's may provide the most economical and flexible power sources for future linear colliders, and that this research on very high frequency radiation sources must proceed in parallel with the development of new high gradient accelerating structures.

Furthermore, one of the most promising novel acceleration schemes is the two-beam accelerator (TBA) based on high-gain amplifiers. In the TBA/FEL scheme, the power provided from a high-efficiency linear accelerator to a first low-energy electron beam, is transferred to a second high-energy electron beam via coherent FEL radiation. This scheme, first proposed by Sessler [1] with an induction linac and long electron bunches, and by Amaldi and Pellegrini [2] with a superconductive linac and short electron bunches, has been very recently reformulated [3] in a way that overcomes or greatly reduces its major difficulties (e.g., the control of the RF phase). Our project can play a basic role in the development of a TBA/FEL with European technologies, along the lines proposed by Amaldi and Pellegrini, because both the operation in the superradiant regime and the velocity control by waveguides should allow for operation in the microwave region even with superconductive accelerator and short electron bunches.

1.1 – Basic Physics

In ELFA the FEL is configured as a single-pass amplifier at $\lambda \simeq 3$mm operating in the high-gain Compton regime described in reference 4,5. In this regime microwave radiation ($\lambda \simeq 9$mm) was obtained at Livermore, with peak power in the 100MW (1GW) range with an untapered [6] (tapered [7]) wiggler, by means of the long-bunch (10ns or 3m), very-high current ($I \simeq 1$KA) electron beams provided by an induction linac. These performances are based on the existence of a collective instability of the system, which leads to electron self-bunching and stimulated emission of radiation with an exponential growth of radiated power up to saturation, with peak power scaling as $I^{4/3}$, where I is the electron current. After saturation the amplification process is replaced by an oscillatory energy exchange between the electrons and the radiation field at the synchrotron frequency. This is the so-called steady-state (SS) regime of a FEL. Only by means of a variable-parameter wiggler [8] saturation has been avoided, with a dramatic improvement of the efficiency of the FEL process.

The ELFA project intends to investigate the high-gain regime of the FEL possibly using European superconducting technologies, i.e., with short-bunch (up to 160 ps) electron beams provided by 352 MHz superconducting LEP–II cavities. The energy and the current should be 10 MeV and 400 A, respectively. Such short bunches set a fundamental problem, namely, the effect of the slippage of radiation over the electrons due to the different velocities of photons and particles. In fact, propagation effects can be dramatic, and the physics quite different from the steady-state (SS) regime, in which slippage is negligible as in the Livermore experiments. In this case the slippage is negligible by the use of very long bunches and of a rather short wiggler. We have shown [9,10] that if slippage is properly taken into account, there are two basically different dynamical regimes of a high-gain FEL, which are defined as short-bunch or long-bunch regimes with respect to a suitably defined cooperation length[11]. The SS regime discussed above is only a limiting case of high-gain FEL dynamics in the long-bunch regime as discussed below.

In the short-bunch regime the radiation emitted by the electrons escapes from the bunch in a time shorter than the synchrotron period, so that the typical steady state saturation can never occur. However, we have predicted exponential growth of radiation [11], with peak power scaling as I^2, i.e. superradiance(SR). The occurrence of this phenomenon demands a properly high current density which allows for self-bunching leading to a new dynamical regime of cooperative spontaneous emission of synchrotron radiation. We call this regime weak SR because the peak power is lower with respect to SS. However, a continuous extraction of energy from the electrons without radiation reabsorption takes place with an untapered wiggler, so that the system exhibits a kind of self-tapering.

In the long-bunch regime there are two different instabilities: one growing from the interaction of the long electron bunch with the long radiation train emitted by the electron beam itself, and this is the so-called steady state instability, and the other growing from the interaction of the very short radiation train emitted by the trailing edge of the bunch slipping over the successive slices of electrons up to the head of the slippage region.

In free space the radiation emitted by an electron beam always slips over it by one wavelength per wiggler period. However, when the electron bunch length L_b is very long compared to this slippage distance — which implies that $N_w \lambda \ll L_b$ i.e. the slippage parameter $S \equiv N_w \lambda / L_b \ll 1$, where N_w is the number of wiggler periods — practically the whole bunch is immersed in the emitted radiation. The Steady State theory holds along almost the whole bunch and the power scales as $I^{4/3}$. However, in the slippage region (a slippage distance from the trailing edge) the slippage turns out to be responsible for the emission of high-power radiation spikes, showing superradiant features (i.e. power scaling as I^2). Due to the fact that this slippage region is a tiny fraction of the bunch length this effect contributes little to the average energy extracted from the electrons, i.e. to the overall efficiency. We call this trailing effect Strong Superradiance. If, instead, $S \gtrsim 1$, that is the slippage distance is long with respect to the bunch length (long wiggler case), the strong superradiant spiking behaviour takes place over the whole bunches. The radiation pulse emitted by the trailing edge runs over all the bunch length, extracting large amounts of energy from the electrons, which radiate in a correlated way to produce superradiant spikes with peak power one-order-of-magnitude higher than at SS.

A beautiful and unique feature of ELFA project is the possibility to test all these three high-gain regimes: the steady-state (up to now observed only with long electron bunches) with short bunches and the yet unobserved weak and strong superradiant regimes with essentially the same experimental setup. In fact, in order to confine radiation in the millimeter range ($\lambda = 3$ mm for ELFA), the FEL process must take place in a suitable waveguide. It follows that the "constraint" of fixed radiation velocity c disappears; on the contrary, we can control both the phase velocity, involved in the FEL resonance condition, and the group velocity, which really controls the slippage in a waveguide. This can be achieved by varying the heigth b (the minor dimension) in a rectangular waveguide of transverse area ab and correspondingly the beam energy, as we shall see in the next Section. Steady-state operation with short bunches has been proposed in Two-beam Acceleration scheme in reference 2.

1.2 – Accelerator and wiggler

The ELFA accelerator will provide a 400 A – 10 Mev electron beam. Its basic components will be a photocathode injector and one superconducting 4–cell LEP II module operating at 352 MHz.

The wiggler will be composed joining up two sections : a hybrid wiggler will be the first section and an Electromagnetic (EM) wiggler the second one. Both wigglers will have canted poles in order to get focussing in both horizontal and vertical planes. The EM wiggler will allow for tapering. A detailed scheme of the experimental apparatus will be discussed elsewhere [14].

2. THE BASIC PARAMETERS AND THE DYNAMICAL REGIMES OF A HIGH-GAIN FEL AMPLIFIER WITH APPLICATION TO ELFA.

In this section we define the fundamental parameters which allow a simple classification of the dynamical regimes of a high-gain FEL amplifier. Also, we shall take into account the effect of the waveguide. A general discussion can be found in ref. 9,12.

ELFA will operate in the high-gain exponential regime, that is with the unsaturated gain per pass

$$G = 4\pi \rho_\circ N_w \left(1 - \frac{X}{2}\right)^{5/6} > 1 \qquad (1)$$

Here $N_w \equiv L_w/\lambda_w$ is the number of wiggler periods and $\rho_\circ$ is the fundamental FEL parameter [5,9]

$$\rho_\circ \simeq 0.136 \frac{B_w^{2/3} \lambda_w^{4/3}}{\gamma_r} \left(\frac{I}{ab/2}\right)^{1/3} f_b^{2/3} \left(\frac{k}{k_\parallel}\right)^{1/3} \qquad \text{(SI)} \qquad (2)$$

In eq. (2), γ_r is the resonant electron energy in rest mass units, B_w the wiggler peak field, I the electron peak current, f_b the usual difference of Bessel functions $f_b(\xi) = J_0(\xi) - J_1(\xi)$, where $\xi = a_w^2/2(1 + a_w^2)$ and a_w is the undulator parameter (for a planar wiggler), defined as $a_w = eB_w^{rms}/mc^2$ and $B_w^{rms} = B_w/\sqrt{2}$. Our analysis is limited to a single TE$_{01}$ mode of a rectangular waveguide of width a and length b, with mode area $ab/2$, dispersion relation $k^2 \equiv (w/c)^2 = k_\parallel^2 + (\pi/b)^2$, phase velocity $v_f = (k/k_\parallel)c$ and group velocity $v_g = (k_\parallel/k)c$; X is the waveguide parameter [9,12]

$$X \equiv \frac{\lambda \lambda_w}{4b^2} \qquad (3)$$

In ELFA $\rho \geq 0.01$, so that the FEL is of the Compton type [13], with negligible plasma effects which on the contrary are relevant in the Raman devices. Note that the resonance condition for the FEL process with a waveguide can be written (for on-axis propagation)

$$\gamma_r^2 = \frac{k}{2k_w}(1 + a_w^2)\left(1 - \frac{X}{2}\right)^{-1} \qquad (4)$$

where k is the wavenumber of the radiation field to be amplified (provided by a magnetron in ELFA). From (4) it follows that $0 < X < 2$.

Let us first introduce the slippage parameter

$$S = \frac{N_w \lambda}{L_b}(1 - X) \tag{5}$$

where $N_w \lambda$ is the slippage length in free space, L_b the bunch length and the factor $(1 - X)$ takes into account waveguide effects.

We classify the long wiggler and the short wiggler regime respectively by the conditions $S > 1$ and $S \ll 1$.

We can now introduce the basic superradiant parameter K

$$K \equiv \frac{S}{G} = \frac{\lambda}{4\pi\rho_\circ L_b} \frac{1 - X}{(1 - \frac{X}{2})^{5/6}} \tag{6}$$

which can also be written as the ratio between the cooperation length

$$L_c = \frac{\lambda}{4\pi\rho_\circ} \frac{1 - X}{(1 - \frac{X}{2})^{5/6}} \tag{7}$$

and the bunch length L_b. Note that K is the slippage when the electrons enter the high-gain regime, i.e. when $N_w = \left[4\pi\rho_\circ(1 - \frac{X}{2})^{5/6}\right]^{-1}$, so that $G = 1$.

The two different dynamical regimes of a high-gain *FEL* are classified as short-bunch regime if $K > 1$ and long-bunch regime if $K \ll 1$. In the former case, $K > 1$, namely for an electron bunch shorter than one cooperation length, we have the phenomenon of weak SR discussed in the previous section.

In the long-bunch case, $L_b \gg L_c$ or $K \ll 1$, we have two different behaviours depending on the wiggler length. In the short-wiggler limit, $S \ll 1$, the slippage is negligible and the system can approximately operate in the SS regime as discussed in the Introduction.

In the opposite long-wiggler limit, $S > 1$, we have both analytical and numerical evidence of the occurrence of the phenomenon of strong SR as described in the previous section.

To sum up, by suitably varying the waveguide heigth b, we can change accordingly the waveguide parameter X ($0 < X < 2$). More precisely:

i) In an oversized waveguide, $X \ll 1$; then, with high-gain ($G \gg 1$) and short-bunch ($K > 1$), which imply strong slippage ($S \gg 1$), we expect to observe weak SR.

ii) By strongly reducing the height b, we can approach the condition $X = 1$ for the suppression of slippage and the observation of the steady-state. In principle, this regime can be observed only this way.

iii) By an intermediate choice of the dimension b such that $0 < X < 1$, we can reduce the slippage length and keep the parameter S greater than one, and eventually observe strong SR.

In conclusion, in ELFA we can observe the three dynamical regimes of a high-gain FEL: steady-state, weak and strong superradiance just tuning the waveguide height and the beam energy.

REFERENCES.

1) A.M. Sessler, in: Laser Acceleration of Particles, American Institute of Physics, Conf. Proc. vol. 91, 154 (1982).
2) U. Amaldi and C. Pellegrini, New Techniques for Future Accelerators, ed. M. Puglisi, S. Stipcich and G. Torelli, Plenum 139–162 (1987).
3) A.M. Sessler, E. Sternbach, J.S. Wurtele, preprint.
4) R. Bonifacio, F. Casagrande and G. Casati, *Optics Comm.*, **40**, 219 (1982).
5) R. Bonifacio, C. Pellegrini and L. Narducci, *Opt. Comm.*, **50**, 373 (1984).
6) T.J. Orzechowski, R.B. Anderson, W.M. Fawley, D. Prosnitz, E.T. Scharlemann, S. Yarema, D. Hopkins, A.C. Paul, A.M. Sessler and J. Wurtele, *Phys. Rev. Lett.*, **54**, 889 (1985).
7) T.J. Orzechowski, R.B. Anderson, J.C. Clark, W.M. Fawley, A.C. Paul, D. Prosnitz, E.T. Scharlemann, S. Yarema, D. Hopkins, A.M. Sessler and J. Wurtele, *Phys. Rev. Lett.*, **57**, 2172 (1986).
8) N.M. Kroll, P.L. Morton and M.N. Rosenbluth, *IEEE J. Quantum Electr.*, **QE–17**, 1436 (1981).
9) R. Bonifacio, F. Casagrande, G. Cerchioni, L. De Salvo Souza and P. Pierini, Proc. INFN Int. School, Varenna, June 1988, to be published.
10) R. Bonifacio, B.W.J. McNeil, *Nucl. Instr. Meth. A*, **272**, 280 (1988).
    R. Bonifacio, B.W.J. McNeil and P. Pierini, submitted to *Phys. Rev. A*.
    R. Bonifacio, C. Maroli and N. Piovella, *Opt. Comm.*, **68**, 369 (1988).
11) R. Bonifacio and F. Casagrande, *Nucl. Instr. Meth. A*, **239**, 36 (1985).
12) R. Bonifacio and L. De Salvo Souza, *Nucl. Instr. Meth. A*, to be published.
13) R. Bonifacio, F. Casagrande and C. Pellegrini, *Opt. Comm.*, **61**, 55 (1987).
14) R. Bonifacio, I. Boscolo, G. Cerchioni, R. Corsini, L. De Salvo Souza, D. Fadini, M. Ferrario, C. Maroli, N. Piovella, Proc. Workshop on FEL, Frascati, Sept. 1988, to be published.

TAPERING AND SELF-TAPERING IN A FREE ELECTRON LASER AMPLIFIER

R.BONIFACIO, F.CASAGRANDE, M.FERRARIO, P.PIERINI and N.PIOVELLA

Dipartimento di Fisica dell'Università, via Celoria 16, 20133 Milano, Italy and Istituto Nazionale di Fisica Nucleare, sezione di Milano.

We consider an FEL amplifier with variable wiggler field, described by a fully Hamiltonian model for operation in the steady-state regime and by a dissipative model for operation in the superradiant regime. In the steady-state regime we present numerical results for different tapering procedures and derive corresponding scaling laws for optimum tapering and for the growth of intensity and efficiency. The radiated power is shown to scale with respect to electron density as $n^{5/3}$, that is half-way between the untapered high-gain steady-state scaling, $n^{4/3}$, and the superradiant n^2 scaling. In the superradiant regime the tapering is shown to have little relevance, contrary to the steady-state case. However, self-tapering with continuous extraction of electron energy occurs; during the growth of the superradiant pulse the resonance is maintained by a shift in the phase of the radiation field which compensates for the decrease of the electron energy.

1. HAMILTONIAN MODEL OF AN FEL WITH TAPERED WIGGLER

Let us consider first an FEL amplifier operating in the steady-state regime, i.e., when the slippage of radiation over the electrons has negligible effects [1]. In the Compton limit the system can be described by the following one-dimensional model Hamiltonian:

$$H = \sum_{j=1}^{N} \left\{ \frac{p_j^2}{2} + i\left[A^* \exp(-i\theta_j) - c.c.\right] \right\} \quad (1)$$

in terms of dimensionless electron energy variation, p, electron phase with respect to the ponderomotive (radiation + wiggler) field, θ, and complex field amplitude A:

$$p = \frac{1}{\rho}\left(\frac{\gamma - \gamma_r}{\gamma_r}\right)$$

$$\theta = (k + k_w)z - ckt$$

$$A = \frac{eE}{\sqrt{\rho\gamma_r}mc\omega_p} \quad (2a)$$

$$|A|^2 = \frac{1}{\rho}\frac{\text{radiation energy}}{\text{electron beam energy}}$$

In eqs. (2a) k (k_w) is the radiation (wiggler) wave number, γ the electron energy in rest mass units, E the complex amplitude of the radiation field; the resonant energy γ_r, plasma frequency ω_p, undulator parameter a_w and FEL parameter ρ are defined as:

$$\omega_p = \left(\frac{4\pi e^2 N}{mV}\right)^{1/2}$$

$$\gamma_r = \left[\frac{k(1+a_w^2)}{2k_w}\right]^{1/2}$$

$$a_w = \frac{eB_w}{k_w mc^2} \quad (2b)$$

$$\rho = \frac{1}{\gamma_r}\left(\frac{a_w\omega_p}{4ck_w}\right)^{2/3}$$

where B_w is the peak magnetostatic field (for a helical wiggler).

Hamiltonian (1) contains no parameters ("universal scaling" [1]) and the FEL dynamic equations are the dimensionless Hamilton equations:

$$\frac{d\vartheta_j}{d\bar{z}} = \frac{\partial H}{\partial p_j} = p_j \quad (3a)$$

$$\frac{dp_j}{d\bar{z}} = -\frac{\partial H}{\partial \theta_j} = -\left(Ae^{i\vartheta_j} + c.c.\right) \quad (3b)$$

$$\frac{dA}{d\bar{z}} = -\frac{i}{N}\frac{\partial H}{\partial A^*} = \langle\exp(-i\theta)\rangle \quad (3c)$$

where $\bar{z}$ is the scaled longitudinal coordinate:

$$\bar{z} = 2k_w\rho z = \frac{4\pi\rho}{\lambda_w}z \quad (4a)$$

and

$$\langle\exp(-i\theta)\rangle = \frac{1}{N}\sum_{j=1}^{N}\exp(-i\theta_j) \equiv b \quad (4b)$$

is the electron bunching parameter which describes longitudinal modulation ($0 \leq |b| \leq 1$). From eqs. (3a–c) one easily verifies that:

$$\langle p \rangle + |A|^2 = const. \tag{5}$$

that is, energy conservation (see definitions (2)).

Now the point is that, due to a collective instability of the system [1], in a high–gain FEL amplifier $|A|^2$ can grow exponentially from initial values $|A|_o^2 \ll 1$ to peak values $|A|_p^2 = O(1)$. Since the radiation energy is extracted from the electron beam (eq.(5)), the mean electron energy $\langle \gamma \rangle$ falls appreciably below the resonant energy γ_r. Thus for fixed radiation and undulator parameters, the FEL process goes out of resonance and saturates: on average, the electrons will stop radiating and will start absorbing energy from the field, until they approach resonance again, and so on. By eq. (2a), $|A|_p^2 = O(1) \gg |A|_o^2$ implies that the efficiency η in the conversion of electron kinetic energy into radiation is:

$$\eta \simeq \rho |A|_p^2 \simeq \rho \tag{6}$$

namely, it is limited to a few percent ($\rho < 0.1$ in a Compton FEL). The efficiency of the FEL process has been dramatically improved [2,3] by application of the concept of variable–parameter or tapered wiggler, suggested by Kroll, Morton and Rosenbluth [4] and other authors [5]. Efficiency up to 40% and peak power in the GW range were observed [2] by suitably decreasing the wiggler field B_w in order to compensate for the decrease of $\langle \gamma \rangle$ and keep the system on resonance [6].

Wiggler field tapering can be included in a straightforward way in the Hamiltonian model [7]. Let us introduce a wiggler field profile function:

$$f_b(z) = \frac{B_w(z)}{\overline{B}_w} = \frac{a_w(z)}{\overline{a}_w} \quad (0 \leq f_b(z) \leq 1) \tag{7}$$

where $\overline{B}_w$ and $\overline{a}_w$ are the untapered values. The resonance energy (with fixed k, k_w) becomes z–dependent and by (2), (7) we obtain:

$$\gamma_r(z) = \left[\frac{k}{2k_w}(1 + a_w^2(z))\right]^{1/2} \equiv \overline{\gamma}_r f_r(z)$$

$$f_r(z) = \left(\frac{1 + \overline{a}_w^2 f_b^2(z)}{1 + \overline{a}_w^2}\right)^{1/2} \tag{8}$$

where $\overline{\gamma}_r$ is again the untapered value.

In the limit:

$$f_b^2(z) \gg \frac{1}{a_w^2} \tag{9}$$

$f_r(z) \simeq f_b(z)$, so that only one tapering function, $f_b(z)$, must be introduced in the model. Furthermore [7], this function appears only in the phase equation, (3a), which becomes:

$$\frac{d\vartheta_j}{d\bar{z}} = \frac{1}{f_b(\bar{z})}[p_j - p_r(\bar{z})] \tag{3a'}$$

where:

$$p_r(\bar{z}) = -\left(\frac{1 - f_b(\bar{z})}{\rho}\right) = -\frac{1}{\rho}\left(\frac{\bar{\gamma}_r - \gamma_r(\bar{z})}{\bar{\gamma}_r}\right) \tag{10}$$

Eqs. (3b),(3c) and the constant of motion (5) are left unchanged.

The dynamic equations (3a'),(3b),(3c) for an FEL with tapered wiggler can be derived as Hamilton equations from a z-dependent Hamiltonian:

$$H(\bar{z}) = \sum_{j=1}^{N}\left\{\frac{1}{2f_b(\bar{z})}[p_j - p_r(\bar{z})]^2 + i(A^*\exp(-i\theta_j) - c.c.)\right\} \tag{1'}$$

An electron with $p = p_r(\bar{z})$ has always the resonant energy $\gamma = \gamma_r(\bar{z})$, while the resonant phase θ_r is defined through eq. (3b) as:

$$\frac{dp_r}{d\bar{z}} = -(A\exp(i\vartheta_r) + c.c.) = -2|A|\cos(\theta_r + \varphi) \tag{11}$$

where we set $A = |A|\exp(i\varphi)$. Hence the resonant particle exists only if $|\cos(\theta_r + \varphi)| \leq 1$, which by (4a), (7), (10) sets a limitation on the tapering gradient:

$$\frac{1}{\bar{B}_w}\left|\frac{dB_w(z)}{dz}\right| \leq \frac{8\pi\rho^2}{\lambda_w}|A(z)| \tag{12}$$

We recall [4,7] that all particles trapped in a "bucket" in the electron phase-space (θ, p) will execute stable synchrotron oscillation around the resonant particle (θ_r, p_r), whereas the untrapped electrons do not interact appreciably with the field. By suitably starting the tapering before saturation and by a proper optimization with respect to both electron deceleration and trapping, all trapped electrons can be continuosly decelerated and forced to radiate coherently with a dramatic improvement of the FEL efficiency.

2. NUMERICAL RESULTS AND SCALING LAWS

We present numerical results and scaling laws for: i) linear tapering, ii) multiple linear tapering, iii) self-consistent continuous tapering. In all cases we start with $|A|_o^2 \ll 1$, $|b|_o^2 \ll 1$, $p_{j,o} = 0$ and fix the untapered undulator parameter $\bar{a}_w = 5$ and the FEL parameter $\rho = 0.01$ such that, by eq. (6), the efficiency $\eta \simeq |A|^2$ %. The tapering begins at $\bar{z} = 3.5$, slightly before the occurrence of the first untapered peak of the bunching parameter.

2.1. Linear tapering

A continuous and linear (on average) growth of intensity (fig.1a, with the untapered result for reference) is obtained via a simple linear tapering (fig.1b):

$$a_w \equiv \bar{a}_w f_b(\bar{z}) = \begin{cases} \bar{a}_w & \bar{z} \leq \bar{z}_T \\ \bar{a}_w - |m|(\bar{z} - \bar{z}_T) & \bar{z} \geq \bar{z}_T \end{cases} \quad (13)$$

The growth of radiation intensity $|A|^2$ turns out to be maximum with a tapering slope $|m|_{opt} = 0.15$. Both this value and the linear growth of intensity or efficiency can be derived from our model.

Actually, from the constant of motion (5) and with initial conditions:

$$\langle p(\bar{z}) \rangle + |A(\bar{z})|^2 = const. \ll 1$$

by approximating:

$$\langle p(\bar{z}) \rangle \simeq n_r p_r(\bar{z}) \quad (14)$$

where n_r is the fraction of trapped or quasi–resonant electrons, ($n_r \simeq |b|$ at saturation), and with the definition (10) of $p_r(\bar{z})$, we find:

$$f_b(\bar{z}) \simeq 1 - \frac{\rho|A(\bar{z})|^2}{n_r}. \quad (15)$$

Now for the linear tapering (13) (for $\bar{z} \geq \bar{z}_T$):

$$f_b(\bar{z}) = 1 - \frac{|m|}{\bar{a}_w}(\bar{z} - \bar{z}_T), \quad (16)$$

which inserted in (15) gives (with $n_r \simeq 0.6$, $\bar{z} = \bar{z}_{peak} = 4.5$, $|A|^2_{peak} \simeq 1.5$):

$$|m|_{opt.} \simeq \frac{\rho \bar{a}_w |A(\bar{z})|^2}{n_r(\bar{z} - \bar{z}_T)} \simeq 0.17,$$

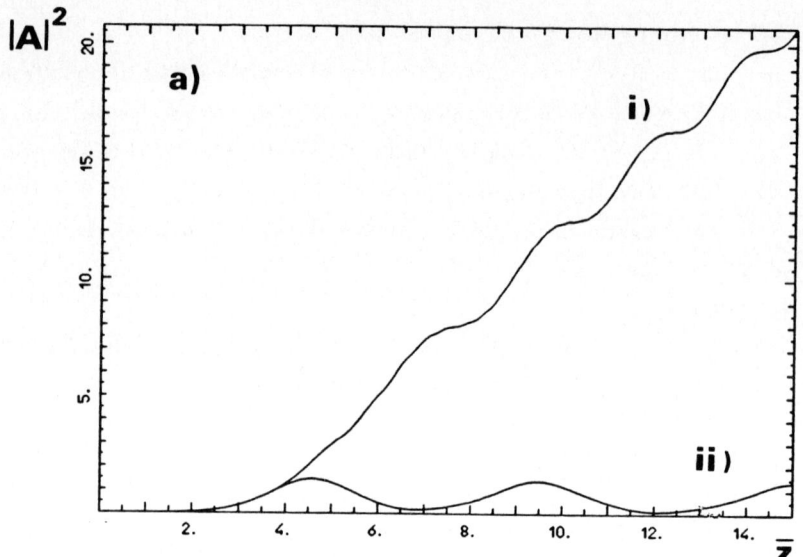

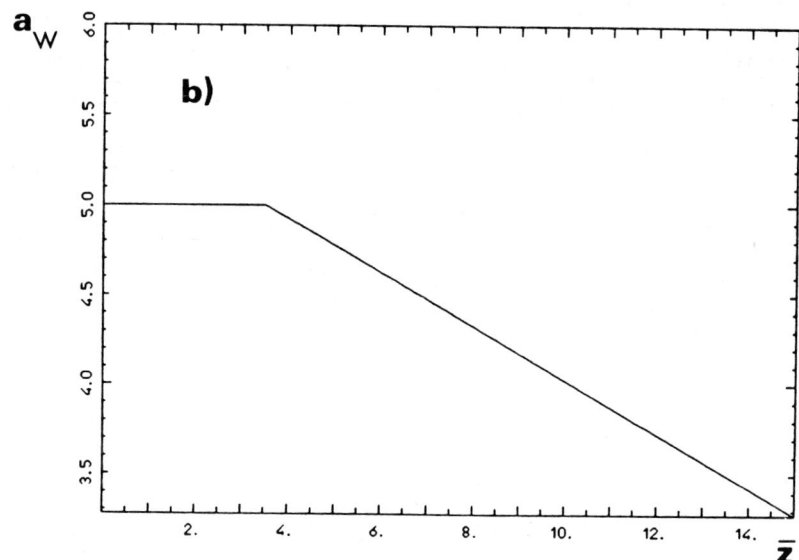

FIGURE 1

Linear tapering of the wiggler field B_w. a) Tapered (i) and untapered (ii) scaled intensity $|A|^2$ vs. scaled longitudinal coordinate $\bar{z}$, from eqs. (3a',b,c). b) Undulator parameter a_w vs. $\bar{z}$. Slope $m = -0.15$ from $\bar{z} = 3.5$. Here and in the following Figs. the parameters are: $\rho = 0.01$, $\bar{a}_w = 5$, and the initial conditions: $|A|_\circ \simeq 0.14$, $|b|_\circ \simeq 10^{-3}$, $(p_j)_\circ = 0$.

that is nearly the numerical optimal slope. On the other hand, again from (15), (16) we obtain a scaling of the average linear growth of intensity after tapering:

$$|A(\bar{z})|^2 \simeq \frac{n_r|m|}{\rho \bar{a}_w}(\bar{z} - \bar{z}_T) \equiv M(\bar{z} - \bar{z}_T) \qquad (17)$$
$$M \simeq 3n_r \simeq 1.8$$

The result $|A(\bar{z})|^2 = O(\bar{z})$ for an FEL with tapered wiggler is to be compared with the result $|A(\bar{z})|^2 = O(1)$ for a high-gain FEL with untapered wiggler. Since $|A|^2 \propto |E|^2/\rho n \propto |E|^2/n^{4/3}$, *the scaling of radiated intensity vs. electron density, which is* $|E|^2 \propto n^{4/3}$ in the untapered case and already indicates a cooperative behaviour, *raises to* $|E|^2 \propto n^{5/3}$ *in the tapered case.*

2.2. Multiple linear tapering

Due to the interplay between deceleration and trapping, a good starting point for a more refined tapering is a slope smaller than the previous one ($|m| = 0.15$), in order to trap more electrons, provided that it does not appreciably lowers the exponential gain.

Actually by starting with $|m| = 0.09$, and then suitably changing slopes other seven times, we have nearly doubled the efficiency η reaching $\eta = 41\%$, with 75% trapped electrons (fig.2a,b). The different slopes and their starting points were chosen via analysis of the electron phase-space [7]. Some windows in phase-space are reported in figs. 3a-c.

2.3. Self-consistent, continuous tapering

Let us set $A = |A| \exp(i\varphi)$ and introduce the total phase $\psi = \theta + \varphi$. The evolution equations (3a',b,c) become:

$$\frac{d\psi_j}{d\bar{z}} = \frac{1}{f_b(\bar{z})}[p_j - p_r(\bar{z})] + \frac{d\varphi}{d\bar{z}} \qquad (18a)$$

$$\frac{dp_j}{d\bar{z}} = -2|A|\cos\psi_j \qquad (18b)$$

$$\frac{d|A|}{d\bar{z}} = \langle\cos\psi\rangle \qquad (18c)$$

$$\frac{d\varphi}{d\bar{z}} = -\frac{1}{|A|}\langle\sin\psi\rangle \qquad (18d)$$

We add to (18) the evolution equation for the resonant particle variable p_r, i.e. eq. (11), and choose a tapering design such that the resonant phase $\psi = \theta + \varphi$ remains constant over the whole tapering region:

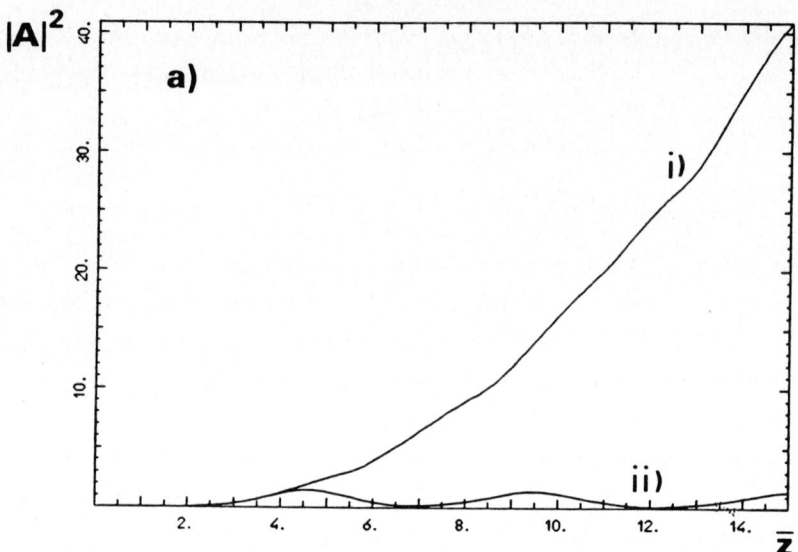

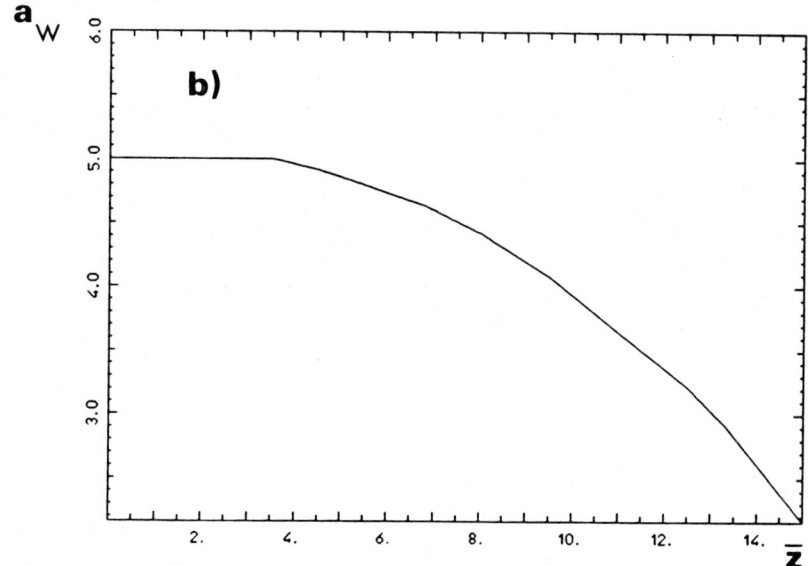

FIGURE 2

Multiple linear tapering of the wiggler field B_w. a) Tapered (i) and untapered (ii) intensity $|A|^2$ vs. longitudinal coordinate $\bar{z}$, from eqs. (3a',b,c). b) Undulator parameter a_w vs. $\bar{z}$. Eight slopes: $|m| = 0.087, 0.112, 0.123, 0.18, 0.227, 0.289, 0.37, 0.45$, starting at $\bar{z} = 3.5, 4.5, 5, 6.8, 8, 9.5, 12.5, 13.3$, respectively.

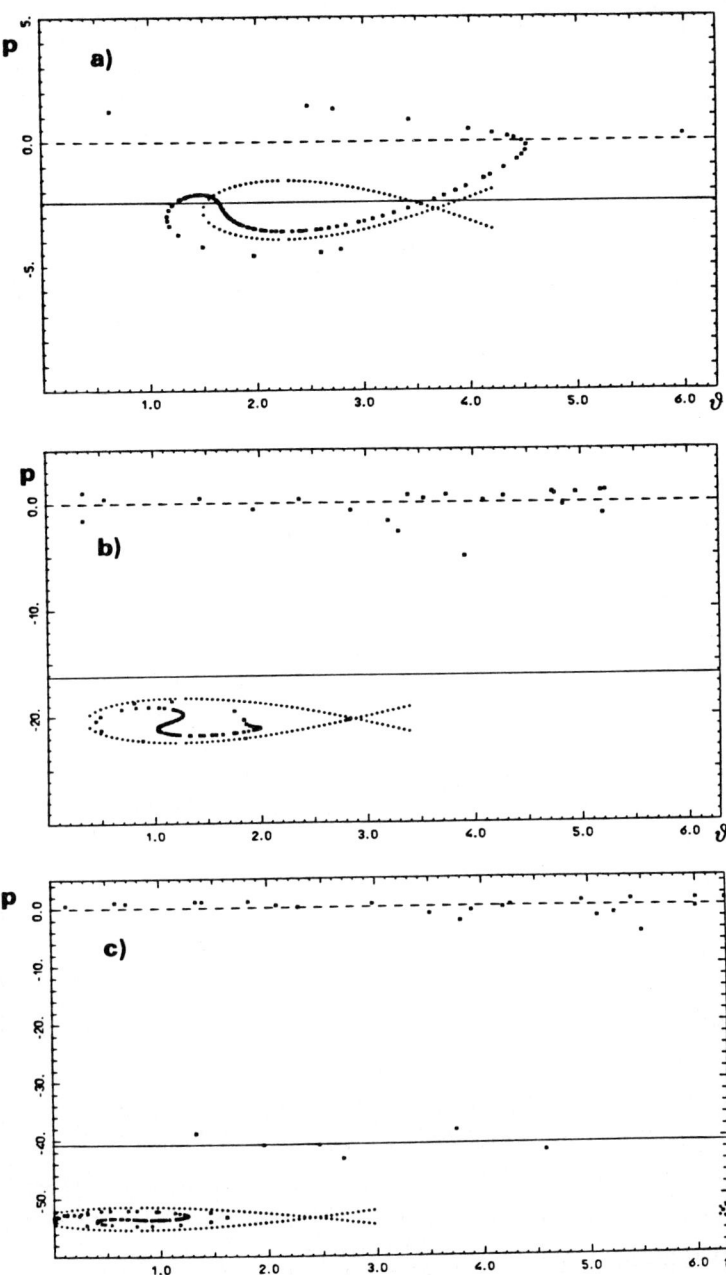

FIGURES

Multiple linear tapering of wiggler field B_w : windows in electron phase–space (θ, p). Circles: the bucket, with circles missing at $\pm p(\theta_r)$; dashed line: initial average $\langle p \rangle_\circ$; solid line: average $\langle p \rangle(\bar{z})$. a) $\bar{z} = 5$, b) $\bar{z} = 10$, c) $\bar{z} = 15$.

$$\frac{dp_r}{d\bar{z}} = -2|A|\cos\psi_r \tag{18e}$$

$$\frac{d\psi_r}{d\bar{z}} = 0 \tag{18f}$$

The choice ($18f$), i.e. $d\theta_r/d\bar{z} = -d\varphi/d\bar{z}$, is such that all particles which are trapped, at the start of tapering, around a phase $\psi = \psi_r$ corresponding to gain ($\cos\psi_r > 0$), remain bunched around that phase until the end of the interaction.

With an optimized choice of the design constant, $\cos\psi_r = 0.67$, we reached an efficiency $\eta = 44\%$, even higher than in the previous case, corresponding to a better trapping, $n_r = 0.8$. The behaviour of $|A|^2$ and a_w (fig. $4a,b$) closely resembles that in figs. $2a,b$; the same remark would apply for the phase–space analysis.

From the model we can derive the scaling law of the intensity and efficiency, and an analytic expression for the optimum continuous tapering function $f_b^{opt.}(\bar{z})$. First of all, from eqs. (18c), (18f) and by an approximation similar to (14), we obtain:

$$\frac{d|A|}{d\bar{z}} = \langle\cos\psi\rangle \simeq n_r \cos\psi_r = const.$$

which gives at once (for $\bar{z} \geq \bar{z}_T$)

$$|A(\bar{z})| \simeq |A(\bar{z}_T)| + (n_r \cos\psi_r)(\bar{z} - \bar{z}_T) \tag{19}$$

Also, from eqs. (15), (19) and with our parameters, we obtain a simple analytical espression for optimum tapering profile, namely:

$$f_b^{opt.}(\bar{z}) \simeq 1 - \rho\left[(\bar{z} - \bar{z}_T) + \frac{1}{3}(\bar{z} - \bar{z}_T)^2\right] \tag{20}$$

Now, eq. (19) seems to predict a quadratic dependence of intensity $|A(\bar{z})|^2$ on $\bar{z}$ for $\bar{z} - \bar{z}_T \gg 1$. By the same reasoning of case 2.1., $|A(\bar{z})|^2 = O(\bar{z}^2)$ would imply $|E|^2 \propto n^2$. However, eq. (20) clearly sets a lower limit on the decrease of $f_b(\bar{z})$, corresponding to the extinction of the wiggler field ($\bar{z} \simeq 19$ in our case from both eqs. (20) and the full system (18a–f); however for $\bar{z} > 15$ the inequality (9) is violated). We can conclude that a parabolic scaling law of intensity, and therefore the superradiant n^2 dependence, can never be reached by tapering the wiggler field in a high–gain Compton FEL amplifier operating in the steady–state regime.

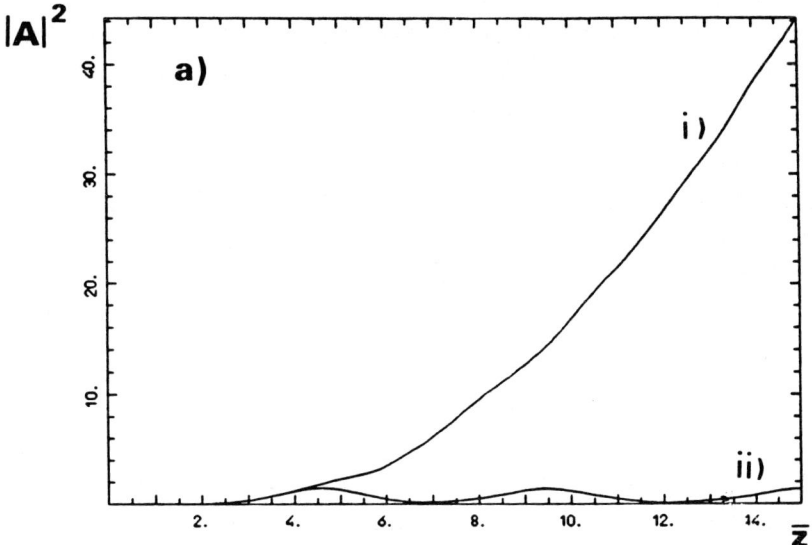

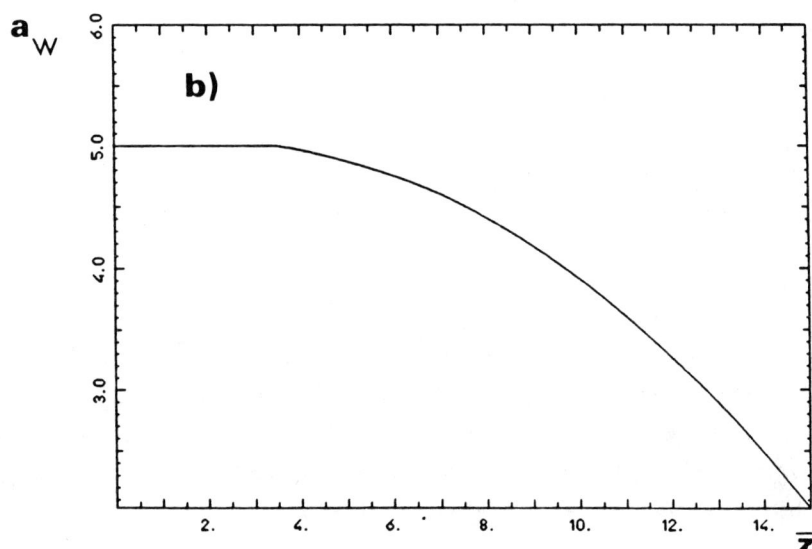

FIGURE 4
Self–consistent continous tapering of the wiggler field B_w a) Tapered (i) and untapered (ii) intensity $|A|^2$ vs. longitudinal coordinate $\bar{z}$, from eqs. (18a–f), with $cos\psi_r = 0.67$. b) Undulator parameter a_w vs. $\bar{z}$.

3. SELF TAPERING IN THE DAMPING MODEL OF A SUPERRADIANT FEL

Some years ago two of us [8] proposed a heuristic model of a superradiant FEL, that is, an FEL in which electrons radiate as n^2. This may occurs if the system starts close to the collective instability but the electron bunch is short enough that the slippage is relevant. In this case the electrons can self–bunch and radiate coherently as at steady–state, but the radiation escapes from the electron bunch before being reabsorbed. Hence the electrons emit collective spontaneous radiation, which is mostly concentrated in a superradiant pulse followed by minor pulses due to electron re–bunching. To model the escape of radiation from the electron bunch we introduced [8] a damping constant K for the field amplitude:

$$K = \frac{\lambda}{4\pi\rho L_e} = \frac{L_C}{L_e} \tag{21}$$

where L_C is the cooperation length and L_e the electron bunch length.

Both in the untapered and in the tapered case the equations of motion for the electrons are still (3a), (3b) or (3a'), (3b), respectively, whereas eq. (3c) is simply modified as follows:

$$\frac{dA}{d\bar{z}} = \langle \exp(-i\theta) \rangle - KA \tag{3c'}$$

The system is no longer Hamiltonian; the conservation law (5) is replaced by the dissipative law:

$$\frac{d}{d\bar{z}}\left(\langle p \rangle + |A|^2\right) = -2K|A|^2 \tag{22}$$

In the short–bunch limit $K \gg 1$ two relevant results can be very simply derived from eqs. (3c') and (22). Actually, in this limit the derivative of the field modulus can be neglected in the l.h.s. of both eqs. (3c'), (22). Hence:

i) from eq. (3c') we obtain the adiabatic relation:

$$A \simeq \frac{\langle \exp(-i\theta) \rangle}{K} = \frac{b}{K} \tag{23}$$

which limits the gain below the steady–state level; however, since $K \propto \rho^{-1}$, for appreciable electron bunching (due to the collective instability)

$$|E|^2 \propto \rho n |A|^2 \propto \rho^3 n \propto n^2 \tag{24}$$

that is *superradiance*;

ii) from eq. (22) we obtain:

$$\frac{d\langle p \rangle}{d\bar{z}} \simeq -2K|A|^2 < 0 \qquad (25)$$

which shows a continuous extraction of energy from the electron beam; in this sense, the system exibits *self-tapering*.

This "damping model" has been now rigorously derived from the propagation equations and turns out to describe nicely the regime of "pure superradiance", as discussed in detail elsewhere in [1,9].

In the tapered damping model, i.e., eqs. (3a'), (3b) and (3c'), the effect of tapering turns out to be of little relevance, as shown in fig. 5 where tapered and untapered superradiant outputs are reported. This result refers to a linear tapering, optimized from $\bar{z}_T = 13.5$ with a slope $|m| = 0.04$. However, this is quite in agreement with the very different nature of the superradiant dynamics with respect to the stady–state (no slippage) regime. Superradiance is a transient, dissipative effect; after the cooperative spontaneous emission of radiation, the electrons are quickly detrapped from the ponderomotive bucket [1], whose height ($\propto |A|^{1/2}$) is strongly reduced due to the adiabatic relation (23). Thus tapering cannot help that much; eventually (see fig. 5), it helps increasing the electron re–bunching which originates the next minor peaks.

However, the *untapered* damping model – i.e. eqs. (3a), (3b) and (3c') – exhibits self–tapering also in the following sense: the variation of the radiation field phase φ is such to compensate exactly for the decrease of the electron energy (eq. (25)), so that the resonance condition is mantained up to the first peak of radiated intensity. Fig. 6 shows the quantity $\langle p \rangle + d\varphi/d\bar{z}$ vs. $\bar{z}$. After a peak which is due to an initial phase variation induced by the smallness of the input field (see eq. (18d)), this quantity remains nearly constant up to the development of the first superradiant peak. This means that after the lethargy in which $\langle p \rangle \simeq d\varphi/d\bar{z} \simeq 0$, the field phase grows in such a way as to balance the decrease of $\langle p \rangle$. Now, by the definitions (2a) and (4a), $\langle p \rangle + d\varphi/d\bar{z} \simeq const.$ means:

$$\frac{\langle \gamma \rangle - \gamma_r}{\gamma_r} + \frac{1}{2k_w}\frac{d\varphi}{dz} \simeq const. \qquad (26)$$

On the other hand, the resonance condition with the variation of the field phase taken into account is:

$$\tilde{\gamma}_r^2 = \frac{k(1+a_w^2)}{2(k_w + d\varphi/dz)} = \frac{\gamma_r^2}{1+(1/k_w)(d\varphi/dz)} \qquad (27)$$

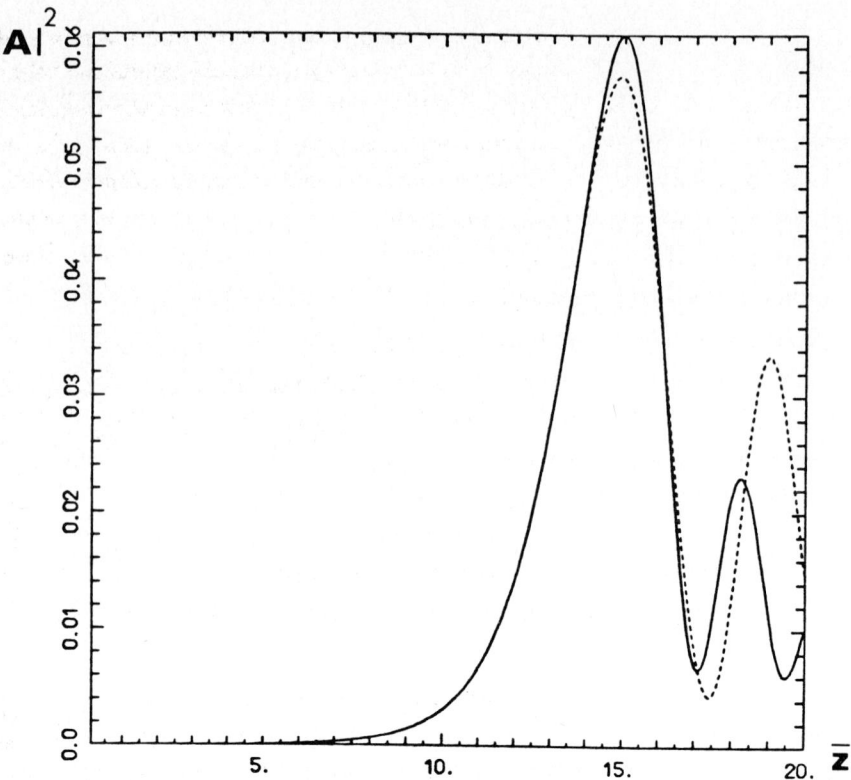

FIGURE 5
Superradiant regime. Tapered (solid line) and untapered (dashed line) intensity $|A|^2$ vs. longitudinal coordinate $\bar{z}$, from the tapered (eqs. (3a', b, c')) or untapered (eqs.(3a, b, c')) damping model. Here and in Fig.6, $K = 3$.

where γ_r is the resonant energy for negligible phase variation (see eq. (2b)). From (27) and (26) it follows that (on average) the electron energy remains close to the resonant value:

$$\frac{\langle\gamma\rangle - \tilde{\gamma}_r}{\tilde{\gamma}_r} \simeq \frac{\langle\gamma\rangle - \gamma_r}{\gamma_r} + \frac{1}{2k_w}\frac{d\varphi}{dz} \simeq const. \qquad (28)$$

This completes the picture of self-tapering in superradiance. While the energy extraction from the electron beam is continuous (eq. (25)) and the radiation growth is limited by the adiabatic relation (23), during the superradiant pulse formation the field phase shift compensates for the electron energy loss. This compensation can occur only when the field has not yet slipped over the electron bunch; after that, the radiation pulse propagates in vacuum.

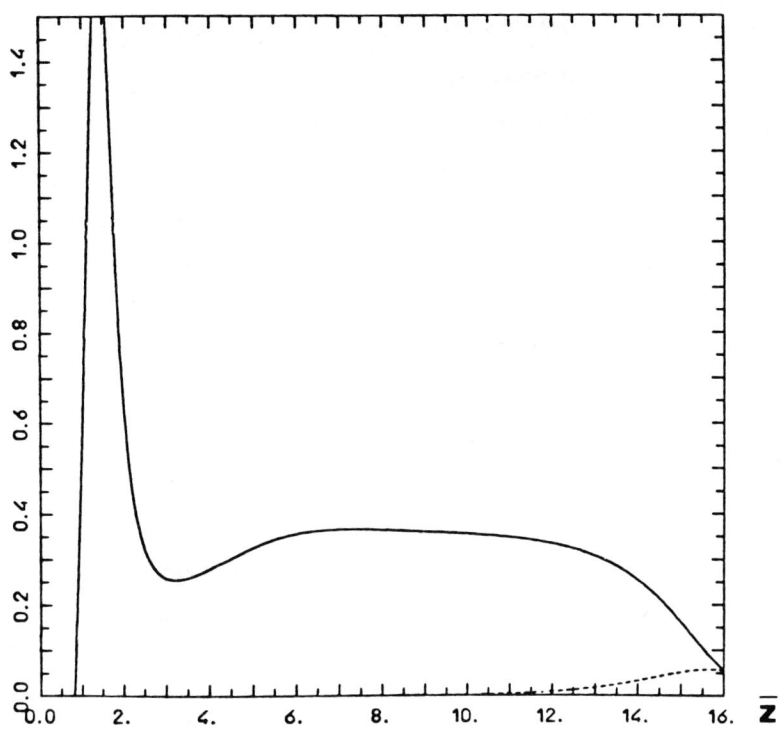

FIGURE 6
Superradiant regime. Solid line: $\langle p \rangle + d\varphi/d\bar{z}$ vs. $\bar{z}$ from the (untapered) damping model (eqs.(3a,b,c')). Dashed line: intensity $|A|^2$ vs. $\bar{z}$.

ACKNOWLEDGEMENT

We are greateful to T. Scharlemann and W. Barletta for helpful discussions.

REFERENCES

1) R. Bonifacio, F. Casagrande, L. De Salvo, and P. Pierini, this volume.
2) T.J. Orzeckowski, B. Anderson, J.C. Clark, W.M. Fawley, A.C. Paul, D. Prosnitz, E.T. Scharlemann, S.M. Yarema, D. Hopkins, A.M. Sessler and J.S. Wurtele, Phys. Rev. Lett. 57 (1986) 2172.
3) J.A. Edighoffer, G.R. Neil, C.E.Hess, T.I. Smith, S.W. Fornaca and H.A. Schwettman, Phys. Rev. Lett. 52 (1984) 344.
4) N.M. Kroll, P.L. Morton and M.N. Rosenbluth, IEEE J. Quantum Electr. QE–17 (1981) 1436.

5) P. Sprangle, C.M. Tang and W.M. Manheimer, Phys. Rev. A21 (1980) 302; D. Prosnitz, A. Szoke and V.K. Neil, Phys. Rev. A24 (1981) 1436.
6) T. Scharlemann, this volume.
7) R. Bonifacio, F. Casagrande, M. Ferrario, P. Pierini and N. Piovella, Optics Commun. 66 (1988) 133.
8) R. Bonifacio and F. Casagrande, Nucl. Instr. and Meth. A239 (1985) 36; in "Complex Systems – Operational Approaches", H. Haken, ed. (Springer, Berlin 1985); J. Opt. Soc. Am. B 2 (1985) 250.
9) R. Bonifacio, C. Maroli and N. Piovella, Optics Comm. 68 (1988) 369; R. Bonifacio and B.W.J. McNeil, Nucl. Instr. and Meth. A272 (1988) 280.

BISTABILITY IN FREE ELECTRON LASERS

R. Bonifacio, F. Castelli, L. De Salvo Souza

Dipartimento di Fisica dell' Universitá, via Celoria 16, Milano, Italy

I.N.F.N. sezione di Milano

We confirm the existence of the bistability phenomenon in a FEL system [1] extending it to a wide range of values for the cavity detuning parameter δ_c. We show that there is a maximum of the hysteresis cycle as a function of δ_c. We show that there is a very good agreement between the single–particle equations and the collective model of ref. 2. Finally we demonstrate well defined invariant properties which are formally identical to these of the mean field model of dispersive optical bistability.

1. Introduction

Bistability is a well–known phenomenon arising in optical systems; it has been theoretically predicted and experimentally observed, both in the absorptive and the dispersive case (for an exhaustive description of the argument see the refs. 3 and 4).

The more simple configuration which can show bistability is an optical resonant ring cavity filled with an atomic absorber; when a coherent driven field is injected, under suitable conditions, the output intensity can assume discontinuously two values, corresponding to an absorbing medium or a saturated medium. It is possible to obtain a hysteresis cycle sweeping the incident intensity from zero to a value beyond the bistable region, and then sweeping back.

Optical bistability, besides the great potential for the construction of optical devices, has also arisen a wide theoretical interest because it is a remarkable example of cooperative behaviour in an open system far from thermal equilibrium [5].

In a free electron laser oscillator with a coherent driving field, bistability can arise as a new physical effect, as first demonstrated by Bonifacio and De Salvo [1]. In this paper we confirm those previous results, and extend our analysis investigating how bistability depends on cavity parameters.

We consider a FEL with long electron bunches in an optical resonant cavity, a configuration which has been successfully operated during the past years [6], and that is close to an optical laser system; the cavity garantees continuous interaction between coherent field and the electrons, provided that the radiation transit time in the resonator is many times shorter than beam duration.

Beam and wiggler parameters are selected in such a way that the FEL behaves like a strong absorber, namely, the energy is transferred from the laser beam to the electrons, as proposed for acceleration of particles with an inverse free electron laser (IFEL) [7]. The electron beam, the cavity and the incident frequency will be properly designed and tuned so that no oscillation is possible over other cavity frequencies.

The FEL is described by a well-known N-electrons model, and the resonant cavity by suitable boundary conditions; the output intensity of radiation is calculated with respect to the incident intensity making use of suitable numerical methods.

We find that this system exhibits a discontinuous bistable behaviour, with a hysteresis cycle like an optical system, over a wide range of cavity parameters. We are able to select the best conditions for the maximization of the hysteresis cycles.

Furthermore we will show that this bistable behaviour in a FEL is reproduced very carefully, both qualitatively and quantitatively, by the collective model developed by Bonifacio, Casagrande and De Salvo [2], that replaces the full N-particle description with three collective variables, and this is a strong argument concerning the validity of this model.

Finally we suggest an interpretation about the nature of the bistability in a FEL, which arises from the study of the behaviour of the hysteresis cycles as a function of cavity parameters and from the analogy with the already widely studied optical bistability.

2. The N-electrons model of Free Electron Laser

In the framework of the Hamiltonian model of Bonifacio, Casagrande and Pellegrini [8], neglecting space charge and radiative effects, the interaction of laser radiation and an electron beam in a helical wiggler is described by the system of $2N+1$ Maxwell–pendulum equations in the adimensional notation [9] :

$$\frac{d\theta_j}{d\bar{z}} = p_j \tag{1.a}$$

$$\frac{dp_j}{d\bar{z}} = -(Ae^{i\theta_j} + c.c.) \tag{1.b}$$

$$\frac{dA}{d\bar{z}} = \langle e^{-i\theta} \rangle \tag{1.c}$$

(the electron index j runs from 1 to N) where A is related to the amplitude of the radiation field E :

$$|A|^2 = \frac{|E|^2 V}{4\pi\rho N \gamma_r m c^2} \tag{2}$$

θ_j is the electron phase with respect to the combined radiation and wiggler fields, and p_j is the electron energy variable (in the near–resonance approximation) :

$$p_j = \frac{\gamma_j - \gamma_r}{\rho \gamma_r} \tag{3}$$

with the energy of jth electron γ_j and the resonance energy γ_r. The symbol $\langle \ \rangle$, here and throughout the paper, means $N^{-1} \sum_{j=1}^{N}$; in particular $\langle e^{-i\theta} \rangle$ is the well-known bunching parameter.

The derivatives are over the scaled wiggler length

$$\bar{z} = \frac{4\pi\rho z}{\lambda_w} \tag{4}$$

where ρ is the fundamental FEL parameter, and λ_w is the wiggler period.

These equations are integrated numerically from $\bar{z}_0 = 0$ to

$$\bar{z}_L = \frac{4\pi\rho L_w}{\lambda_w} = 4\pi\rho N_w,$$

where N_w is the total number of the wiggler periods, with the initial conditions:

$$\langle e^{-i\theta} \rangle_0 = 0, \quad p_j(0) = \delta = \frac{\langle \gamma \rangle_0 - \gamma_r}{\rho \gamma_r}, \quad A(0) = A_0 \tag{5}$$

which means that the electrons are initially unbunched and with the same momentum (cold beam).

For various values of scaled wiggler length $\bar{z}_L$ and initial energy detuning with respect to resonance δ, the equations (1) allow to compute the output field A_L as a nonlinear (complex) function of the input field A_0:

$$A_L = \mathcal{F}(A_0, \delta, \bar{z}_L) \tag{6}$$

Calculating the small signal gain, defined by:

$$G(\delta, \bar{z}_L) = \frac{|A_L|^2 - |A_0|^2}{|A_0|^2}$$

with a small value of $|A_0|$ (for example $|A_0| = 10^{-4}$) we find for $\bar{z}_L \ll 1$ the well known Madey's gain curve [10] which is an odd function of δ, with $G(0, \bar{z}_L) = 0$; for $\bar{z}_L > 1$ the symmetry of the gain curve is broken (see fig. 1), and $G(0, \bar{z}_L)$ is no more zero but starts increasing (exponential gain region).

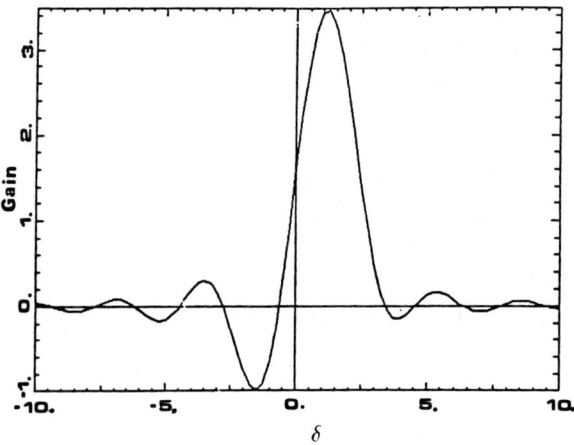

Fig. 1 : Plot of the gain G versus δ for $\bar{z}_L = 2$.

We center our attention over the values of the parameters:

$$\delta = -1.5; \qquad \bar{z}_L = 2. \tag{7}$$

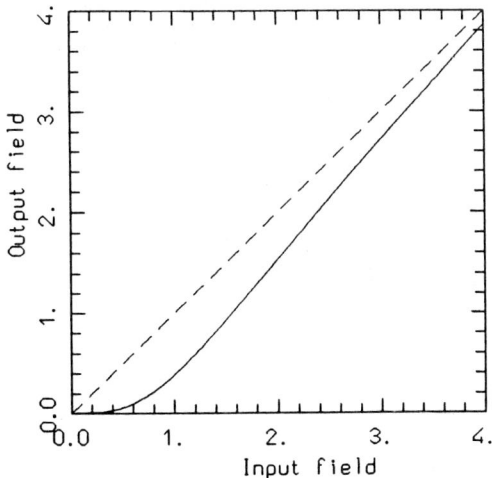

Fig. 2 : Plot of the output field $|A_L|$ versus input field $|A_0|$

where $G \simeq -1$ (exactly $G = -0.99$), *i.e.* there is a 99% absorption of the incident field. Solving eq. (1) for different values of $|A_0|$ we look for a transmission characteristic of this FEL system, obtaining a curve $|A_L| = \mathcal{F}(|A_0|)$ as in fig. 2.

We note that there is strong absorption for $|A_0| < 1$, as already seen from the small signal gain regime, and tendency to saturation for $|A_0| \gg 1$, so that the system becomes transparent ($|A_L| = |A_0|$).

We conclude that with this choice of the parameters, a relativistic beam of electrons in a wiggler behaves like a nonlinear atomic absorber with saturation; hence we expect that this system inside a resonant cavity behaves like a bistable one.

3. The FEL in a resonant cavity

Let's consider now a FEL system in a ring cavity (fig. 3) with an incident external coherent field A_I ; mirrors 1 and 2 have reflectivity R and mirrors 3 and 4 have 100% reflectivity.

The assumption of the ring cavity is just to have strictly one directional propagation, even if a Fabry–Perot configuration can be used without particular complications [1].

A more important assumption is that the ring cavity is supposed designed in such a configuration that there is only one proper mode which can interact with the radiation field of the FEL in the gain region.

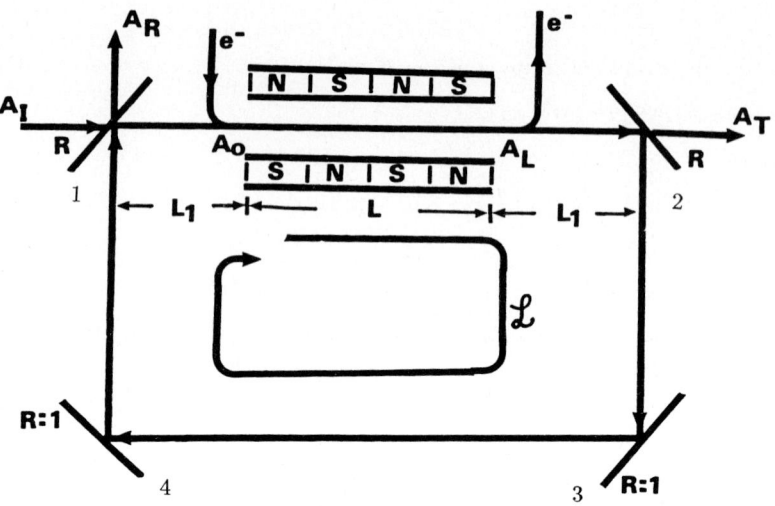

Fig. 3 : Plan of the FEL and an optical ring cavity.

The input field A_0 is the sum of two contributions, the incident field A_I directly transmitted from the mirror, and the feedback field coming from the output A_L after three reflections.

This imposes the boundary conditions :

$$A_0 = \sqrt{T} A_I + R e^{-i\delta_c} A_L \tag{8.a}$$

$$A_T = \sqrt{T} A_L \tag{8.b}$$

where $T = 1 - R$ (transmittivity coefficient), and $\delta_c = (\omega_c - \omega_l)\mathcal{L}/c$ (cavity detuning of the proper mode), with :

$\omega_c = $ proper frequency of the cavity

$\omega_l = $ frequency of the incident field.

The equations (1) and these boundary conditions form a closed system, from which in principle one can obtain the internal fields A_0 and A_L, and hence the transmitted field A_T as a function of the external fixed A_I, namely :

$$A_T = \mathcal{F}(A_I) \tag{9}$$

A_T can be a discontinuous function showing virtually instantaneous jumps from different stationary states, whereas (6) is always a continuous function.

General analitical solutions of the $2N+1$ equations (1) are not known, hence for solving simultaneously (1) and (8) we use a recurrence method (derived from Newton method for zeros of functions) which simulates the experimental situation in which the incident field propagates with the electrons along the wiggler.

The scheme of this method is illustrated below; from the eqs. (6) and (8.a) we consider :

$$A_L^{(n)} = \mathcal{F}(A_0^{(n)}) \tag{10.a}$$

$$A_0^{(n+1)} = \sqrt{T} A_I + R e^{-i\delta_c} A_L^{(n)} \tag{10.b}$$

for $n = 0, 1, 2, ...$; in the first step $A_0^{(0)} = \sqrt{T} A_I$ or same other value if the field is already present inside the cavity.

The recurrence scheme converges if $A_0^{(n+1)} = A_0^{(n)} = \tilde{A}_0$, for n large enough (n simulates the number of round trips in the cavity). Obviously $\tilde{A}_0$ and $\tilde{A}_L$ are simultaneous solutions of propagation equations and boundary conditions, and we get $A_T = \sqrt{T} \tilde{A}_L$ for the given value of A_I.

Varying A_I from 0 to a maximum value, and then going back to 0, it is possible to construct a hysteresis cycle (depending on the cavity parameters δ_c and R), showing that $|A_T|$ is a discontinuous and multivalued function of $|A_I|$. Several examples of these cycles are plotted in fig. 4 for $R = 0.99$, and in fig. 5 for $R = 0.95$.

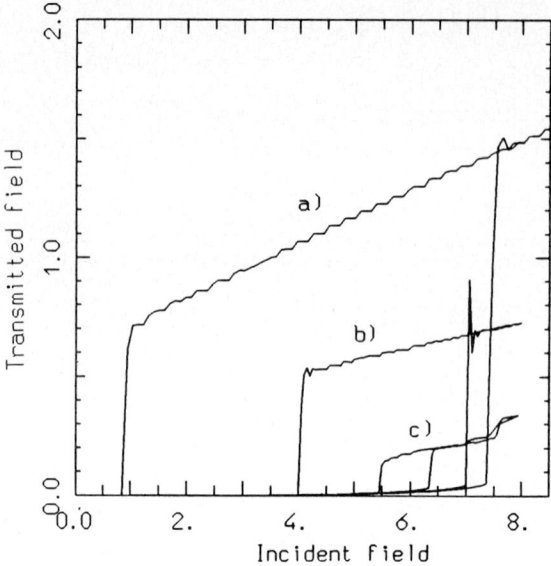

Fig. 4 : Hysteresis cycles of $|A_T|$ versus $|A_I|$ for $R = 0.99$ and
a) $\delta_c = 0.1$; b) $\delta_c = 0.2$. ; c) $\delta_c = 0.5$

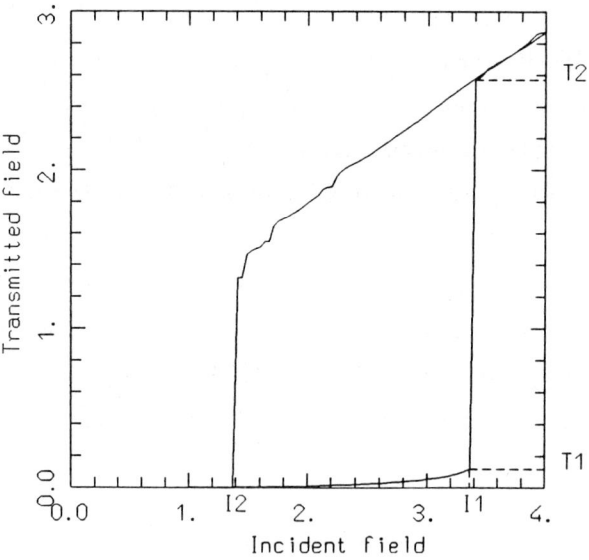

Fig. 5 : Hysteresis cycles of $|A_T|$ versus $|A_I|$ for $R = 0.95$ and
$\delta_c = 0.1$; the symbols I1, I2, T1, T2 refer to section 4

4. Bistability conditions for FEL

The occurrence of the hysteresis cycles in the transmitted field is a normal behaviour of the FEL system for R sufficiently close to 1 (*i.e.* a good cavity), and over a wide range of δ_c, as we have shown in the last paragraph. In this section we discuss in more detail these phenomena.

For incident field increasing from 0 the system is absorbing, and the transmitted field A_T is very small (lower branch of the curve of the hysteresis cycle), until a threshold value of A_I (called I1 in fig. 5). Above this value, this kind of solution of eqs. (10) becames unstable, and A_T jumps suddenly towards high values (straight line from T1 to T2 in fig. 5), namely, the absorber behaves like a saturated one.

Then, decreasing A_I, the system remains in a high transmission state (upper branch of the curve), until a second threshold on A_I is reached (I2 in fig. 5), beyond which A_T falls versus zero and the system returns in the initial absorbing state. This second threshold is always lower and well-separated from the first one.

Hence, over a wide range of the incident coherent field, there are two stable solutions of the FEL system in a resonant cavity, and in essence this is the bistable behaviour.

We have analyzed the influence of the cavity parameters on the appearance of a hysteresis cycle; the phenomenon is particularly evident (*i.e.* the hysteresis cycle is larger) if the cavity frequency is detuned from resonance, as one can observe from the two threshold values of A_I on the lower and upper branch, which are plotted in fig. 6 versus the cavity detuning δ_c for the case $R = 0.95$.

The width of the hysteresis cycle is larger for $\delta_c \simeq 0.1$, while bistability disappears for $\delta_c < -0.1$ or $\delta_c > 0.7$.

In fig. 7 we have plotted the transmitted field A_T corresponding to values of A_I below and above the threshold on the lower branch. Around the previous noticeable value of cavity detuning ($\delta_c \simeq 0.1$), the transmitted field over the threshold (namely, in the higher branch) tends to the incident field, in other words the FEL system becomes nearly transparent.

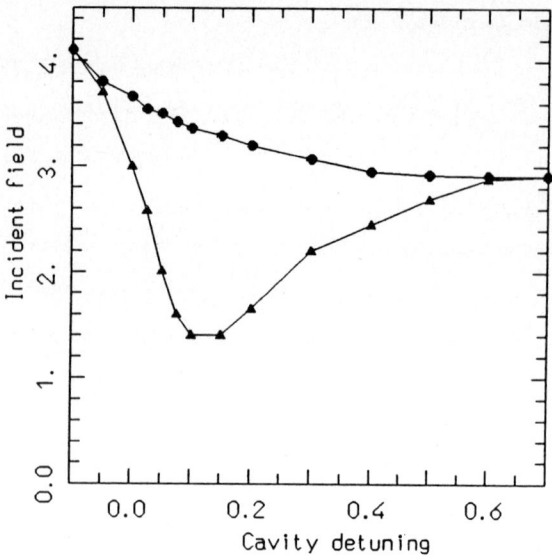

Fig. 6 : Plot of the threshold values of $|A_I|$ versus δ_c for $R = 0.95$;
1) circles : threshold on the lower branch (I1);
2) triangles : threshold on the upper branch (I2).

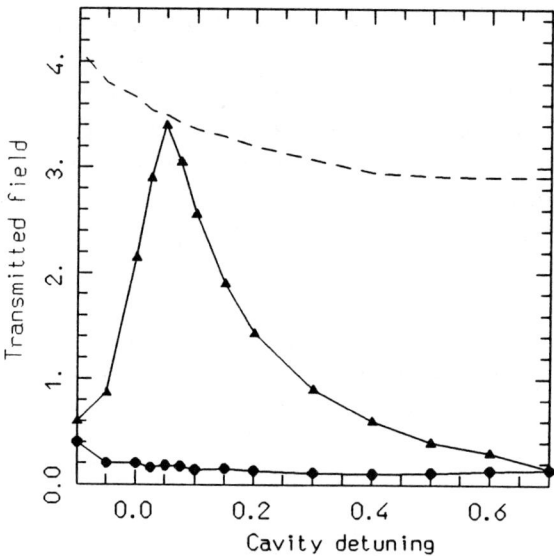

Fig. 7 : Plot of the transmitted field $|A_T|$ versus δ_c for $R = 0.95$;
1) circles : if A_I is below the threshold on the lower branch (T1);
2) triangles : if A_I is above the same threshold (T2);
3) dashed line : values of $|A_I|$ for the case 2.

5. Bistability and collective model of FEL

The previous results have been reproduced also with the collective model of a free electron laser developed in ref. 2. In this model the FEL is represented by the three complex variables:

A : amplitude of radiation field (as in eq. (2))

$b = \langle e^{-i\theta} \rangle$: the bunching parameter

$P = \langle p e^{-i\theta} \rangle$: the phase–momentum average

which obey the system of differential equations :

$$\frac{dA}{d\bar{z}} = b \tag{11.a}$$

$$\frac{db}{d\bar{z}} = -iP \tag{11.b}$$

$$\frac{dP}{d\bar{z}} = -A - i\langle p^2 e^{-i\theta}\rangle - A^* \langle e^{-2i\theta}\rangle \tag{11.c}$$

which can be obtained from eqs. (1).

Using the same procedure as in ref. 2, with the initial conditions

$$b(0) = 0, \qquad \langle p \rangle_0 = \delta, \qquad \langle p^2 \rangle_0 = \delta^2 \tag{12}$$

(the same as in eq. (5)), we rewrite the following set of equations for the three collective variables A, b, P , that allows to study the cavity problem without solving the full system (1) :

$$\frac{dA}{d\bar{z}} = b \tag{13.a}$$

$$\frac{db}{d\bar{z}} = -iP \tag{13.b}$$

$$\frac{dP}{d\bar{z}} = -A - 2b(A^*b - Ab^*) - ib\delta^2 + 2iPF + 2ibF^2 \tag{13.c}$$

where we have defined :

$$F = |A|^2 - |A_0|^2 - \delta.$$

As in section 3, these equations are integrated from $\bar{z} = 0$ to $\bar{z} = 2$ with the initial conditions (from (5) and (7)) :

$$b(0) = 0, \qquad P(0) = 0, \qquad \delta = -1.5 \tag{14}$$

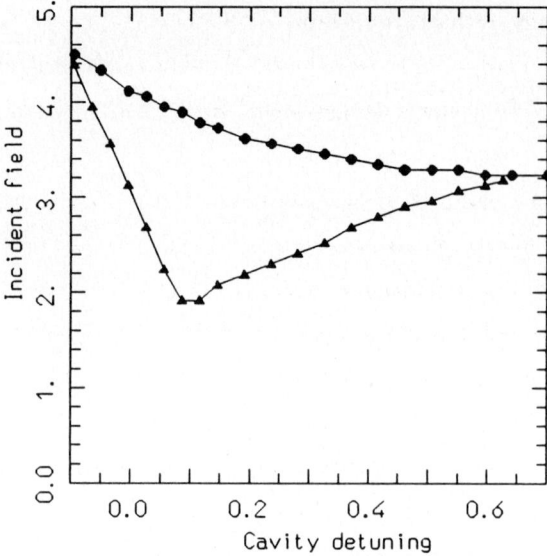

Fig. 8 : Plot of the threshold values of $|A_I|$ versus δ_c calculated with the collective model for $R = 0.95$;
1) circles : threshold on the lower branch (I1);
2) triangles : threshold on the upper branch (I2).

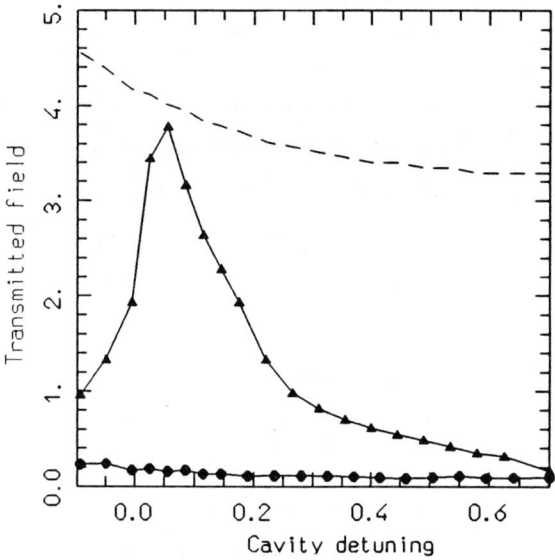

Fig. 9 : Plot of the transmitted field $|A_T|$ versus δ_c calculated with the collective model for $R = 0.95$;
1) circles : if A_I is below the threshold on the lower branch (T1);
2) triangles : if A_I is above the same threshold (T2);
3) dashed line : values of $|A_I|$ for the case 2.

and A_0 assigned by the boundary conditions (8.a). With the same recurrence method used for the N–electron system, we are able to solve simultaneously the collective propagation equations (12) and the boundary conditions (with the same parameters) obtaining surprisingly the same hysteresis cycles.

This shows that the collective model reproduces very well the bistable behaviour found with the complete system. As one can see comparing fig. 8, in which we plot the threshold values of A_I obtained with the collective model, with the correspondent fig. 6, there is full qualitative and good quantitative agreement with the previous results; the same happens comparing fig. 9 and fig. 7, which shows the transmitted field around the threshold on the lower branch.

In particular, the range of values of δ_c over which the system displays bistable behaviour is the same; moreover for $\delta_c \simeq 0.1$ the hysteresis cycle has the maximum width, as we already pointed out in section 4.

This analysis confirm the validity of the collective model, which is capable to reproduce all the interesting features of the FEL in a resonant cavity; it allows also for further theoretical and computational investigations, which are now in progress.

6. Nature of bistability phenomenon

One evident feature of the hysteresis cycles, showed in fig. 6 and in fig. 8, is the asymmetry with respect to the sign of δ_c.

As we infer from analogy with the well–known optical bistability [3,4,11], this suggests the preponderance of the dispersive component over the absorptive one in the nonlinearity of the medium.

We have checked this hypotesis investigating the behaviour of the hysteresis cycle as a function of the mirror transmittivity coefficient T, with both the N–electrons and the collective models.

In this analysis we consider, instead of A_T and A_I, the 'normalized' field variables $A_{eff} = A_I\sqrt{T}$, that is the effective incident field inside the cavity, and $A_L = A_T/\sqrt{T}$, that is the output field after the interaction in the wiggler (see fig. 3).

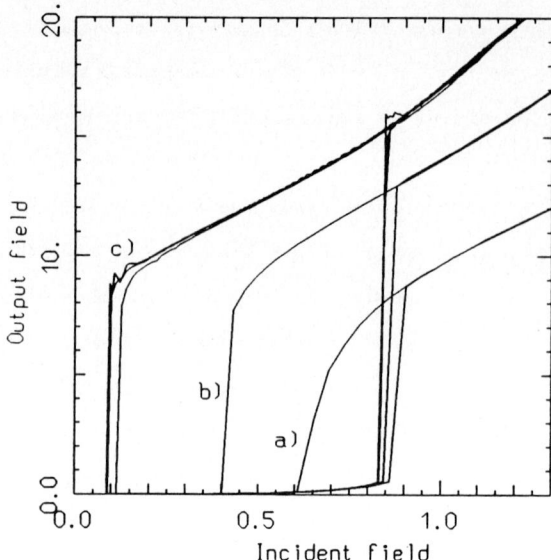

Fig. 10 : Hysteresis cycles of $|A_L|$ versus $|A_{eff}|$ calculated with the collective model for different T ; a) $T = 0.1$; b) $T = 0.05$; c) $T = 0.01$, $T = 0.005$ and $T = 0.002$.

Plotting the hysteresis cycles in these new variables, for $\delta_c = 0.1$ and several small values of T (fig. 10), we see that for $T \leq 0.01$ the hysteresis cycle is practically independent on T.

This behaviour is typical of dispersive bistability. In fact, let's consider the general approximate relation for a nonlinear medium in a resonant cavity :

$$\frac{|A_I|^2}{T} = |A_0|^2 \left[(1 + \frac{\alpha_2}{T})^2 + (\frac{\delta_c - \alpha_1}{T})^2\right] \qquad (15)$$

derived in ref. 1 by setting :

$$A_L = A_0 e^{i\alpha}; \qquad \alpha = \alpha_1 + i\alpha_2.$$

Here α_1 describes dispersion and α_2 absorption, and in general they are nonlinear function of the field; for example in optical bistability one has :

$$\alpha_i = \frac{a_i}{1 + b_i|A_L|^2}; \qquad i = 1, 2$$

(however the following is independent from this particular statement).

In purely dispersive case $\alpha_2 = 0$ and $\delta_c \gg T$, so one has $|A_0|^2 = |A_L|^2$ and the dispersive term in eq. 15 dominates the absorptive term; hence we obtain :

$$\frac{|A_I|^2}{T} = |A_L|^2 \left(\frac{\delta_c - \alpha_1}{T}\right)^2$$

Multiplying this formula by T^2 we get, in normalized field variables :

$$|A_{eff}|^2 = |A_L|^2 (\delta_c - \alpha_1)^2 \tag{16}$$

which is independent of T, as expected. This shows that in the dispersive case, *i.e.* small T, the relation between the normalized fields depends only on the cavity detuning.

Concluding, we stress that these observations clearly indicate the predominant dispersive nature of the bistable behaviour in a FEL with a resonant cavity.

References

1) R. Bonifacio and L. De Salvo, Optical Bistability in a relativistic electron beam, in: 'Instabilities and chaos in quantum optics II ' (Plenum Press), in press.

2) R. Bonifacio, F. Casagrande and L. De Salvo, Phys. Rev. A **33** (1986), 2836.

3) R. Bonifacio and L. A. Lugiato, Theory of Optical Bistability, in: 'Dissipative Systems in Quantum Optics ', Topics in Current Physics, ed. R. Bonifacio (Springer–Verlag, 1982).

4) L. A. Lugiato, Theory of Optical Bistability, in: Progress in Optics, vol XXI, ed. E.Wolf (North–Holland Physics Publishing, 1984).

5) H. Haken, 'Synergetics, An Introduction ' (Springer, Berlin 1977); G. Nicolis and I. Prigogine, 'Self–Organization in Non–Equilibrium Systems ' (Wiley, New York, 1977).

6) J. Mathew and J. A. Pasour, Phys. Rew. Lett. **56** (1986) 1805

7) E. D. Courant, C. Pellegrini and W. Zakowicz, Phys. Rev. A **32** (1985) 2813.

8) R. Bonifacio, F. Casagrande and C. Pellegrini, Opt. Comm. **61** (1987), 55.

9) R. Bonifacio, C. Pellegrini and L. Narducci, Opt. Comm. **50** (1984), 373.

10) J. M. J. Madey, Jour. Appl. Phys. **42** (1971), 1906.

11) R. Bonifacio, M. Gronchi, L. A. Lugiato, Nuovo Cim. **53B** (1979), 311.

SLIPPAGE AND SUPERRADIANCE IN A HIGH–GAIN FEL : LINEAR THEORY *

Rodolfo BONIFACIO, Cesare MAROLI and Nicola PIOVELLA

Dipartimento di Fisica dell'Università, via Celoria 16, 20133 Milano, Italy and Istituto Nazionale di Fisica Nucleare, Sezione di Milano, Italy

We describe the linear regime of a Free–Electron–Laser Amplifier taking into account propagation effects (slippage). We demonstrate analytically, (for a simple case) the existence of two different solutions of the pulse propagation equations, which in suitable limits describe the Steady–State and the Superradiant regimes.

1. INTRODUCTION

The Free Electron Laser (FEL) uses a relativistic high–current electron beam to amplify a copropagating electromagnetic wave. The optical wave and the electrons are coupled as they pass through a periodic, transverse magnetic undulator field.

A theoretical analysis [1] based on the equations of the traveling–wave–tube type, shows the existence of two completely different dynamical regimes of a Compton FEL, namely, a stable regime in which one finds an interference gain (the well known Small–Gain regime), and an unstable cooperative regime, experimentally observed [3], in which the gain grows exponentially up to saturation (High–Gain regime). This regime, in which the FEL is operating in a single–pass configuration, has been the object of detailed analysis in previous works [1,2]; it is named the *Amplified Spontaneous Emission (ASE)* if the lasing process is developed from noise.

Most previous theories describing the ASE regime, for a long electron pulse, neglect the relative slippage of the electron and radiation pulses (which arises as the electron velocity $v_\| < c$). The amount of the slippage, after $N_\circ$ periods of the undulator, is $N_\circ \lambda_r$, where λ_r is the radiation wavelength. This neglect of the slippage is incorrect in the High–Gain regime if the electron pulse is sufficiently short. More precisely, as discussed elsewhere in this volume [4], introducing a dimensionless parameter K [5,6], which is the measure of

---

* Work supported by Istituto Nazionale di Fisica Nucleare (INFN)

the slippage in a gain length, we define the long pulse limit by $K \ll 1$, and the short pulse limit by $K \geq 1$. Only in the first case the slippage may be neglected if the undulator is sufficiently short; in this limit, the ASE regime can be described by the steady–state theory (i.e. without slippage), with peak intensity proportional to $n_e^{4/3}$ (where n_e is the electron beam density).

In the short–bunch limit ($K \geq 1$) we find a completely different dynamical regime, with peak intensity proportional to n_e^2 (Superradiant regime).

In Ref. [5] the slippage effect was modelled by introducing a loss term in the Maxwell's equation; from this simple dissipative model, the Superradiant limit, (in which the field variables can be adiabatically eliminated), and the condition under which a Superradiant emission should be observed, were defined.

In this paper, we linearize the equations describing the one dimensional evolution of the radiation and the electrons, taking into account the slippage effects. After defining the initial and the boundary conditions, we describe analytically the evolution of the field and the electron distribution inside an electron pulse of finite length, passing through the undulator. Assuming for the sake of simplicity an initial square density distribution, we find that, in the electron pulse, two regions can be identified, where the field and electron variables evolve in two completely different ways.

In the first region, the radiation is time–independent, and every particle emits radiation in the same way, giving a steady–state behaviour. This is the ASE regime, in which the electrons can reabsorb the emitted radiation. We re–obtain the ASE solution of the equations without slippage effects [1,2].

In the second region, near the trailing edge of the electron pulse, the field is space and time dependent, and the electrons do not evolve identically; this occurs because the electron pulse experiences less radiation reabsorbtion than in the first 'steady–state' region, so that slippage inhibits saturation.

In the Superradiant limit ($K \geq 1$) the steady–state part disappears when the electron pulse enters the High–Gain regime, so that only the Superradiant trailing edge part is observable (undergoing an exponential gain).

It is important to note that, even in the long–pulse limit, there is always a region of the pulse, near the trailing edge, where the behaviour is Superradiant. In this 'slippage' region (as it is demonstrated in Ref. [6]), the radiation exhibits spiking behaviour, with peak intensity reaching many times the saturated intensity predicted in the Steady–State regime.

2. LINEAR PROPAGATION EQUATIONS

The equations describing the one–dimensional evolution of the electrons variables and the complex amplitude of the radiation field, as derived in detail elsewhere in this volume [4] , are:

$$\frac{d\theta_j}{d\bar{z}} = p_j \tag{1a}$$

$$\frac{dp_j}{d\bar{z}} = -\left(Ae^{i\theta_j} + c.c.\right) \tag{1b}$$

$$\left(\frac{\partial}{\partial \bar{z}} + \frac{\partial}{\partial \bar{t}}\right) A = \chi_b \langle e^{-i\theta}\rangle + i\delta A \tag{1c}$$

($j = 1,\ldots, N_e$), where

$$\theta_j = \frac{2\pi}{\lambda_\circ} z + \frac{2\pi}{\lambda_r}(z - ct_j) - \delta \bar{z}$$

$$p_j = \frac{1}{\rho}\frac{\gamma_j - \langle\gamma\rangle_\circ}{\langle\gamma\rangle_\circ}$$

$$A = \frac{E_\circ}{\sqrt{4\pi\rho n_e \langle\gamma\rangle_\circ mc^2}}\exp(i\delta\bar{z}) \tag{2a}$$

$$\delta = \frac{1}{\rho}\frac{\langle\gamma\rangle_\circ - \gamma_R}{\langle\gamma\rangle_\circ}$$

$$\bar{z} = \frac{4\pi\rho}{\lambda_\circ}z \qquad \bar{t} = \frac{4\pi\rho}{\lambda_\circ}ct \tag{2b}$$

$$\rho = \frac{1}{\langle\gamma\rangle_\circ}\left(\frac{a_\circ}{4}\frac{\omega_P \lambda_\circ}{2\pi c}\right)^{2/3} \tag{2c}$$

$$a_\circ = \frac{eB_\circ \lambda_\circ}{2\pi mc^2}$$

and $\langle\ldots\rangle$ is the average $\frac{1}{N_e}\sum_{j=1}^{N_e}(\ldots)$.

The meaning of the dimensionless variables and parameters in eqs. (1a) – (1c) is the following: θ_j and p_j are the phase and the energy variation, respectively, of the j th electron; A is the scaled complex amplitude of the radiation field, $\bar{z}$ and $\bar{t}$ are the scaled position and time, δ is the detuning parameter. In eqs. (2) these quantities are expressed in terms of the complex amplitude E_o and the wavelength λ_r of the radiation field, the magnetostatic amplitude B_o and the period λ_o of the undulator whose length is $L_o = N_o \lambda_o$ (N_o is the number of periods); ρ is the fundamental FEL parameter, ω_P is the plasma frequency and a_o is the undulator parameter; γ_j is the energy of the j th electron in rest mass units, $\langle \gamma \rangle_o = N_e^{-1} \sum_{j=1}^{N_e} \gamma_j(0)$ and γ_R is the resonance energy, defined as:

$$\gamma_R{}^2 \equiv \frac{\lambda_o}{2\lambda_r}(1 + a_o^2) \tag{3}$$

In (1c) $\chi_b(\bar{z} - \beta_\| \bar{t})$ is the macroscopic electron density function ($\beta_\| = v_\|/c$, where $v_\|$ is the mean longitudinal electron velocity). We assume χ_b to be a rectangular function normalized to one.

We linearize eqs.(1) around the initial condition, which is an equilibrium condition,

$$A^\circ = 0 \qquad p_j{}^\circ = 0 \qquad \langle e^{-i\theta^\circ} \rangle = 0$$

(no field and monokinetic, unbunched electron beam).

By introducing the phase deviation $\delta\theta_j = \theta_j - \theta_j{}^\circ$, and considering $\delta\theta_j$, p_j and A in (1) as infinitesimal quantities, the linearized system can be written as

$$\left(\frac{\partial}{\partial \bar{z}} + \frac{1}{\beta_\|}\frac{\partial}{\partial \bar{t}}\right) b = -i\mathcal{P} \tag{4a}$$

$$\left(\frac{\partial}{\partial \bar{z}} + \frac{1}{\beta_\|}\frac{\partial}{\partial \bar{t}}\right) \mathcal{P} = -A \tag{4b}$$

$$\left(\frac{\partial}{\partial \bar{z}} + \frac{\partial}{\partial \bar{t}}\right) A = \chi_b b + i\delta A \tag{4c}$$

in terms of the electron collective variables [1] b, $\mathcal{P}$:

$$\begin{aligned} b &\equiv -i\langle e^{-i\theta^\circ} \delta\theta \rangle \\ \mathcal{P} &\equiv \langle e^{-i\theta^\circ} p \rangle \end{aligned} \tag{5}$$

In eqs.(4a,b) we have introduced a fluid description for the electron variables by

$$\frac{d}{d\bar{z}} = \frac{\partial}{\partial \bar{z}} + \frac{1}{\beta_\|}\frac{\partial}{\partial \bar{t}}.$$

The eqs.(4) are defined for $\bar{z} \geq 0$ (inside the undulator) and only in the region occupied by the electrons. We assume that the electrons are continuously distributed in a finite bunch length L_b. So, the eqs.(4) are defined in the moving frame with the velocity $v_\parallel$, in the region of space occupied by the bunch; hence, in eq.(4c), $\chi_b = 1$ for $0 < z - v_\parallel t < L_b$; outside the bunch, the field propagates in the vacuum according to eq.(4c) with $\chi_b = 0$.

We introduce

$$\begin{cases} z' = \bar{z} = \dfrac{4\pi\rho}{\lambda_\circ}z \\ \zeta = \bar{z} - \beta_\parallel \bar{t} = \dfrac{4\pi\rho}{\lambda_\circ}(z - v_\parallel t) \end{cases} \quad (6)$$

so that the eqs.(4) become :

$$\frac{\partial b}{\partial z'} = -i\mathcal{P} \quad (7a)$$

$$\frac{\partial \mathcal{P}}{\partial z'} = -A \quad (7b)$$

$$\frac{\partial A}{\partial z'} + (1 - \beta_\parallel)\frac{\partial A}{\partial \zeta} = b + i\delta A \quad (7c)$$

We set $t = 0$ when the trailing edge of the bunch is crossing the position $z = 0$, i.e. the beginning of the undulator. Hence eqs.(7) are defined for $z' \geq 0$ and for

$$0 \leq \zeta \leq \zeta_\circ \equiv \frac{4\pi\rho}{\lambda_\circ}L_b \quad (8)$$

with the initial condition at $z' = 0$

$$A(0,\zeta) = A_\circ(\zeta)$$
$$\mathcal{P}(0,\zeta) = 0 \quad (9)$$
$$b(0,\zeta) = b_\circ(\zeta)$$

and with

$$A(z',0) = A_\circ(0)e^{i\delta z'} \quad (10)$$

in the trailing edge of the bunch. $A_\circ(\zeta)$ and $b_\circ(\zeta)$ are the known initial field and electron distribution inside the bunch. The boundary condition (10) is a consequence of the S.V.E.A. approximation in the Maxwell equation: considering only the first order derivatives in (1c), we assume that the electron are emitting only in the forward direction, so that the field does not evolve in the trailing edge of the electron bunch.

The solution of eqs.(7) proceeds by introducing the Laplace transform in z' :

$$\overline{X}(\omega,\zeta) = \int_0^\infty dz' e^{i\omega z'} X(z',\zeta) \qquad (Im\,\omega > 0) \quad (11)$$

Eqs.(7) become:

$$-i\omega\overline{b}(\omega,\zeta) - b_\circ(\zeta) = -i\overline{P}(\omega,\zeta)$$
$$-i\omega\overline{P}(\omega,\zeta) = -\overline{A}(\omega,\zeta) \qquad (12)$$
$$-i\omega\overline{A}(\omega,\zeta) + (1-\beta_\parallel)\frac{d\overline{A}(\omega,\zeta)}{d\zeta} - A_\circ(\zeta) = \overline{b} + i\delta\overline{A}(\omega,\zeta)$$

This yields the ordinary first order differential equation

$$\frac{d\overline{A}(\omega,\zeta)}{d\zeta} - i\frac{\Delta(\omega)}{\omega^2(1-\beta_\parallel)}\overline{A}(\omega,\zeta) = \frac{1}{1-\beta_\parallel}\left\{A_\circ(\zeta) + i\frac{b_\circ(\zeta)}{\omega}\right\} \qquad (13)$$

with the initial condition $\overline{A}(\omega,0) = i\frac{A_\circ(0)}{\omega+\delta}$, and where

$$\Delta(\omega) = \omega^3 + \delta\omega^2 - 1 \qquad (14)$$

Eq.(13) has the solution

$$\begin{aligned}\overline{A}(\omega,\zeta) =& i\frac{A_\circ(0)}{\omega+\delta}\exp\left\{i\frac{\Delta(\omega)}{\omega^2(1-\beta_\parallel)}\zeta\right\} \\ &+ \frac{1}{1-\beta_\parallel}\int_0^\zeta d\zeta'\left\{A_\circ(\zeta') + i\frac{b_\circ(\zeta')}{\omega}\right\}\exp\left\{i\frac{\Delta(\omega)}{\omega^2(1-\beta_\parallel)}(\zeta-\zeta')\right\}\end{aligned} \qquad (15)$$

3. THE STEADY-STATE AND THE SUPERRADIANT SOLUTIONS

We assume for the sake of simplicity $A_\circ = 0$ (no field excitation) and we invert the Laplace transform (15), after integrating over ζ' with a constant initial bunching $b_\circ$. Physically, it corresponds to consider no laser pump and some noise $b_\circ$ due to the non uniform distribution of the electrons (electron shot–noise). We obtain:

$$A(z',\zeta) = -\frac{b_\circ}{2\pi}\int_{-\infty+ic}^{\infty+ic}d\omega\frac{\omega e^{-i\omega z'}}{\Delta(\omega)}\left\{1-\exp\left[i\frac{\Delta(\omega)\zeta}{\omega^2(1-\beta_\parallel)}\right]\right\} \qquad (16)$$

where ζ is ranging in the interval (8) and the real number c is chosen so that $\omega = ic$ lies above all the singularities of the integrand function.

The field amplitude A is the sum of two terms. The first, only space dependent, is given by the residues at the three simple poles ω_j, obtained by solving the dispersion equation $\Delta(\omega) = \omega^3 + \delta\omega^2 - 1 = 0$; it is the well known cubic equation, obtained in the 'steady–state' linear analysis; setting $\omega' = -\omega$, it can be written as in Ref. [1] :

$$\omega'^3 - \delta\omega'^2 + 1 = 0.$$

It follows that the first term of (16) is the sum of three exponential modes:

$$A_S = ib_\circ \sum_{j=1}^{3} \frac{\omega_j e^{-i\omega_j \bar{z}}}{(\omega_j - \omega_k)(\omega_j - \omega_m)} \qquad (j \neq k \neq m) \qquad (17)$$

The second term of (16) (z' and ζ dependent) is the sum of the residues at the same three simple poles ω_j and the residue at the essential singularity $\omega = 0$; we note that, evaluating the residue at the simple poles, the exponential in ζ is one (because $\Delta = 0$), so this contribution cancels exactly the first steady–state term A_S , and only the contribution from the essential singularity remains.

Writing (16) in the original coordinates $\bar{z}$ and $\bar{t}$ and using (14), we obtain:

$$A = A_S(\bar{z}) + A_1(\bar{z}, \bar{t}) \qquad (18)$$

where

$$A_1 = \frac{b_\circ}{2\pi} e^{i\delta z_1} \int_{-\infty+ic}^{\infty+ic} d\omega \frac{\omega}{\Delta(\omega)} \exp\left\{-i\left(\omega z_2 + \frac{z_1}{\omega^2}\right)\right\} \qquad (19)$$

In (19) we have introduced, as in Ref. [4,6] , the following coordinates:

$$z_1 = \frac{\zeta}{1-\beta_\parallel} = \frac{z - v_\parallel t}{\beta_\parallel l_C} \qquad (20a)$$

$$z_2 = \beta_\parallel \frac{\bar{t} - \bar{z}}{1 - \beta_\parallel} = \frac{ct - z}{l_C} \qquad (20b)$$

with

$$z_1 + z_2 = \bar{z} \qquad (21)$$

and where

$$l_C = \frac{1 - \beta_\parallel}{\beta_\parallel} \frac{\lambda_\circ}{4\pi\rho} = \frac{\lambda_S}{4\pi\rho} \qquad (22)$$

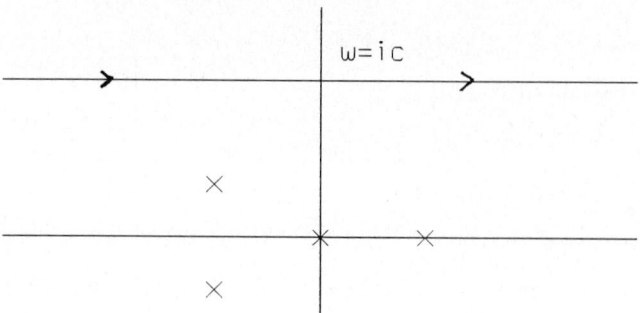

is the cooperation length [4,6] ; in (22) $\lambda_S = (1 - \beta_\|) \lambda_\circ/\beta_\| \simeq \lambda_\circ (1 + a_\circ^2)/2\langle\gamma\rangle_\circ^2$ is the wavelength of the spontaneous emission (at resonance $\lambda_S = \lambda_r$).

We note that the limitation (8) becomes:

$$0 \leq z_1 \leq \frac{1}{K} \qquad (23)$$

where (assuming $\beta_\| \simeq 1$):

$$K = \frac{1 - \beta_\|}{\zeta_\circ} = \frac{l_C}{L_b} \qquad (24)$$

is the *Superradiant* parameter. In (23), $z_1 = 0$ and $z_1 = 1/K$ refer respectively to the trailing and leading edges of the electron bunch.

From (23,24) we introduce the *short* and *long bunch limit* respectively defined for $K \geq 1$ (i.e. $L_b \leq l_C$) and $K \ll 1$ (i.e. $L_b \gg l_C$).

We discuss the integral in (19) in more detail. The integration in (19) is to be performed along a line $\omega = ic$ in the complex plane (see the figure above); the real number c is chosen so that $\omega = ic$ lies above all the singularities (the crosses in the figure). For $z_2 < 0$ the contour integral can be closed in the upper half plane without affecting the value of the integral; since the integrand is regular inside the closed contour the integral vanishes. Hence for $z_2 < 0$ (i.e. $ct < z$) $A_1 = 0$ and A is the steady-state solution A_S of eq.(17). Otherwise, for $z_2 > 0$ the contour is closed in the lower half plane and the integral is given by $-2\pi i$ times the residues at the singularities of $(\omega/\Delta)\exp\{-i(z_2\omega + z_1/\omega^2)\}$; these are the same simple poles of A_S (the roots of $\Delta(\omega)$) and the essential singularity $\omega = 0$. For the first contribution (simple poles) it follows from (14) that

$$\frac{1}{\omega^2} = \omega + \delta$$

so that

$$\exp\left\{-i\left(\omega z_2 + \frac{z_1}{\omega^2}\right)\right\} = \exp\left\{-i\delta z_1 - i\omega(z_1+z_2)\right\} = \exp\left\{-i\delta z_1 - i\omega \bar{z}\right\}$$

because $z_1+z_2=\bar{z}$. Hence the residue of A_1 at the simple poles cancels the steady-state contribution A_S, so that only the residue at the essential singularity remains. Hence for $z_2>0$ (i.e. $ct>z$) $A_1 = -A_S + A_{SR}$ and, from (18) $A = A_{SR}$, where:

$$A_{SR} = -\frac{b_o e^{i\delta z_1}}{2\pi} \oint_{C_o} \frac{d\omega\, \omega}{\Delta(\omega)} \exp\left\{-i\left(\omega z_2 + \frac{z_1}{\omega^2}\right)\right\} \quad (25)$$

In (25) C_o is a counterclockwise circular path enclosing the singularity $\omega=0$, of radius R sufficiently small, so that the simple poles, roots of Δ, are excluded from the path.

Hence, there are two regions inside the bunch, where the field evolves in two completely different ways.

In the first region (for $z_2 \leq 0$ i.e. for $\bar{z} \leq z_1 \leq 1/K$), $A = A_S(\bar{z})$ and the radiation is uniform over the bunch (*steady-state region*); in the High-Gain regime, A_S is growing exponentially [1] as $e^{|Im\omega_j|\bar{z}}$, where $\bar{z} = \frac{4\pi\rho}{\lambda_o}z \equiv G$ is the unsaturated gain; in this region of the pulse there are not slippage effects.

The second region of the bunch, where $A = A_{SR}(\bar{z},\bar{t})$, is ranging from the trailing edge $z_1=0$ to $z_2=0$ (i.e. in $0 \leq z_1 < \bar{z}$); $z_2=0$ (i.e. $ct=z$) corresponds to the point of the bunch reached by the radiation emitted from the trailing edge. In this *slippage region* the radiation is not uniform over the bunch: here the electrons cannot reabsorb the emitted radiation at the same rate that in the 'steady-state region', because there are no more electrons behind them and so there is less radiation propagating into this region.

We observe that, from (23), the first steady-state region $\bar{z} \leq z_1 \leq 1/K$ can exist only for $\bar{z} < 1/K$, or equivalently for $S<1$, where:

$$S = K\bar{z} = \frac{l_S}{L_b} \quad (26)$$

is the Slippage parameter. In (26) $l_S = (1-\beta_\|)z/\beta_\|$ is the slippage length: after a distance z covered by the electrons, the radiation pulse has slipped over the electron bunch for a distance l_S. If the electrons have passed N_o periods of the undulator, l_S is equal to $N_o \lambda_S$.

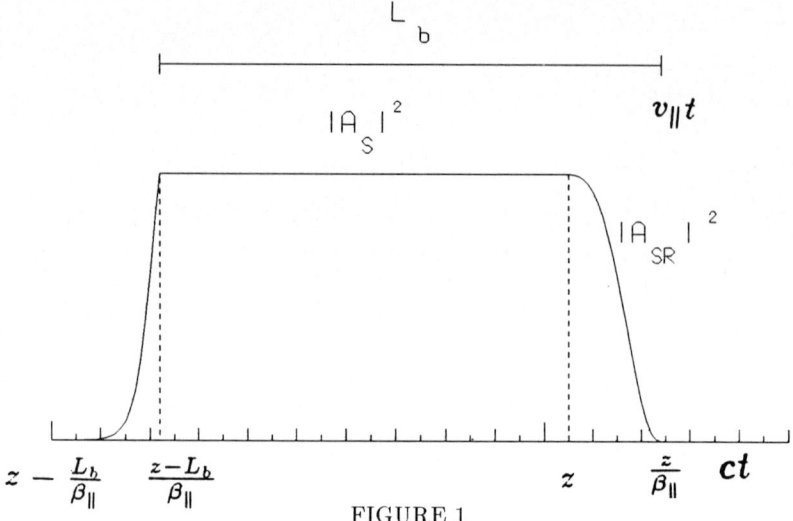

FIGURE 1

$|A|^2$ as a function of ct for a fixed z and $S < 1$. $|A_S|^2$ is the steady-state intensity (17), $|A_{SR}|^2$ is the intensity (25); the electron pulse is crossing z in the interval $z - L_b < v_\| t < z$.

Hence for $S > 1$ the steady-state region does not exist; in this case, the slippage length is greater than the electron bunch length, and the radiation has completely slipped over the electron bunch.

In Fig.1 $|A|^2$ is plotted as function of ct, for a fixed z such that the slippage length l_S is smaller than L_b ($S < 1$); the electron pulse is crossing z in the interval $z - L_b < v_\| t < z$. The steady-state solution $|A_S|^2$ (17) and the solution $|A_{SR}|^2$ (25) are plotted, respectively in the intervals $(z - L_b)/\beta_\| < ct < z$ ('steady-state region') and $z < ct < z/\beta_\|$ ('slippage region' of width $c\Delta t = l_S$); for $z - L_b/\beta_\| < ct < (z - L_b)/\beta_\|$, $|A|^2$ is plotted as it escapes the leading edge of the electron bunch.

If, on the other hand, the slippage length is greater than L_b (i.e. $S > 1$), the 'steady-state region' does not exist : in this case of complete slippage of the radiation pulse over the electrons, as show in Fig.2, the only contribute to the radiation intensity comes from the solution (25) A_{SR}.

Calculating the bunching parameter $b(\bar z, \bar t)$ following the same steps as for $A(\bar z, \bar t)$, we obtain, for $z_2 < 0$ (i.e. $ct < z$):

$$b_S = b_\circ \sum_{j=1}^{3} \frac{e^{-i\omega_j \bar z}}{\omega_j(\omega_j - \omega_k)(\omega_j - \omega_m)} \qquad (j \neq k \neq m) \qquad (27)$$

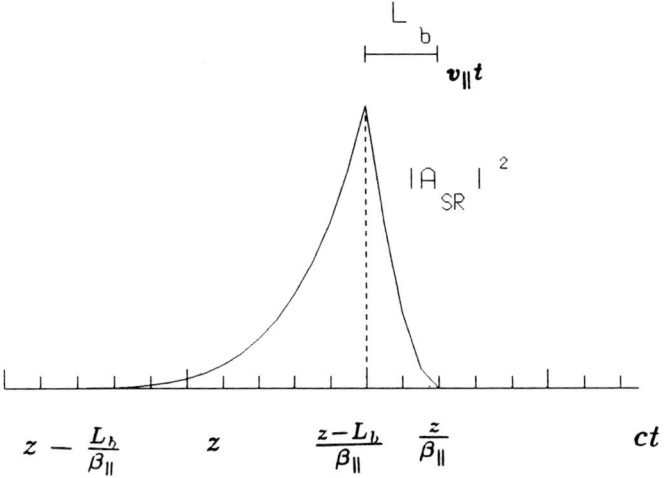

FIGURE 2

$|A|^2$ versus ct for $S > 1$. In this case no steady-state part exists.

whereas for $z_2 > 0$ (i.e. $ct > z$):

$$b_{SR} = \frac{ib_\circ e^{i\delta z_1}}{2\pi} \oint_{C_\circ} \frac{d\omega}{\omega \Delta(\omega)} \exp\left\{-i\left(\omega z_2 + \frac{z_1}{\omega^2}\right)\right\} \tag{28}$$

We can write (25,28) in a more convenient form, by defining $\xi \equiv \omega \left(\frac{z_2}{z_1}\right)^{1/3}$ and $y = \sqrt{z_1 z_2}$:

$$A_{SR} = \frac{b_\circ}{2\pi} \frac{z_1}{y^{2/3}} e^{i\delta z_1} \oint_{C_\circ} \frac{d\xi\, \xi}{1 - \delta\left(\frac{z_1}{z_2}\right)^{2/3} \xi^2 - \frac{z_1}{z_2}\xi^3} \exp\{-iy^{2/3}\Phi(\xi)\} \tag{29a}$$

$$b_{SR} = \frac{b_\circ}{2\pi i} e^{i\delta z_1} \oint_{C_\circ} \frac{d\xi}{\xi} \frac{1}{1 - \delta\left(\frac{z_1}{z_2}\right)^{2/3} \xi^2 - \frac{z_1}{z_2}\xi^3} \exp\{-iy^{2/3}\Phi(\xi)\} \tag{29b}$$

where $\Phi(\xi) = \xi + \frac{1}{\xi^2}$.

The integrals in eqs.(29a,b) can be evaluated at resonance ($\delta = 0$) as a power series by expanding each of the exponentials $\exp(-iy^{2/3}\xi)$ and $\exp(-iy^{2/3}/\xi^2)$ and the rational function $1/(1 - \frac{z_1}{z_2}\xi^3)$ in a power series and looking for simple poles. We obtain:

$$A_{SR} = b_\circ z_1 \sum_{m=0}^{\infty} \left(\frac{z_1}{z_2}\right)^m F_m^{(1)}(y) \tag{30}$$

$$b_{SR} = b_\circ \sum_{m=0}^{\infty} \left(\frac{z_1}{z_2}\right)^m F_m^{(0)}(y) \qquad (31)$$

where

$$F_m^{(j)} = \sum_{n=[\frac{m+1}{2}]}^{\infty} \frac{i^n y^{2n}}{(n+m+j)!(2n-m)!} \qquad (j=0,1) \qquad (32)$$

and $[\frac{m+1}{2}]$ is the largest integer smaller or equal than $\frac{m+1}{2}$.

Finally, we comment on the phase factor $\exp(i\delta z_1)$ appearing in the solution (25) in the slippage region.

We demonstrate that this factor shifts the carrier wavelength of the radiation field $E_\circ(z,t)\exp\{ik_r(z-ct)\}$ from the value $\lambda_r = 2\pi/k_r$ to the wavelength $\lambda_S = 2\pi/k_S = \lambda_\circ(1+a_\circ^2)/2\langle\gamma\rangle_\circ^2$ of the spontaneous emission. In fact, from (2a), (21) and (25), it follows that the radiation phase of the plane wave is $k_r(z-ct) - \delta z_2$; from (20b) and since

$$\delta = \frac{1}{\rho}\frac{\langle\gamma\rangle_\circ - \gamma_R}{\langle\gamma\rangle_\circ} \simeq \frac{1}{2\rho}\frac{k_S - k_r}{k_S} \qquad (33)$$

(assuming $\langle\gamma\rangle_\circ \simeq \gamma_R$), then $\delta z_2 = (k_r - k_S)(z-ct)$, so that the radiation phase is $k_S(z-ct)$.

Hence we conclude that in the slippage region the electrons emit *spontaneous* radiation due only to their wiggling motion in the undulator.

4. THE SUPERRADIANT REGIME

As stated elsewhere in this volume [4] on the base of numerical analysis, the Superradiant regime take place in the short–bunch limit $K \geq 1$. From (26), where $\bar{z}$ is the gain G, we have:

$$K = \frac{S}{G} \qquad (34)$$

K measures the slippage in one gain length (when $G=1$ or equivalently $z = \lambda_\circ/4\pi\rho$) and it does not depend on the undulator length.

Now we can obtain an analytical understanding of the Superradiant regime.

As we have seen, the Steady–state region in the bunch exists only for $S<1$. Now, in the High–Gain regime ($G>1$), if we have a short bunch ($K \geq 1$) the steady-state regime can not exist, since necessarily $S = GK > 1$: the steady–state solution (17) is observable only in the Small–Gain regime and it disappears when the electron pulse enters the High–Gain regime, so that only the solution of the 'slippage region' A_{SR} (25) is observable.

We now demonstrate that a short bunch in a High Gain FEL implies radiation intensity scaling as n_e^2, where n_e is the electron density.

For $K \geq 1$, in the linear High–Gain regime, the field and the bunching parameter are respectively (25) and (28) or equivalently (29a) and (29b). From (21) and (23) it follows that $z_2/z_1 = \bar{z}/z_1 - 1 > S - 1$. Hence for $S \gg 1$ (strong slippage) then $z_1 \ll z_2$ and we can neglect the terms dependent on z_1/z_2 in (29a,b), so that:

$$A_{SR} \simeq \frac{b_\circ}{2\pi} \frac{z_1}{y^{2/3}} e^{i\delta z_1} \oint_{C_\circ} d\xi \xi \exp\left\{-iy^{2/3}\Phi(\xi)\right\} \qquad (35)$$

$$b_{SR} \simeq \frac{b_\circ}{2\pi i} e^{i\delta z_1} \oint_{C_\circ} \frac{d\xi}{\xi} \exp\left\{-iy^{2/3}\Phi(\xi)\right\} \qquad (36)$$

These yield esplicitly:

$$A_{SR} \simeq b_\circ z_1 e^{i\delta z_1} \sum_{n=0}^{\infty} \frac{i^n y^{2n}}{(n+1)!2n!} \qquad (37)$$

$$b_{SR} \simeq b_\circ e^{i\delta z_1} \sum_{n=0}^{\infty} \frac{i^n y^{2n}}{n!2n!} \qquad (38)$$

(we note that (37,38) corresponds to consider only the terms $m = 0$ in the sums of the exact espressions (30,31)).

We observe that in the limit $z_1 \ll z_2$, $A_{SR} \exp(-i\delta z_1)/z_1$ and $b_{SR} \exp(-i\delta z_1)$ are functions of y only. If we assume in the exact propagation equations (1) $\theta \equiv \theta_1(y) - \delta z_1$, $p \equiv \sqrt{z_1} p_1(y)$ and $A \equiv z_1 \exp(i\delta z_1) A_1(y)$, we derive the following one variable equations:

$$\frac{d\theta_1}{dy} = p_1 \qquad (39a)$$

$$\frac{dp_1}{dy} = -\left(A_1 e^{i\theta_1} + c.c.\right) \qquad (39b)$$

$$\frac{y}{2} \frac{dA_1}{dy} + A_1 = \langle e^{-i\theta_1} \rangle \equiv b_1 \qquad (39c)$$

It is possible to demonstrate that the solution of the *linearized* equations (39) is exactly $A_{SR} \exp(-i\delta z_1)/z_1$ and $b_{SR} \exp(-i\delta z_1)$ from (37,38).

It is important to note that eqs.(39) do not depend on z_1 and δ; hence the Superradiant field $|A_{SR}| = z_1|A_1(y)|$ scales as z_1. As in the leading edge of the electron bunch $z_1 = 1/K$, then $|A_{SR}|^2 \propto 1/K^2 \propto \rho^2$; from (2), it follows that the emitted intensity is proportional to n_e^2, instead of the $n_e^{4/3}$ dependence in the Steady-State regime.

Setting $x = y^2$, (39c) becomes $\frac{d(xA_1)}{dx} = b_1$ and, after integration:

$$A_1 = \frac{1}{x}\int_0^x dx' b_1(x') \equiv \bar{b}_1 \tag{40}$$

where $\bar{b}_1$ is the average on x of the bunching parameter; in the head of the electron pulse (i.e. for $z_1 = 1/K$) it follows that:

$$A_{SR} = \frac{\bar{b}_{SR}}{K} \simeq \frac{b_{SR}}{K} \tag{41}$$

and we obtain so the adiabatic relation of Ref. [5].

The importance of (41) is better understood if we consider the equation for the mean electron energy, from (39b):

$$\frac{d\langle p_1 \rangle}{dy} = -\left(A_1 \langle e^{i\theta_1}\rangle + c.c.\right) = -\left(\bar{b}_1 b_1^* + c.c.\right) \simeq -2|b_1|^2 < 0 \tag{42}$$

Hence in the Superradiant regime, the electron energy is continuosly decreasing: this occurs because the emitted radiation escapes from the bunch before to experience reabsorbtion by the electrons.

Following Ref. [7], we can derive an asymptotic expression for A_{SR} and b_{SR}, doing a stationary-phase evaluation of the contour integral (35,36) at large values of y. There are three points of stationary phase, namely the roots of the eq. $\Phi'(\xi) = 0$, corrisponding to an amplifying, an oscillatory and a decaying exponential. By keeping only the amplifying exponential, we get:

$$A_{SR} \simeq \frac{b_\circ}{\sqrt{3\pi}} \frac{z_1}{y} \exp\left[3e^{i\pi/6}\left(\frac{y}{2}\right)^{2/3} + i\delta z_1 - i\pi/4\right] \tag{43}$$

$$b_{SR} \simeq \frac{1}{z_1}\left(\frac{y}{2}\right)^{2/3} e^{i\pi/6} A_{SR} \tag{44}$$

If we observe the field amplitude in the head of the electron pulse (for $z_1 = 1/K$), then $y = (\bar{z} - 1/K)/\sqrt{K}$ and the absolute values of (43) and (44) are:

$$|A_{SR}| \simeq \frac{b_\circ}{\sqrt{3\pi}K} \frac{\sqrt{K}}{\bar{z} - 1/K} \exp\left[3\frac{\sqrt{3}}{2}\left(\frac{\bar{z} - 1/K}{2\sqrt{K}}\right)^{2/3}\right] \tag{45}$$

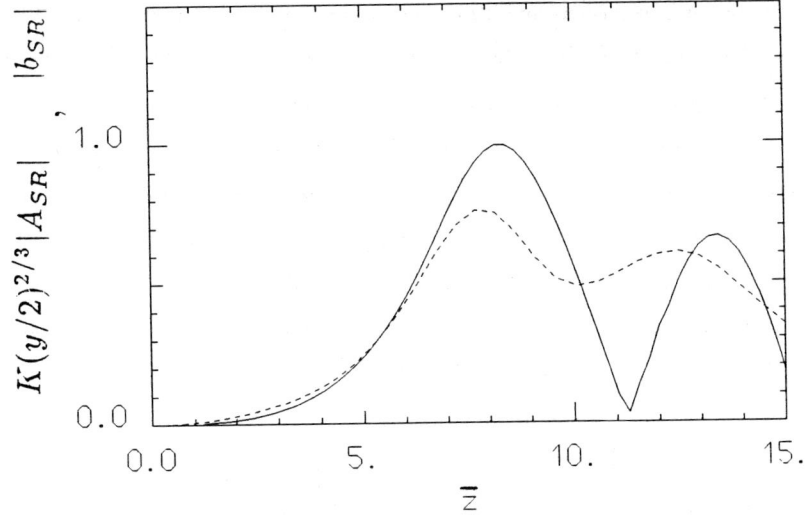

FIGURE 3

$K(y/2)^{2/3}|A_{SR}|$ (solid line) and $|b_{SR}|$ (dashed line) vs. $\bar{z}$, for $y = (\bar{z} - 1/K)/\sqrt{K}$, $K = 1$, $\delta = 0$, $b_\circ = 0$, $A_\circ = \sqrt{2} \times 10^{-2}$.

$$|b_{SR}| \simeq K \left(\frac{\bar{z} - 1/K}{2\sqrt{K}} \right)^{2/3} |A_{SR}| \qquad (46)$$

Hence for $\bar{z} \gg \sqrt{K}$, the field amplitude and the bunching parameter grow exponentially, with a rate $1/\sqrt{K}$, as derived in Ref. [5] from the dissipative model; however, the exponent is proportional to $\bar{z}^{2/3}$ [7]. The validity of (46) in linear regime is shown in Fig.3, where the numerical values (from the exact equations (1), integrated as discussed in Ref. [6] and also in this volume) of $K(y/2)^{2/3}|A_{SR}|$ (solid line) and $|b_{SR}|$ (dashed line) are plotted as function of $\bar{z}$, for $K = 1$, $\delta = 0$, $b_\circ = 0$, $A_\circ = \sqrt{2} \times 10^{-2}$ and $y = (\bar{z} - 1/K)/\sqrt{K}$. We see as (46) is in agreement with the numerical solution in the linear stage of the emission, before the saturation occurs.

5. CONCLUSIONS

We have described analytically the linear regime of the pulse propagation in a high-gain FEL.

We have demontrated the existence of two regions of the electron bunch, where the field evolves in two completely different ways. Whereas in the region of the leading edge the slippage has no effect and a steady-state regime occurs, in the region of the trailing edge the evolution is dominated by the slippage and is completely different from the steady-state. In the short-bunch limit $K \geq 1$ the steady-state exponential gain regime cannot exist; however a Superradiant regime with peak power $\propto n_e^2$ occurs.

The analytical expression and the scaling as n_e^2 of the Superradiant solution are obtained and an asymptotic evaluation of the exponential growth of the Superradiant field is derived. The Superradiant solution turns out to be in a qualitative agreement with the results of the simple damping model of Ref. [5].

Finally, we observe that also in the long-pulse limit ($K \ll 1$), (where the steady-state behaviour is dominant for $S \ll 1$ [4]), there is always a trailing region of the electron pulse in which the FEL dynamics is Superradiant. In this region the slippage effects can give rise, as demonstrated in Ref. [6], to a spiking emission. This unspected powerful emission will be the object of our further investigation.

REFERENCES

1) R. Bonifacio, C. Pellegrini and L. Narducci, Optics Comm. 50 (1984) 373.
2) R. Bonifacio, F. Casagrande and C. Pellegrini, Optics Comm. 61 (1987) 55.
3) T.J. Orzeckowsky, R.B. Anderson, J.C. Clark, W.M. Fawley, D. Prosnitz, E.T. Scharlemann, S. Yarema, D. Hopkins, A.C. Paul, A.M. Sessler, Phys. Rew. Lett. 54 (1985) 889.
4) R. Bonifacio, F. Casagrande, G. Cerchioni, L.De Salvo and P. Pierini, this volume.
5) R. Bonifacio and F. Casagrande, Nucl. Instr. and Meth. in Phys. Res. A239 (1985) 36.
6) R. Bonifacio and B.W.J. Mc Neil, Nucl. Instr. and Meth. in Phys. Res. A272 (1988) 280.
7) G. Moore, Nucl. Instr. and Meth.in Phy. Res. A239 (1985) 19.

THREE-DIMENSIONAL EFFECTS BY BEAT WAVE EXCITATION IN MAGNETOACTIVE PLASMAS

F. Esposito[1], R. Fedele[1,2], G. Miano[3] and V.G. Vaccaro[1,2]

[1] Dipartimento di Scienze Fisiche, Università di Napoli, Italy
[2] I.N.F.N. Sezione di Napoli, Italy
[3] Dipartimento di Ingegneria Elettrica, Università di Napoli. Italy

ABSTRACT
The excitation of plasma waves in magnetoactive plasmas by the beating between two circularly polarized external high-frequency electromagnetic waves is analyzed in the linear fluid theory by taking into account the radial profile of the amplitudes. A low frequency resonance is found which depends, other than on the plasma density and the radial profile, also on the electron cyclotron frequency. This circumstance allows to tune the system with a given plasma density, a fixed beat wave frequency and a given radial profile by adjusting the intensity of the external magnetic field. In such a way the resonance condition can be easily attained. The relevance of this result concerns plasma accelerator physics as well as the generation of coherent radiation by means of plasma wigglers.

1. INTRODUCTION

The study of three-dimensional effects in the plasma beat wave (PBW) excitation is important for ultra-high gradient particles acceleration [1] (100 GeV/m) as well as strong beam focusing [2] and for the generation of coherent radiation with small wavelengths [3] (plasma wigglers).

In the orginal formulation of the PBW excitation scheme [4] two collinear e.m. waves (pumps) with very near frequencies ω_1 and ω_2 (with $\omega_1, \omega_2 >> \omega_p$) are launched in a plasma. They create a beat-wave which generates a ponderomotive force [5] along the direction of propagation in such a way to produce a charge separation between ions and electrons. If the difference $\omega_1 - \omega_2$ is equal to ω_p (the plasma frequency), the ponderomotive effect is resonant with the plasma and it drives the linear growth of the amplitude plasma wave (0.1-0.01 times $\frac{m\omega_p c}{e}$) up to the relativistic saturation time, because of the frequency mismatch between the driver frequency $(\omega_1 - \omega_2)$ and the effective plasma frequency (ω_p/γ) due to the relativistic electron mass variation.

When a spatial profile for the e.m. wave amplitudes in the transverse plane is taken into account, there is also a transverse ponderomotive force which generates transverse plasma oscillations.

The three-dimensional effects in unmagnetized cold plasmas have been recently studied for a radial gaussian pump profile and applied to the PBW accelerator [6].

The three-dimensional effects in the PBW excitation have been studied in the microwave open resonator [7] and suggested for a relatively easier (with respect to laser regime) experimental set-up to produce and study large amplitude waves in plasmas [8].

In this paper we are concerned with three-dimensional effects in the PBW excitation when a constant, uniform magnetic field $\mathbf{B}_o$ is present in the plasma, along the direction of propagation of the pumps. Thus the expression for the ponderomotive force must be changed [9] and in general it is not possible to take the electrostatic approximation ($\nabla \times \mathbf{E} = 0$). We find a low-frequency resonance ($\sim \omega_p$), depending both on the pump profile width and on external magnetic field. This circumstance allows to tune a system with a given plasma density, a fixed beat wave frequency and a given radial profile by adjusting the intensity of the external magnetic field. In such a way the resonance condition can be easily attained. The relevance of this result concerns plasma accelerators as well as the generation of coherent radiation (for example Free Electron Laser) by means of the plasma wigglers.

In section 2 we give the general formulation of the problem, while in section 3 we find the resonance frequency for a particular pump radial profile. Finally, in section 4, we discuss the results and remark their importance for the plasma wigglers.

2. LINEAR THEORY MODEL

Our model consists of a homogeneous and cold plasma with infinitely massive ions, in the presence of an external uniform magnetic field $\mathbf{B}_o$. Ignoring nonlinear effects (relativistic mass variation, convective effects and nonlinear current effects) the linearized model equations of our interest are:

$$m_o \frac{\partial \mathbf{v}}{\partial t} = -e\mathbf{E} - e\frac{\mathbf{v}}{c} \times \mathbf{B}_o + \mathbf{F}_{NL} , \qquad (1)$$

$$\nabla \times \mathbf{E} = -\frac{1}{c}\frac{\partial \mathbf{B}}{\partial t} \qquad (2)$$

$$\nabla \times \mathbf{B} = \frac{1}{c}\frac{\partial \mathbf{E}}{\partial t} - \frac{4\pi e n_o \mathbf{v}}{c} \qquad (3)$$

where $\mathbf{v}$, $\mathbf{E}$ and $\mathbf{B}$ are the first-order electron velocity, electric field and magnetic field in the perturbed plasma, while n_o is the unperturbed plasma number density and $\mathbf{F}_{NL}$ is the ponderomotive force on the electrons due to the beat of two transverse electromagnetic waves of frequency ω_1 and ω_2 and wave numbers k_1 and k_2. Those high frequency electromagnetic waves have transverse spatial variations in order to take into account the finite transverse size of the pumps; therefore the ponderomotive force $\mathbf{F}_{NL}$ has both transverse and longitudinal components. The perturbed plasma density n is calculated from the linearized continuity equation

$$\frac{\partial n}{\partial t} + n_o \nabla \cdot \mathbf{v} = 0 \qquad (4)$$

once the first order velocity $\mathbf{v}$ is found.

We assume a transverse pump profile with axial symmetry so that we refer to a cylindrical coordinate system (r, ϕ, z) with z axis along the direction of the external magnetic field.

We consider the case of the beat propagation parallel to the magnetic field. For two circularly polarized pump waves the ponderomotive force $\mathbf{F}_{NL}$ has the following expression [9]:

$$\mathbf{F}_{NL} = F_o \left[-\Phi'(r) \cos(kz - \omega t)\hat{\mathbf{e}}_r + \frac{\omega_o \pm \omega_c}{\omega_o} k\Phi(r) \cdot \sin(kz - \omega t)\hat{\mathbf{e}}_z \right] \qquad (5)$$

where

$$\omega \equiv \omega_1 - \omega_2,\, k \equiv k_1 - k_2,\, F_o \equiv \frac{1}{2}\frac{e^2 E_{\text{pump}}^2}{m_o}/(\omega_o \pm \omega_c)^2,\, \omega_o = \frac{\omega_1 + \omega_2}{2} \qquad (6)$$

and $\Phi = \Phi(r)$ describes the radial profile of the beat; E_{pump} is the electric field amplitude of the pumps. The sign + and − indicate whether the waves are right or left polarized. Thus the expression of the ponderomotive force can be rewritten in the following form:

$$\mathbf{F}_{NL} = e\,\text{Re}\left\{\mathbf{E}_o(r)e^{i\omega(t - z/\beta_{ph}c)}\right\}\,, \qquad (7)$$

where

$$\mathbf{E}_o(r) = \frac{F_o}{e}\left[-\Phi'(r)\hat{\mathbf{e}}_r + ik\Phi(r)\hat{\mathbf{e}}_z\right]\,, \qquad (8)$$

$$\beta_{ph} = \frac{1}{c}\frac{\omega}{k} = \frac{1}{c}\frac{\omega_1 - \omega_2}{k_1 - k_2} \qquad (9)$$

Our aim is to study the sinusoidal regime behaviours and resonance conditions; then we assume a periodic time variation for the electric field $\mathbf{E}$, the magnetic field $\mathbf{B}$ and the velocity $\mathbf{v}$ in the form:

$$\mathbf{E} = \text{Re}\{\tilde{\mathbf{E}}e^{-j\omega t}\}\,, \qquad (10)$$

$$\mathbf{B} = \text{Re}\{\tilde{\mathbf{B}}e^{-j\omega t}\}\,, \qquad (11)$$

$$\mathbf{v} = \text{Re}\{\tilde{\mathbf{v}}e^{-j\omega t}\}\,, \qquad (12)$$

Inserting (4) and (10)-(12) into the model equations (1)-(3), we easily obtain:

$$\nabla \times \tilde{\mathbf{E}} = -\frac{i\omega}{c}\tilde{\mathbf{B}}\,, \qquad (13)$$

$$\nabla \times \tilde{\mathbf{B}} = \frac{i\omega}{c}\underline{\underline{\varepsilon}} \cdot \tilde{\mathbf{E}} - \frac{i\omega}{e}\underline{\underline{\gamma}} \cdot \tilde{\mathbf{E}}_o e^{-i\omega z/\beta_{ph}c}\,, \qquad (14)$$

where $\underline{\underline{\varepsilon}}$ is the well known magnetized plasma dielectric dyad represented by

$$\underline{\underline{\varepsilon}} = \begin{pmatrix} \varepsilon_1 & i\varepsilon_2 & 0 \\ -i\varepsilon_2 & \varepsilon_1 & 0 \\ 0 & 0 & \varepsilon_3 \end{pmatrix} , \qquad (15)$$

and

$$\varepsilon_1 = 1 + \frac{\omega_p^2}{\omega_c^2 - \omega^2} , \quad \varepsilon_2 = \frac{\omega_c}{\omega} \frac{\omega_p^2}{\omega_c^2 - \omega^2} , \quad \varepsilon_3 = 1 - \frac{\omega_p^2}{\omega^2} , \qquad (16)$$

$$\underline{\underline{\gamma}} = \underline{\underline{\varepsilon}} - \underline{\underline{I}} , \quad \omega_p^2 = \frac{4\pi e^2 n_o}{m_o} , \quad \omega_c = \frac{eB_o}{mc} \qquad (17)$$

When $B_o \to 0$, ($\omega_c \to 0$, $\varepsilon_1, \varepsilon_3 \to 1 - \frac{\omega_p^2}{\omega^2}$ and $\varepsilon_2 \to 0$) the equations (13) and (14) become:

$$\nabla \times \tilde{\mathbf{E}} = -\frac{i\omega}{c}\tilde{\mathbf{B}} , \qquad (18)$$

$$\nabla \times \tilde{\mathbf{B}} = \frac{i\omega}{c} \left[\varepsilon_3 \tilde{\mathbf{E}} - (\varepsilon_3 - 1)\tilde{\mathbf{E}}_o e^{i\omega z/\beta_{ph}c} \right] , \qquad (19)$$

Since the field $\tilde{\mathbf{E}}(r)e^{i\omega z/\beta_{ph}c}$ is curl-free it follows:

$$\tilde{\mathbf{E}} = \frac{\varepsilon_3 - 1}{\varepsilon_3}\tilde{\mathbf{E}}(r)e^{i\omega z/\beta_{ph}c} \qquad (20)$$

which is the solution of the problem; this is the electric field generated by the beat wave with transverse profile for an unmagnetized plasma. In this case the resonance is obtained when $\omega = \omega_p$ [6].

In order to solve equations (13) and (14) when $B_o \neq 0$ we look for a solution of the following kind

$$\tilde{\mathbf{E}}(r,z) = (\tilde{E}_r(r), \tilde{E}_\phi(r), \tilde{E}_z(r))e^{-i\omega z/\beta_{ph}c} \qquad (21)$$
$$\tilde{\mathbf{B}}(r,z) = (\tilde{B}_r(r), \tilde{B}_\phi(r), \tilde{B}_z(r))e^{-i\omega z/\beta_{ph}c} \qquad (22)$$

and we express the r and z components of the fields in terms of the ϕ-components.

To reach this goal, we project equations (13) and (14) along r and z-directions getting

$$\tilde{E}_r = -i\frac{\varepsilon_2}{\varepsilon_1}\tilde{E}_\phi + \frac{1}{\beta_{ph}\varepsilon_1}\tilde{B}_\phi + (1 - 1/\varepsilon_1)E_{or} , \qquad (23)$$

$$\tilde{B}_r = -\frac{1}{\beta_{ph}}\tilde{E}_\phi , \qquad (24)$$

$$\tilde{E}_z = -i\frac{c}{\omega\varepsilon_3}\frac{1}{r}\frac{d}{dr}(r\tilde{B}_\phi) + (1 - 1/\varepsilon_3)E_{oz} , \qquad (25)$$

$$\tilde{B}_z = i\frac{c}{\omega}\frac{1}{r}\frac{d}{dr}(r\tilde{E}_\phi) , \qquad (26)$$

Projecting equations (13) and (14) along the ϕ-direction and using equations (23)-(26), we finally obtain two coupled equations for $\tilde{E}_\phi$ and $\tilde{B}_\phi$

$$\varepsilon_1 \mathcal{L}(\tilde{B}_\phi) + C_{11}\tilde{B}_\phi + C_{12}\tilde{E}_\phi = S_1 \tag{27}$$

$$\varepsilon_1 \mathcal{L}(\tilde{E}_\phi) + C_{21}\tilde{B}_\phi + C_{22}\tilde{E}_\phi = S_2 \tag{28}$$

where

$$C_{11} = \varepsilon_3 \frac{\omega^2}{c^2}(\varepsilon_1 - 1/\beta_{ph}^2) \; , \quad C_{12} = i\frac{\omega^2}{c^2}\frac{\varepsilon_2 \varepsilon_3}{\beta_{ph}} \tag{29}$$

$$C_{21} = -i\frac{\omega^2}{c^2}\frac{\varepsilon_2}{\beta_{ph}} \; , \quad C_{22} = \frac{\omega^2}{c^2}\left[\varepsilon_1^2 - \varepsilon_2^2 - \frac{\varepsilon_1}{\beta_{ph}^2}\right] \tag{30}$$

and

$$S_1 = \frac{\omega^2}{c}\frac{\varepsilon_3(\varepsilon_1 - 1)}{\beta_{ph}}E_{or} - i\frac{\omega}{c}\varepsilon_1(\varepsilon_3 - 1)\frac{d}{dr}E_{oz} \; , \tag{31}$$

$$S_2 = -i\frac{\omega^2}{c^2}\varepsilon_2 E_{or} \tag{32}$$

The differential operator $\mathcal{L}$, defined as

$$\mathcal{L}(f) = \frac{d}{dr}\left[\frac{1}{r}\frac{d}{dr}(rf)\right] \; , \tag{33}$$

it is the well known Bessel operator.

3. THE RESONANCE FREQUENCY

In this section we investigate the existence of some resonance from equations (27) and (28) when the radial profile for the ponderomotive potential is assumed proportional to $Z_o(q_o r)$, where Z_o is a zero-order Bessel function with q_o imaginary or real. According to the definition of S_1 and S_2, we thus get (see equations (27) and (28)):

$$S_1 = S_{o1}\mathcal{E}_o Z_1(q_o z) \tag{34}$$

$$S_2 = S_{o2}\mathcal{E}_o Z_1(q_o z) \tag{35}$$

where

$$S_{01} = \frac{\omega_c}{c}\frac{\omega}{c}\frac{\varepsilon_2}{\beta_{ph}} \mp \frac{\omega_c}{\omega}\frac{\omega^2}{c^2}\frac{\varepsilon_3 - 1}{\beta_{ph}}\varepsilon_1 \; , \tag{36}$$

$$S_{02} = -i\frac{\omega^2}{c^2}\varepsilon_2 \; , \tag{37}$$

$$\mathcal{E}_o = \frac{e}{m_o}\frac{q_o}{(\omega_o \pm \omega_c)^2}E_{pump}^2 \; , \tag{38}$$

and $Z_1(q_o r)$ is the first-order Bessel function.

Introducing the vectors

$$\mathbf{A} = \begin{pmatrix} \tilde{B}_\phi \\ \tilde{E}_\phi \end{pmatrix} \quad \mathbf{S} = \begin{pmatrix} S_1 \\ S_2 \end{pmatrix}, \tag{39}$$

and the matrix

$$\underline{\underline{C}} = \begin{pmatrix} C_{11} & C_{12} \\ C_{21} & C_{22} \end{pmatrix} \tag{40}$$

it is convenient to write the equations for $\tilde{E}_\phi$ and $\tilde{B}_\phi$ in the following matrix form:

$$\varepsilon_1 \mathcal{L}(\mathbf{A}) + \underline{\underline{C}} \cdot \mathbf{A} = \mathbf{S} \tag{41}$$

We look for a solution like

$$\mathbf{A} = \mathbf{A}_o \mathcal{E}_o Z_1(q_o r) \tag{42}$$

So, by substituting (42) in (41) and by using the definition of C_{ij} given by (29)-(30) we solve for $\mathbf{A}_o$:

$$\mathbf{A}_o = \frac{\underline{\underline{G}} \cdot \mathbf{S}}{\varepsilon_1 f(\omega, q_o)} \tag{43}$$

where

$$\underline{\underline{G}} = \begin{pmatrix} \left[\frac{\omega^2}{c^2}(1+\varepsilon_3) - (q_o^2 + \frac{\omega^2}{c^2}\beta_{ph}^{-2})\right]\varepsilon_1 - \frac{\omega^2}{c^2}\varepsilon_3 & -i\frac{\omega^2}{c^2}\frac{\varepsilon_2 \varepsilon_3}{\beta_{ph}} \\ i\frac{\omega^2}{c^2}\frac{\varepsilon_2}{\beta_{ph}} & \left[\frac{\omega^2}{c^2}\varepsilon_3 - q_o^2\right] - \frac{\omega^2}{c^2}\frac{\varepsilon_3}{\beta_{ph}^2} \end{pmatrix} \tag{44}$$

and

$$\mathbf{S}_o = \begin{pmatrix} S_{o1} \\ S_{o2} \end{pmatrix} \tag{45}$$

$$f(\omega, q_o) = \varepsilon_1 q_o^4 - T q_o^2 + \Delta_o \tag{46}$$

$$\Delta_o = \frac{1}{\varepsilon_1} \det(\underline{\underline{C}}) = \frac{\omega^4}{c^4}\varepsilon_3[(1+\eta^2)^2 - 2(1+\eta^2)\varepsilon_1 + \varepsilon_1\varepsilon_3 + (\varepsilon_1 - \varepsilon_3)] \tag{47}$$

$$T = \text{tr}(\underline{\underline{C}}) = -\frac{\omega^2}{c^2}[\eta^2(\varepsilon_1 + \varepsilon_3) - 2\varepsilon_3(\varepsilon_1 - 1)] \tag{48}$$

$$\eta^2 = \beta_{ph}^{-2} - 1 \tag{49}$$

It is well known that, when the ponderomotive force is an ideal plane wave (no radial amplitude profile) travelling along z, three natural modes can be excited; one is longitudinal (Langmuir oscillations), corresponding to $\varepsilon_3 = 0$ (i.e. $\omega = \omega_p$), and the other ones are transverse circularly polarized electromagnetic corresponding to $\frac{k^2 c^2}{\omega^2} = \varepsilon_1 \pm \varepsilon_2$

In the low-frequency plasma response, when the ponderomotive force has a radial profile, more natural modes can be excited and we have to start our investigation from (41)-(43) and (23)-(26). The explicit presence of ε_1 and ε_3 in the denominator of these equations may suggest the existence of resonance for $\varepsilon_1 = 0$ and $\varepsilon_3 = 0$, which would correspond to the resonance $\omega = (\omega_c^2 + \omega_p^2)^{1/2} \equiv \omega_H$ (upper hybrid frequency) and $\omega = \omega_p$, respectively. Nevertheless, it can be seen that any resonance for $\varepsilon_1 = 0$ and $\varepsilon_3 = 0$ doesn't exist, because also the numerator of (43) vanishes, but there is a low-frequency resonance $(\omega \sim \omega_p, \omega_c)$.

According to (49) the values for η^2 are given by $\eta^2 \simeq \frac{c^2}{v_g^2} - 1$, where

$$v_g = \frac{d\omega}{dk} \simeq \frac{\omega_1 - \omega_2}{k_1 - k_2} = \frac{\omega}{k} \qquad (50)$$

The resonances are found putting

$$f(\omega, q_o) = 0 \qquad (51)$$

Consequently, in the limit for $\eta^2 \to 0$ ($v_g \to c$) the (51) gives the solution:

$$\omega = \left[\omega_p^2 + \omega_c^2 \frac{\Omega_0^4}{(\omega_p^2 \pm \Omega_0^2)^2}\right]^{1/2} \equiv \omega_s \qquad (52)$$

where $\Omega_o = c|q_o|$. The signs $+$ and $-$ correspond to q_o real or imaginary, respectively. Since $|q_o|^{-1}$ is of the order of few pump wavelengths, $\Omega_o \sim \omega_p$, thus by choosing the sign $+$ ω_s corrisponds to a low-frequency resonant mode $(\omega_s \leq \omega_H)$.

The excitation of the resonant mode implies that the resonance condition (52) occurs when the group velocity v_g of the beat-wave is equal to the phase velocity v_{ph} of the low-frequency mode (see (50)).

In the limit $\Omega_o \to \infty$, this resonance reduces to $\varepsilon_1 = 0$ (upper hybrid resonance), while for $B_o = 0$ it reduces to $\varepsilon_3 = 0$. Consequently, every resonant solution for the fields $(\omega = \omega_s)$ has the same linear growth rate

$$\gamma = \omega_s^{-1} \qquad (53)$$

For $B_o = 0$ ($\varepsilon_2 = \omega_c = 0$) the resonance becomes $\varepsilon_3 = 0$ and B_ϕ, E_ϕ, B_r and B_z vanish. Physically that means the equations of our model are in agreement with the electrostatic approximation ($\nabla \times \mathbf{E} = 0$); as previously seen, the only possible fields are those generated by the charge separation between ions and electrons. In fact, with no external magnetic field, the ponderomotive force acts only in the radial and longitudinal direction while for $B_o \neq 0$ the transverse oscillations are coupled with $\mathbf{B_o}$, so in general $\nabla \times \mathbf{E} \neq 0$.

4. COMMENTS AND CONCLUSIONS

Recently the PWB excitation has been proposed for plasma accelerator physics as well as the plasma wigglers (for example FEL). In the plasma accelerator physics the large amplitude electric fields can be obtained. In plasma wigglers both high fields and very small plasma wavelength can be obtained.

According to ref. [3] the frequency ω_R of the coherent radiation emitted in the FEL with plasma wiggler is given by:

$$\omega_R = \alpha \gamma_o^2 (\omega_1 - \omega_2)$$

where $\alpha = 2$ or 4, depending on the scheme used.

We remark the system would be able to resonate at the beat-wave frequency $\omega_1 - \omega_2$. The unavoidable density fluctuations represent an experimental difficulty to attain the resonance condition. The result found in the previous section suggests the possibility to make a fine tune by varying the external magnetic field B_o. In order to get a significant dependence for ω_s on B_o, ω_c should be of the order of ω_p, assuming $\Omega_o \sim \omega_p \sim \omega_1 - \omega_2$. Thus $\omega_1 - \omega_2 \sim \omega_c$. Consequently, assuming $B_o = 10^5$ gauss ($f_c = \frac{2\pi}{\omega_c} = 280$ GHz) the radiation wavelength is:

$$\lambda_R \sim \frac{1 \text{ mm}}{\alpha \gamma_o^2}$$

This result is competing with the conventional wigglers.

ACKNOWLEDGEMENTS

The authors thank Dr. P.K. Shukla of the University of Bochum (FRG) for stimulating this work and for the useful discussions.

REFERENCES

1) T. Tajima and J.M. Dawson, Phys. Rev. Lett. <u>43</u>, (1979) 267.
2) F.F. Chen, "Laser Accelerators", PPG-1107 U.C.L.A. report, Sept. 1987, to be published in the Handbook of Plasma Physics, ed. by R.Z. Sagdeev and M.N. Rosenbluth, Vol.4, "Physics of Laser Plasma", ed. by A. Rubenchik and S. Witkowski, North-Holland, Amsterdam.
3) C. Joshi, T. Katsouleas, J.M. Dawson, Y.T. Yan and J.M. Slater, IEEE Journal of Quantum Electronics, vol.QE-23, (1987) 1571.
4) M.N. Rosenbluth and C.S. Liu, Phys. Rev. Lett. <u>29</u>, (1972) 701.
5) L.D. Landau and E.M. Lifshitz, "Electrodynamics of continuous media", Pergamon Press, (1963) 68.
6) R. Fedele, U. de Angelis and T. Katsouleas, Phys. Rev. <u>A</u>33, (1986) 4412.
7) U. de Angelis, R. Fedele, G. Miano and C. Nappi, Plasma Phys. and Contr. Fus. <u>29</u>, (1987) 789.
8) U. de Angelis, L. De Menna, R. Fedele, G. Miano, C. Nappi and V.G. Vaccaro, IEEE Trans. on Plasma Science <u>15</u>, (1987) 179.
9) H. Washimi and V.I. Karpman, Sov. Phys. JETP <u>44</u>, (1976) 528; J. Vaclavick, M.L. Sawley and F. Anderegg, LRP 261/85 Ecole Polytechnique Federale de Lausanne, April 1985.

Propagation of a short RF pulse train in an iris-loaded waveguide

L.Ferrucci, C.Pagani and L.Serafini

Istituto Nazionale di Fisica Nucleare and Universitá di Milano, Milano, Italy

In this paper we present some numerical results on the propagation of RF pulses inside an accelerating structure of the iris loaded, constant impedance, type.
The dispersion curve for a SLAC-type structure (S-band, 2856 MHz) fitting with good approximation the experimental data for the phase shift and the attenuation (within the passband) is given by [1]:

$$-\cosh \Gamma(\omega) = 1 + \left[\frac{\alpha_0}{\sqrt{2}} + \frac{\sqrt{2}}{\omega_c}i(\omega - \omega_0)\right]^2 \quad (1)$$

where:
- $\Gamma(\omega)$ is the complex propagation constant for the fundamental space harmonic in the accelerating pass-band
- ω_0 is the mid-band frequency of the pass-band
- ω_c is the half pass-band width
- α_0 is the attenuation coefficient of the structure at mid-band frequency.

Following the method by J.E. Leiss [1] the electric field $E_q(t)$ at a certain position q along the waveguide, as a function of time, is given by the convolution integral:

$$E_q(t) = \int_0^t E_0(t-\tau)G_q(\tau)d\tau \quad (2)$$

where:
- $E_0(t)$ is an arbitrary input waveform (train of rectangular RF pulses or train of Gaussian-shaped RF pulses etc.)
- $G_q(t)$ is the response, at a position q, to a delta-function input at the point $q = 0$. This response completely determines the pulse transmission inside the waveguide.

For an accelerating structure with a dispersion curve as described, the expression for the function $G_q(t)$ is:

$$G_q(t) = 2q\frac{J_{2q}(\omega_c t)}{t}e^{i(\omega_0 t - \pi q)}e^{-\frac{\alpha_0}{2}\omega_c t} \quad (3)$$

where $2q$ is the position in number of cells.

From the literature we know that this analysis allows to obtain analytical results for the propagation of a semi-infinite RF pulse into a waveguide. Our aim is to study the propagation of a train of short RF pulses with a specific shape; in this case we have numerically integrated eq. (2), in order to be able to look at the pulses propagation through the waveguide.

The main goal is to evaluate the possibility to fill an accelerating structure with a train of short RF pulses (shorter than the filling-time τ_f), instead of a single RF pulse with a lenght of the order of the filling-time.

In fig. 1 a to d and in fig. 2 we present some numerical results on the propagation in a SLAC-type accelerating section of a train of four rectangular RF pulses with unity amplitude, a lenght of 3.5 ns (10 RF cycles) and a spacing of 0.2 μs corresponding to about 1/4 of the structure filling- time .

In particular, fig. 1 gives the normalized amplitude of the electric field at some given position, as a function of time, after the entrance of the first pulse into the waveguide. The strong spreading of the pulses due to the dispersion, is visible. Fig. 2 gives the normalized amplitude and the real part of the electric field, for the same train of pulses as before, at a given time ($\omega_c t = 82\ rad$), corresponding to about 4/5 of the filling time, as a function of the longitudinal position $2q$ in number of cells. From the plot of the real part of the electric field is possible to see that the $2/3\ \pi$ phase relationship between adiacent cells is not always satisfied, expecially at the head-tail region of each pulse.

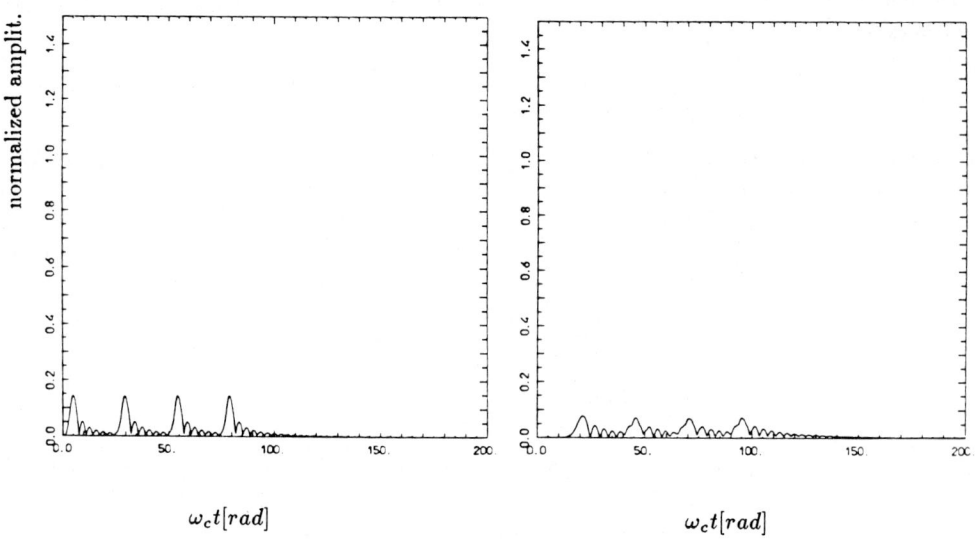

Fig. 1 a and b : normalized amplitude of electric field at 5th cell and at 20th cell as a function of the time for a train of rectangular RF pulses in a SLAC-type structure. (see text for details)

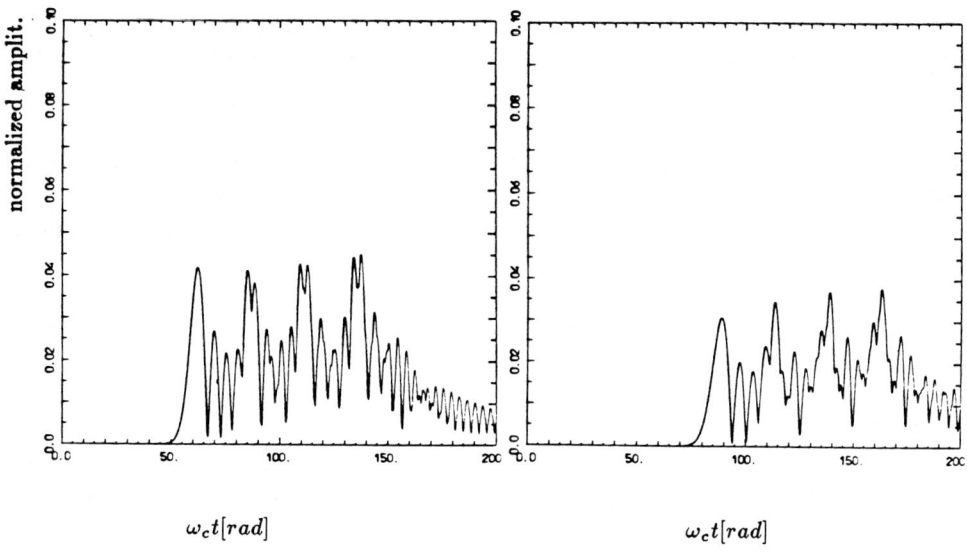

Fig. 1 c and d: normalized amplitude of electric field at 60th cell and at 86th cell as a function of the time for a train of rectangular RF pulses in a SLAC-type structure. (see text for details)

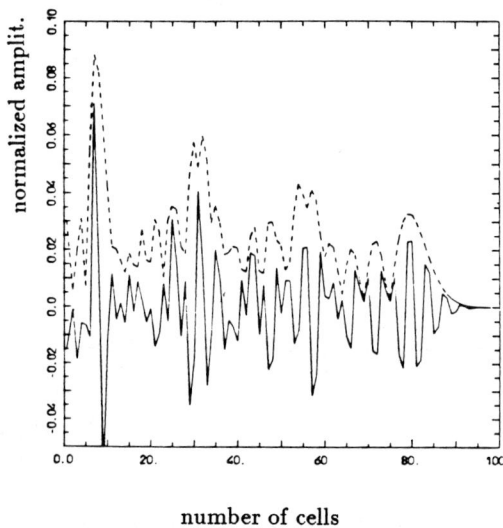

Fig. 2: normalized electric field amplitude (- - -) and real part (——) at a given time $\omega_c t = 82\ rad$, as a function of the longitudinal position in number of cells, for a train of four RF rectangular pulses in a SLAC-type structure. (see text for details)

The question now arising is: what about the energy gain for a relativistic electron passing throughout such a field?
Following again the method in ref. 1 we have numerically integrated the equation for the energy gain $W(t)$ of a fully relativistic electron (v=c) passing through the waveguide, without beam-loading effect, as a function of the injection time, measured from the entrance of the first pulse into the waveguide:

$$W(t) = \int_0^{q_m} E_q(t + q\frac{(\pi + \psi_0)}{\omega_a})dq \qquad (4)$$

where:
- q_m is equal to the half of the total number of cells
- $E_q(t + q\frac{(\pi+\psi_0)}{\omega_a})$ is the electric field seen by the syncronous electron and has the same expression as in eq. (2):

$$E_q(t + q\frac{(\pi + \psi_0)}{\omega_a}) = \int_0^{t+q\frac{(\pi+\psi_0)}{\omega_a}} E_0(t + q\frac{(\pi + \psi_0)}{\omega_a} - \tau)G_q(\tau)d\tau \qquad (5)$$

- $q\frac{(\pi+\psi_0)}{\omega_a}$ is the transit time for the electron to point q. ψ_0 is a phase which depends on the operating mode of the accelerating structure and in this example is equal to $\frac{\pi}{3}$.

In fig. 3 we present, only as a reference, the energy gain for a syncronous electron travelling in a 30 cells SLAC-type structure, filled with a semi-infinite RF pulse of unity-amplitude-per-cell, as a function of the injection time after the turn-on of the RF generator.

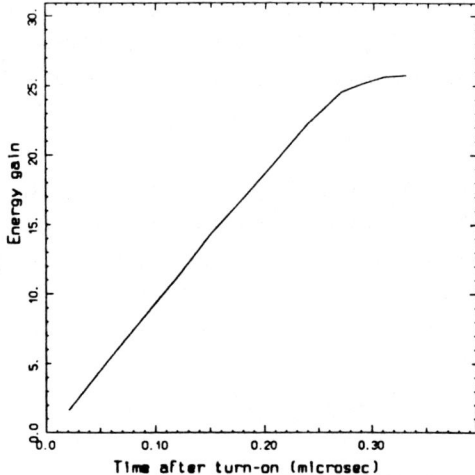

Fig. 3 : energy gain with a semi-infinite RF pulse for a v=c electron travelling in a 30 cells SLAC-type accelerating structure. (see text for details)

The energy gain is in arbitrary units such that if the electric field seen by syncronous particle is equal to 1-per-cell in every cell, the total energy gain is equal to the total number of cells. The plot is in complete agreement with the one published in ref. 1, scaled for a 30 cells SLAC-type structure, and this confirms the validity of our numerical integration method.

Some new results are those in Fig. 4 a to h, that refer to a situation where the 30 cells SLAC-type structure is filled with a train of short RF pulses, much shorter than the filling time.
A train of four rectangular, unity-amplitude-per-cell RF pulses, 10 RF cycles width and with a spacing of 70 nsec. (about 1/4 of the filling time for the structure) is shown in the first four figures. The quantity plotted is the electric field seen by the syncronous electron travelling down to the waveguide, for some given times after the entrance of the first pulse into the structure.
Similar plots are shown in fig. 4 e to h, but for the case of a train of four Gaussian-shaped RF pulses. The Gaussians peak-value has been choosen to have same area as the one of the rectangular pulses train, having fixed the Gaussian-width equal to that one of the rectangular pulses.
The energy gain for the two cases is similar, as can be seen in fig. 5 a and b.

Let us now compare the energy gain for an electron with a RF pulse having a lenght equal to the filling time and with a train of very short RF pulses, to have an idea of how much of the RF energy put into the waveguide with short pulses is useful for acceleration.
Consider first a rectangular unity-amplitude-per-cell RF pulse with a lenght equal to the filling time for a 30 cells SLAC-type structure (300 ns): the energy gain W_{rect} for a v=c electron that enters the waveguide at a time equal to the filling time is: $W_{rect} = 26.2$, measured in the same arbitrary units as previously described. Consider also a train of four rectangular RF pulses with a lenght of 10 RF cycles (3.5 ns), a spacing of 70 ns and an amplitude E_{train} choosen to have the same total area of the rectangular unity-amplitude pulse:

$$E_{train} = \frac{300}{4 \times 3.5} = 21.43$$

The energy gain W_{train} for a v=c electron that enters the waveguide at a time equal to the filling time is: $W_{train} = 23.5$. This value is less than the energy gain with a single pulse by only about 11%, this means that the loss in efficiency due to the dispersion effects for the 30 cells accelerating strucure filled with short pulses is quite low.
The strong difference is between the amount of peak power P_p that the RF generator must supply for the two cases; that is proportional to the square of the amplitude of the RF pulse:

$$\frac{P_{p\ pulse}}{P_{p\ train}} = \frac{1}{(21.43)^2} = \frac{1}{460}$$

The RF energy put into the waveguide is equal to the RF power times the time width of the pulse, so that the ratio between the RF energy in the waveguide for the two cases is:

$$\frac{P_{p\ pulse} \times \tau_{pulse}}{P_{p\ train} \times \tau_{train}} = \frac{1}{460 \times (300/14)} = 0.047$$

A "figure of merit" M_t for the train of pulses can be introduced, that tells us how much is the accelerating efficiency with a certain train of pulses, in respect of a single RF pulse with a length equal to the filling time.

$$M_t = \frac{W_{train}}{P_{p\ train} \times \tau_{train}} \bigg/ \frac{W_{pulse}}{P_{p\ pulse} \times \tau_{pulse}}$$

In our case is: $M_t = 0.042$.

We can conclude that the accelerating efficiency for a train of RF pulses as described is much less than the efficiency with a single RF pulse, with a length of the order of the filling time. The dispersion of the waveguide affect only a little this efficiency and the loss is about equal to the duty cycle of the pulse train. The use of accelerating structure with higher bandwidth, as those described in ref. 2 and in ref. 3, reduces further the loss due to the dispersion.

It is also straightforward that with short RF pulses the accelerating structure must have an higher peak power handling capacity.

References

1) J.E. Leiss, Beam Loading and Transient Behaviour in Travelling Wave Electron Linear Accelerators, in: Linear Accelerators, eds. P. Lapostolle and A. Septier (North-Holland, Amsterdam, 1969), pp. 147-172

2) G. Geshonke, Distorsion of a short RF pulse in travelling-wave accelerating structures, CERN 87-11, ECFA 87/110, 12 October 1987, pp. 181-187.

3) J.P. Boiteux et al., Studies of RF accelerating structures for an electron linear collider, CERN-LEP-RF/87-25 and CLIC NOTE 36, March 1987.

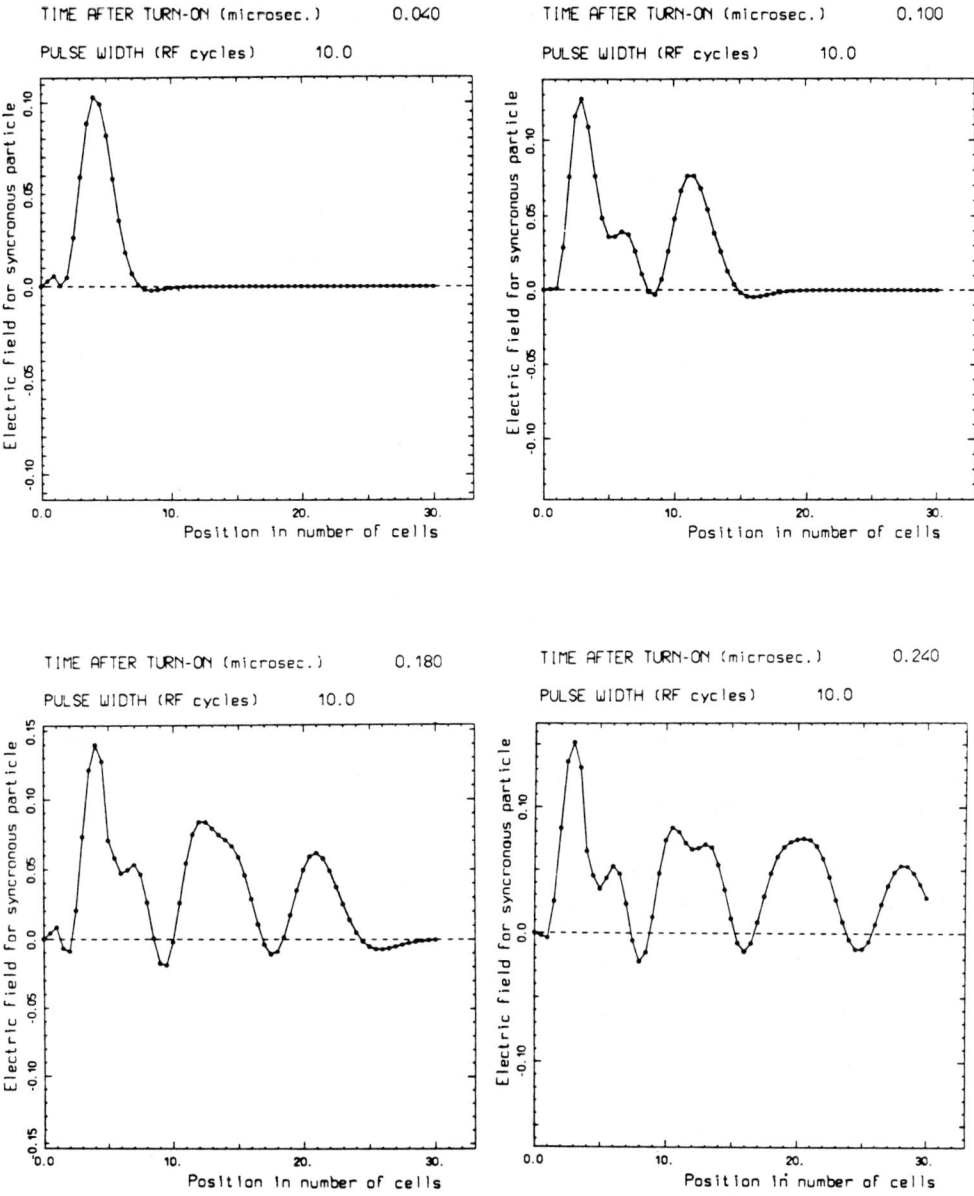

Fig. 4 *a* to *d* : normalized electric field seen by syncronous electron travelling down to the waveguide for a train of four rectangular RF pulses in a SLAC-type accelerating structure for some given times after the entrance of the first pulse into the structure. (see text for details)

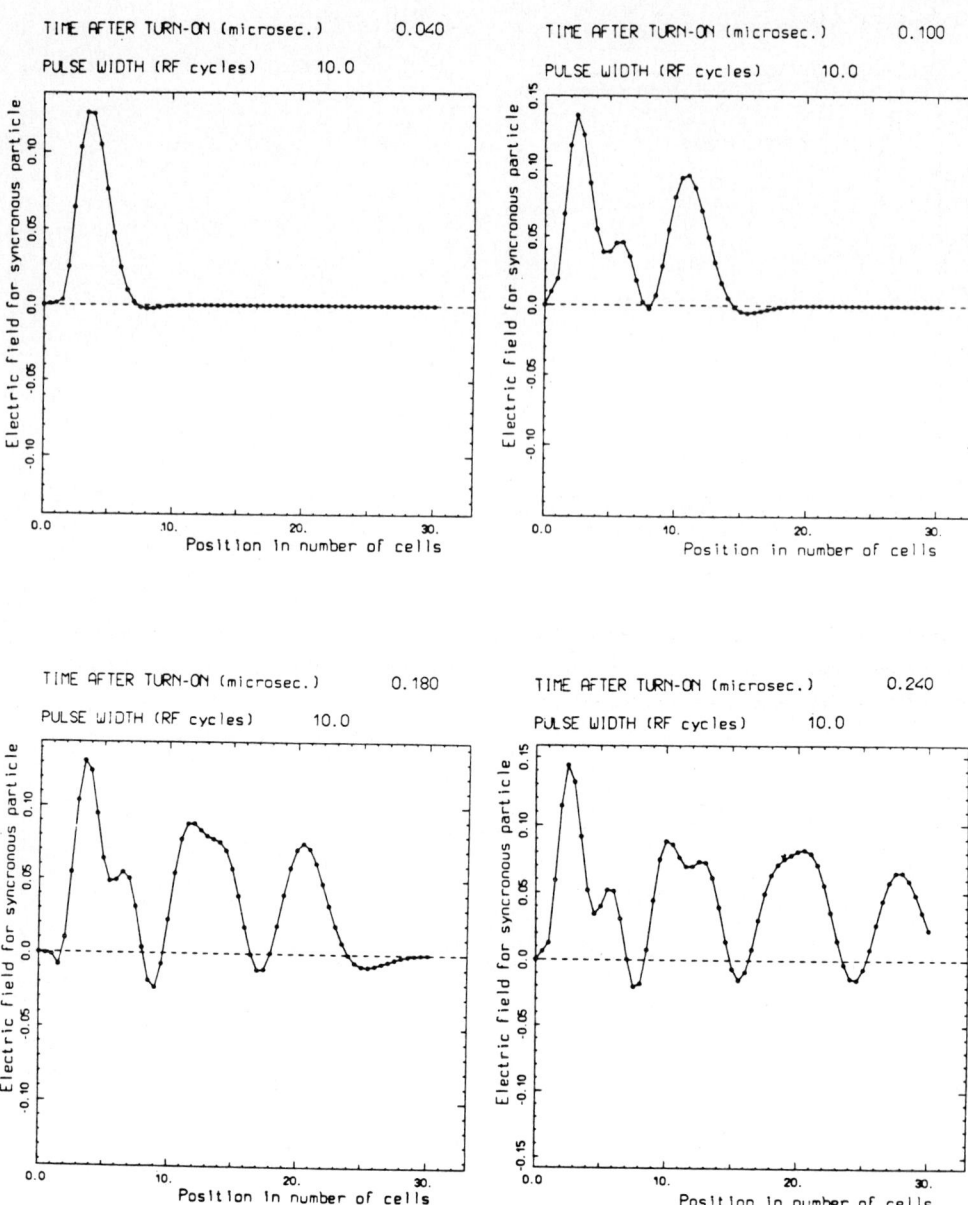

Fig. 4 *e* to *h* : normalized electric field seen by syncronous electron travelling down to the waveguide for a train of four Gaussian RF pulses in a SLAC-type accelerating structure for some given times after the entrance of the first pulse into the structure (see text for details)

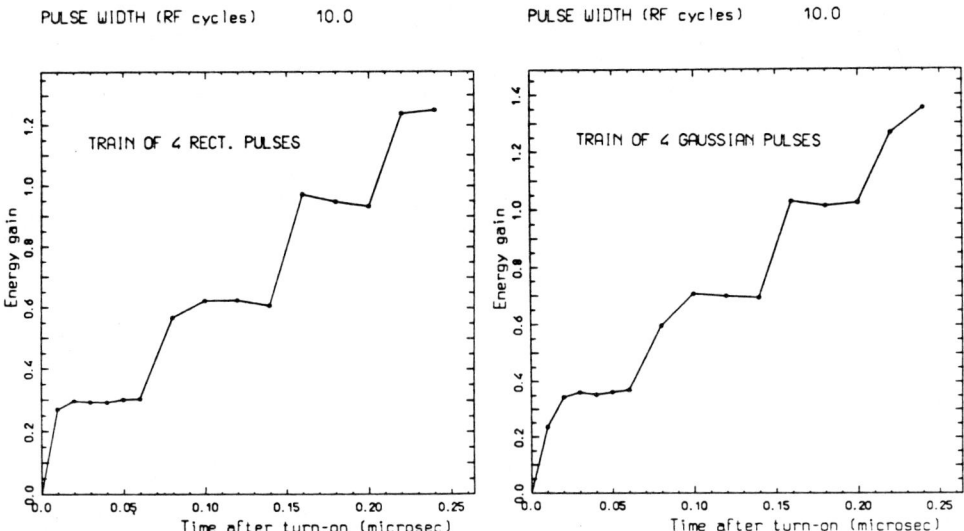

Fig. 5 a and b : energy gain with a train of four rectangular RF pulses and with a train of four Gaussian pulses of same area for a v=c electron travelling in a 30 cells SLAC-type accelerating structure. (see text for details)

NUMERICAL INTEGRATION OF TRANSIENT
PARTICLE AND FIELD EQUATIONS IN AXI-SYMMETRICAL CAVITIES

L. Serafini, C. Pagani
L. Ferrucci, L. Muda and A. Peretti.
Istituto Nazionale di Fisica Nucleare and Universita' di Milano, Milano, Italy

INTRODUCTION

A method for the numerical simulation of the interaction between high charged bunches and axi-symmetric fields inside a cylindrical structure (RF cavity or wave guide) is presented, together with a computer code named ITACA (Integration of Transients in Axi-symmetrical Cavities for Accelerators), which has been written on the basis of this method.

ITACA is an electromagnetic particle-in cell code able to study both the radial and axial motion of a bunch of particles moving through a cavity, and the propagation of the wake-field, excited by the bunch itself, inside the cavity. The code can inject the particles of the bunch into the resonant field of the accelerating mode eventually present inside the cavity. This makes possible to study the dynamics of the bunch in presence of all the relevant forces acting on it: the accelerating field, the space charge forces due to the self-field of the bunch and the wake-field excited by the interaction between the self-field of the bunch and the boundary condition imposed by the cavity surface.

The computer code ITACA can presently handle axi-symmetrical fields and bunches, allowing to study monopole wake fields and their effects on the particles dynamics. The code has been designed in order to represent with accuracy the behavior of the fields and the dynamics of the particles. As an example of the results which can be obtained, we present the integration of the motion of low and medium energy (1÷100 MeV) high charged (.1-1 µC) bunches moving through a LEP 4-cells SC cavity, with and without the resonant field of the accelerating mode.

TRANSIENTS AXI-SYMMETRICAL FIELDS INTEGRATION METHOD

It is well known that axi-symmetrical fields in a cylindrical cavity can be expressed as functions of a scalar potential, which is given by $\Phi_h = r \cdot H_\phi$ for fields which can be expanded in a sum of TM_{0np} modes (i.e. TM-like fields), and is given by $\Phi_e = r \cdot E_\phi$ for field which can be expanded in a sum of TE_{0np} modes (i.e. TE-like fields).

Writing down the wave equations for axi-symmetrical **E** and **H** fields in a system of cylindrical coordinates, it can be seen at a glance that, in case of a cylindrical driving current, having only radial and axial components, the two set of fields (the TM-like, specified by H_ϕ, E_r, E_z and the TE-like, specified by E_ϕ, H_r, H_z) are independent. Moreover, the driving current excites only the TM-like fields. Since a particle moving in a TM_{0np} field experiences only a radial and an axial force (but no azimuthal one), it produces, inside a cavity, a driving current which can couple only to TM-like fields. Then, the interaction between a bunch of particles, being accelerated through an axi-symmetrical cavity, and the field inside the cavity can be fully described solving the wave equation for the $\Phi = r \cdot H_\phi$ potential, as a function of r,z,t, in presence of a driving current $J(r,z,t) = (J_r(r,z,t), J_z(r,z,t))$:

1)
$$\mathcal{L}\Phi - \frac{1}{c^2}\frac{\partial^2 \Phi}{\partial t^2} = r\frac{\partial J_z}{\partial r} - r\frac{\partial J_r}{\partial z}$$

$$\mathcal{L} = \frac{\partial^2}{\partial z^2} + \frac{\partial^2}{\partial r^2} - \frac{1}{r}\frac{\partial}{\partial r}$$

together with the boundary condition $\partial \Phi / \partial n = 0$ on the cavity surface.

This typical hyperbolic equation gives, starting from some initial condition of the field inside the cavity (typically the field of a resonating mode) at t=0, the time evolution of the fields, given that the driving current is known inside the cavity at each instant.

To solve this equation we adopt the standard technique of the FDM, discretizing the fields over a regular rectangular mesh covering all the cavity section in the r-z plane. Using the explicit scheme for the discretization of time, the value of the field at each point of the mesh is given, for the next time-instant, as a function of both the field-values in the same point, at the present time and at the previous one, and the field-values in the neighborly mesh points, at the present time. The numerical stability of this well established technique is assured if the time integration step ($\tau=c t$) is below a threshold given by [2]:

$$\tau < 1/\sqrt{\frac{1}{\epsilon^2} + \frac{1}{h^2}}$$

where ϵ and h are the mesh-steps in r,z respectively.

A special treatment for curved boundary has been developed. At the location of all the mesh points close to the boundary, the field has been expanded in a

Taylor series up to the II order and the boundary condition has been imposed on the intersections between the mesh lines and the boundary. This procedure allows to handle all the special points near to the boundary with the same equation as for the normal ones, making possible to describe the field over a regular mesh with the same accuracy of an irregular one. The total amount of memory needed is reduced: in fact the gain in memory given by the regular mesh overcomes the lost due to the special boundary points handling. We think moreover that a regular mesh becomes recommended for short high charged bunches, because the field contains high order spatial harmonics (of wavelength comparable to the mesh step) which are excited by the bunch and propagate through the cavity. These harmonics, eventually excited in some point of an irregular mesh, couldn't propagate through other regions of the mesh with larger discretization steps.

The algorithm for the field integration has been tested starting the computation at t=0 from the field distribution of the fundamental accelerating π-mode of the LEP SC cavities: during three RF periods (ν=352 MHz) of integration, the field distribution follows the harmonic evolution in time with a great accuracy, reproducing at the end of the third period the starting condition with a maximum error of a few parts per thousand (in the case of a mesh with 20000 points, i.e. a mesh step of 6 mm). Other tests have been performed against the two typical analytical cases of the pill-box and of the spherical resonators, giving similar results with meshes having a few thousands of points.

PARTICLES AND FIELD COUPLING

The driving current is produced by the bunch of particles which moves around the axis of the cavity.
Since the driving current must be axi-symmetrical, the particles of the bunch are free to move only in the r-z plane, so that the current and charge densities associated to them are the same as the ones produced by rings of charge centered on-axis, free to move axially and to expand radially. In order to derive, from the distributions of both particles and their velocities, the current density distribution, we adopt the Gaussian assignment algorithm [3], which treats the bunch as constituted by Gaussian axi-symmetrical sub-bunches. The charge density associated to a mesh point close to the i-th particle is given by:

$$\rho_i = \frac{q}{(2\pi)^{\frac{3}{2}} \sigma_z \sigma_r^2} \frac{e^{\left[\frac{(r-r_i)^2}{2\sigma_r^2} + \frac{(z-z_i)^2}{2\sigma_z^2}\right]}}{e^{-\frac{r_i^2}{2\sigma_r^2}} + \sqrt{2\pi}\frac{r_i}{\sigma_r} errf\left(\frac{r_i}{\sigma_r}\right)}$$

where r_i, z_i are the particle coordinates and r,z are the mesh point coordinates.

Due to the fast decrease of the Gaussian, only the mesh points closer then three times the Gaussian-width have assigned a density value. Choosing a value for the Gaussian width close to the mesh step, it can be seen that the unphysical fluctuations in the charge density (and in the current density) distribution, due to the assignment algorithm, are quite low (less than one percent for a uniform distribution)[3]. This property of the assignment algorithm is very important: since each fluctuation in the driving current distribution excites a spatial harmonic of the field with the same wavelength as the one of the fluctuations itself, one must avoid to produce unphysical fluctuations in the driving current which, if smaller than the mesh step, can cause instability in the field integration [4] and in general produce unphysical feedback effects on the dynamics of the particles.

Once known the charge density associated to a particle, the current density distribution is computed, at each point in the mesh, simply by summing over the contributions of all the particles which are closer to that point more than three times the Gaussian width:

$$J_r = c\Sigma_{i=1}^n \beta_{i\perp}\rho_i \qquad J_z = c\Sigma_{i=1}^n \beta_{i\|}\rho_i$$

Recalculating at each integration time the current densities, the inhomogeneous field equation 1) can be used to compute the field at the next integration time.

THE COMPUTER CODE ITACA

The computer code ITACA solves simultaneously the equation 1) for the field propagation and the equations 2) for the particles motion:

2)
$$\begin{cases} \dfrac{d\beta_{i\|}}{d\tau} = \dfrac{\Omega}{\gamma_i}[E_z^i(1-\beta_{i\|}^2) - E_r^i\beta_{i\|}\beta_{i\perp} + H_\phi^i\beta_{i\perp}] \\ \dfrac{d\beta_{i\perp}}{d\tau} = \dfrac{\Omega}{\gamma_i}[E_r^i(1-\beta_{i\perp}^2) - E_z^i\beta_{i\|}\beta_{i\perp} - H_\phi^i\beta_{i\|}] \\ \gamma_i = \dfrac{1}{\sqrt{1-(\beta_{i\|}^2+\beta_{i\perp}^2)}} \end{cases} \quad i=1,...,n$$

$$\begin{cases} \dfrac{dz_i}{d\tau} = \beta_{i\|} \\ \dfrac{dr_i}{d\tau} = \beta_{i\perp} \end{cases} \qquad \Omega = \dfrac{q\mu_o}{m_o c} \qquad \tau = ct$$

where $\beta_{i\perp}$ and $\beta_{i\|}$ are the transverse and parallel beta's, and E_r^i E_z^i H_ϕ^i are the interpolated field values on the mesh at the location of the i-th particle. These equations are integrated by a standard R.K. procedure (note that the electric field is expressed, for convenience, in the same unit of H).

The electric field components are derived integrating versus the time one of the Maxwell equation, over a reduced mesh which covers the region around the axis where the bunch is expected to move (in the examples here presented this reduced mesh extends up to the radius r=98 mm, i.e. it covers the volume occupied by the acceleration tube).

$$\begin{cases} \dfrac{\partial E_r}{\partial \tau} = -\dfrac{\partial H_\phi}{\partial z} - J_r \\ \dfrac{\partial E_z}{\partial \tau} = \dfrac{1}{r}\dfrac{\partial (rH_\phi)}{\partial r} - J_z \end{cases}$$

Knowing the field H_ϕ at each point of the mesh (both at the present time and at the next integration time), this equation can be integrated with respect to the time with a standard R.K. procedure.

During the integration of the whole set of equations the program monitors, at some locations of the bunch in the cavity, the energies stored in the bunch and in the e.m. field. The computation of the total e.m. field energy requires an integration of the Poynting vector flux across the cylindrical surface separating the reduced mesh (where both the E and the H fields are known) and the outer part of the cavity.

The code is able to handle mesh with up to 200,000 points for the H_ϕ field inside the cavity, requiring a core memory of about 3.5 Mbytes; the CPU time needed for the case presented in the next section (1,000 particles in the bunch traced over 2.5 m, 20,000 mesh points) is about 6 hours of VAX-8600. The CPU time T scales like $(n+.02N)\sqrt{N}$, where n is the number of particles and N is the number of mesh points (the factor $\sqrt{N}$ is due to the numerical stability criterion, which states that the integration time step must scale like the mesh step, i.e. like the inverse of $\sqrt{N}$).

EXAMPLES OF THE RESULTS OBTAINABLE WITH THE CODE ITACA

As a reference case, we chose a bunch of 1µC charge and 10 MeV energy injected in the empty LEP SC cavity. The bunch is initially Gaussian in the r-z plane and semi-Gaussian in the two phase spaces (Gaussian in r and uniform in r', Gaussian in z and uniform in Δγ), with a normalized emittance of $4 \cdot 10^{-4}$ m·rad and an initial radius (rms) of 20 mm. The bunch length at injection is 40 mm (rms). The wake-field excited in the cavity by the bunch passage, is shown in

Fig.1 at some integration times. The excitation and propagation of the higher order modes is evident: the driving current, shown in Fig.2 at the same bunch position, is initially Gaussian, but it exhibits an increasing splitting while the bunch approaches the last two cells, due to the explosion of the bunch itself under the strong effects of its self-field. The split driving current starts to excite higher and higher frequencies, which propagate more as in free space than in a cavity: these frequencies are at the limits of the mesh capability of propagating high order modes, but the numerical stability is not yet violated, as can be deduced from the fact that these modes propagate through the mesh and that the total energy (e.m. energy plus bunch energy) still stays constant until the bunch exits from the cavity. At this time the bunch has lost 1.02 joule of its initially 10 joule energy in the cavity: the computation of the e.m. field energy when the bunch has left the cavity gives .98 joules stored in the field, with an error of a few percent in the energy exchange.

The electric field on axis, shown in Fig.3, reflects evidently the wake associated to the bunch: a peak field larger than 10 MV/m propagates just behind the bunch, and a great fluctuation of the field inside the bunch induces the explosion of the particles off the axis while the bunch propagates in the cavity.

The strong dependence of the beam dynamics from the bunch shape at the input and from its energy is evident comparing Fig. 1 to 3 with Fig. 4 to 6, the last three being referred to a case similar to the previous one, but with a higher injection energy and a more compact bunch. Particularly in this example a bunch of 1 µC charge and 100 MeV (instead of 10 MeV) energy is injected in the LEP SC cavity, having a similar normalized emittance ($4.8 \cdot 10^{-4}$ instead of $4 \cdot 10^{-4}$ m rad) and the same shape, but a smaller dimension: 12 mm and 24 mm (instead of 20 mm and 40 mm) respectively for the rms radius and length.

In Fig. 7 (left and right) we present the longitudinal phase space of the bunch in the three positions specified respectively in Fig. 1 and Fig. 4. The effect of the very high wake field in these extreme examples is summarized by the fact that the electrons of the bunch tail are strongly decelerated.

To conclude, we present also the acceleration of a 1 MeV 100 nC bunch injected in the same cavity ($1.5 \cdot 10^{-4}$ of normalized emittance and same dimensions as the first case), when a 115 joule of e.m. energy is stored in the fundamental accelerating mode (which corresponds to an accelerating field of ~6.5 MV/m on the active length). In this case the accelerating field is much higher than the wake-field excited by the bunch, as can be seen from Fig. 8, where the wake field is visible only at those phases of the accelerating field where the H_ϕ field is close to zero. Here it is important to note the reliability of the field integration algorithm, which after three RF periods

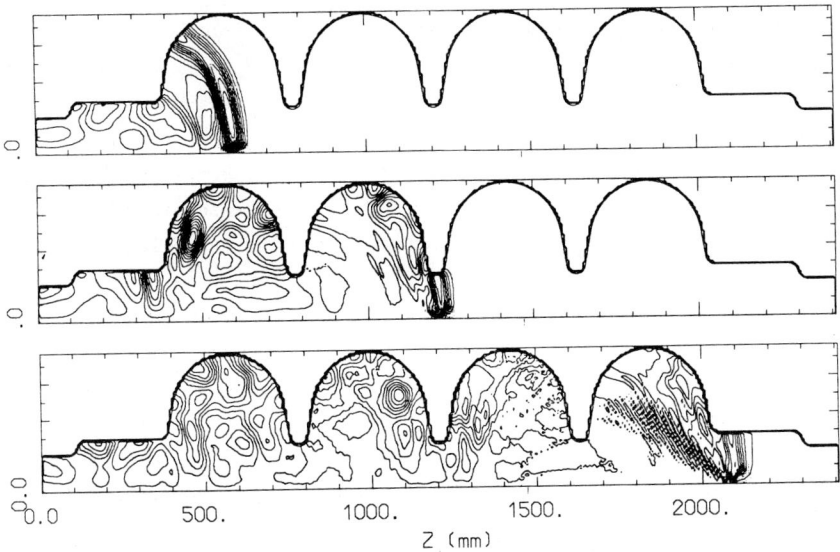

Fig.1 - Motion of a 10 MeV, 1 µC, electron-bunch and wake fields excitation in an initially empty LEP SC cavity. $r \cdot H_\phi$ = constant lines are shown.

reproduces the starting spatial distribution of the H_ϕ field with a great accuracy. The bunch is focused during the acceleration by the radial component of the electric field, (see Fig. 9) but it emerges from the cavity with a large energy spread due both to the wake field excitation and to its large phase spread (about 40° RF). In this case the driving current associated to the bunch stays fairly Gaussian, with some distortions, all over the acceleration process (see Fig.10). The bunch exits from the cavity at an average energy of 11.12 MeV: the total energy, given by the sum of the e.m. energy and the bunch energy, remains constant near its initial value of 115.1 Joule with a maximum oscillation of 0.05 joule.

The final energy gained by the bunch is 1.02 J, against an energy lost by the e.m. field of 1.07 J. The final normalized emittance of the bunch is $4.6 \cdot 10^{-3}$ m·rad, 30 time larger than the input one.

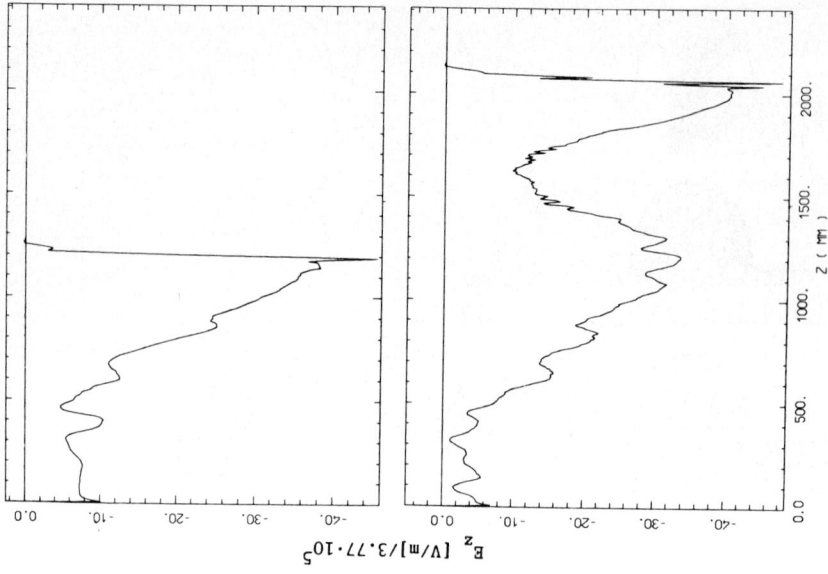

Fig.3 - Electric field on the axis produced by the bunch of Fig.1 at the last two positions.

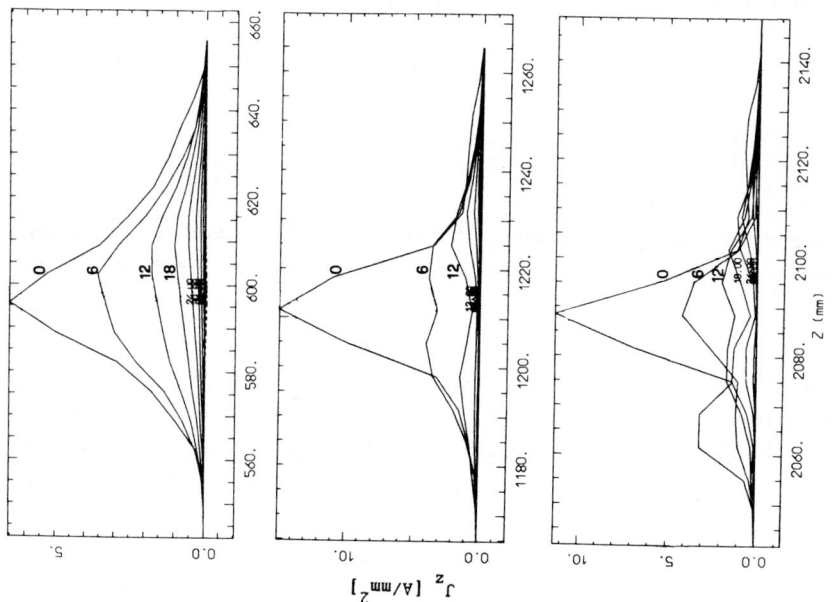

Fig.2 - Current density J_z as a function of z (at some indicated radii) for the bunch shown in Fig.1.

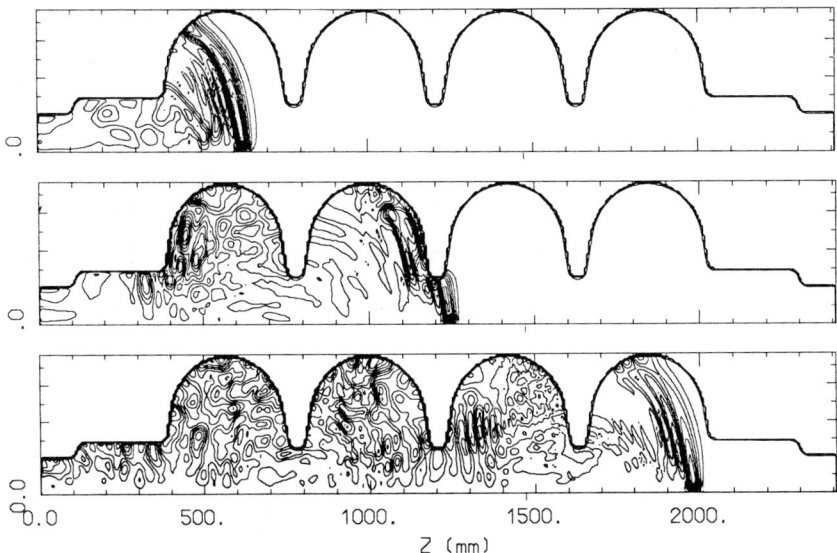

Fig.4 - Motion and wake fields excitation (like Fig.1) for a bunch having the same charge, but more energy and a small dimension (see text for details).

CONCLUSIONS

In this paper we presented a summary of the ITACA code, together with a few extreme examples to show the program performances. In particular we want to point out that the results were obtained for a mesh with a number of points lower by a factor of ten with respect to the full capability of the code (which can handle up to 200,000 points). The very good stability of the algorithm (checked also versus the sphere and the pill-box) ensures that the contributions to the total field given by the wake fields with frequency content up to 22 GHz can be taken into account for the LEP cavity, using a mesh with just 20,000 points.

In the near future a more complete presentation of ITACA will be published, together with a description of the overall package (which includes a resonating modes finder and a graphic post-processor).

We assume that, once fully tested, ITACA will be ready for external users by the end of this year, in a version compatible for vector computation.

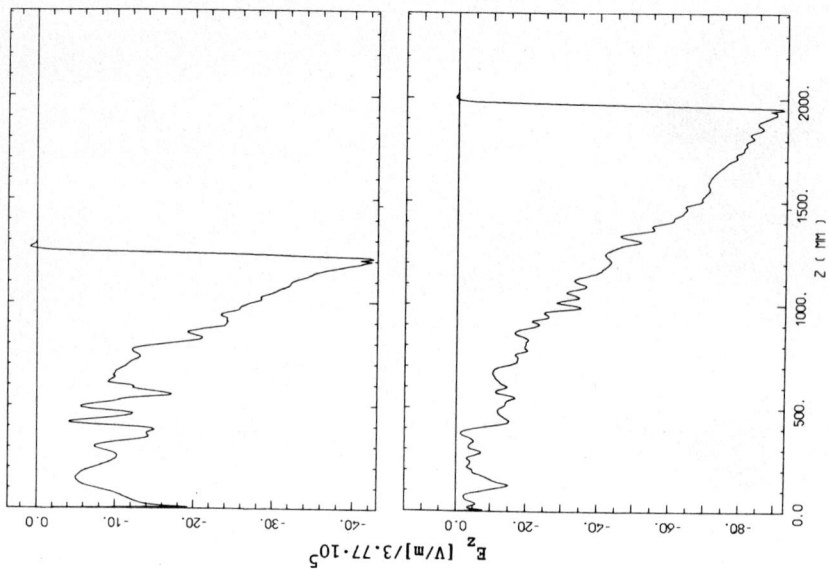

Fig.6 - Electric field on the axis produced by the bunch of Fig.4 at the last two positions.

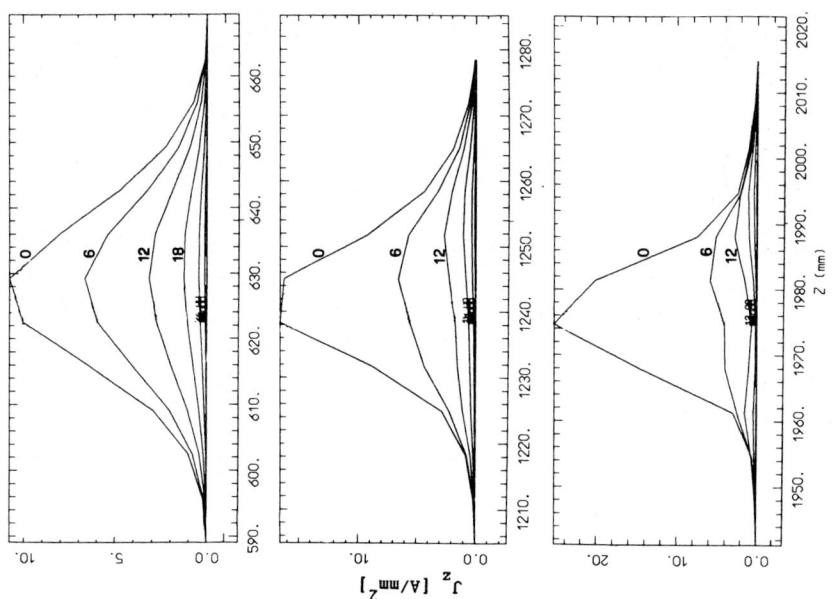

Fig.5 - Current density J_z as a function of z (at some indicated radii) for the bunch shown in Fig.4.

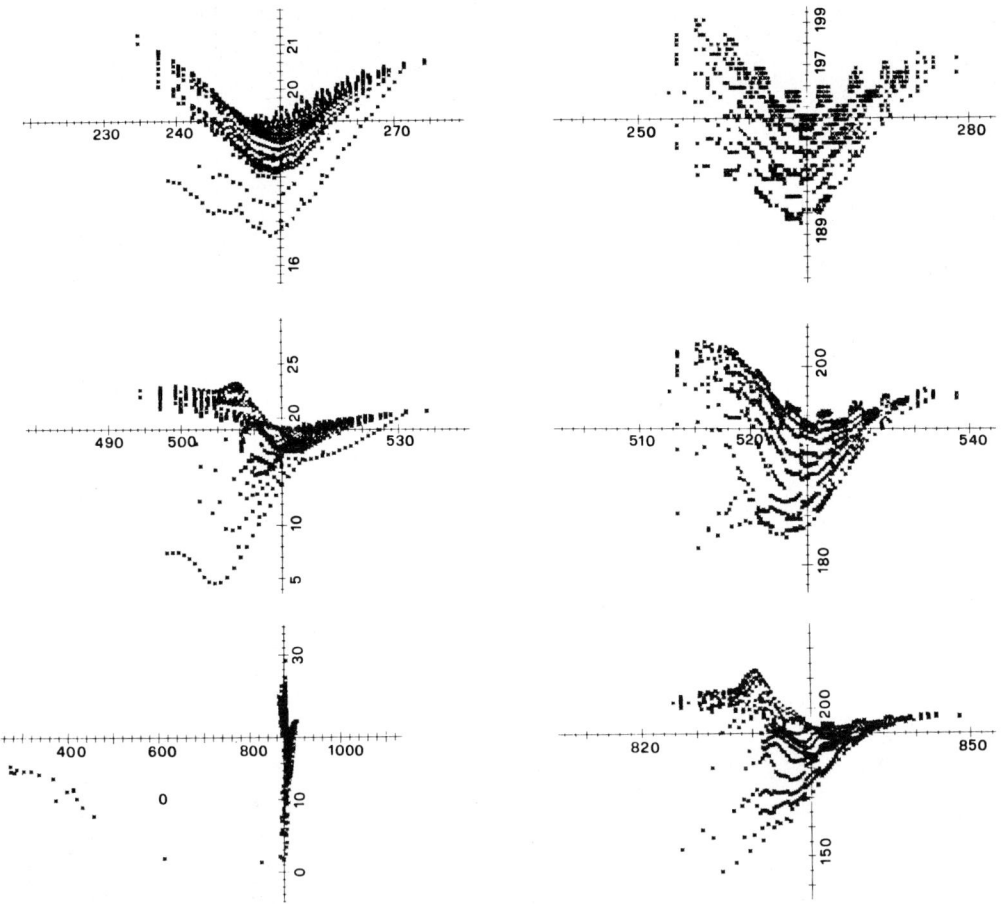

Fig.7 - Longitudinal phase space of the bunch in the three positions specified respectively in Fig.1 (left) and Fig.4 (right). The effect of the very high wake field in these estreme examples is summarized by the fact that the electrons of the bunch tail are strongly decelerated.

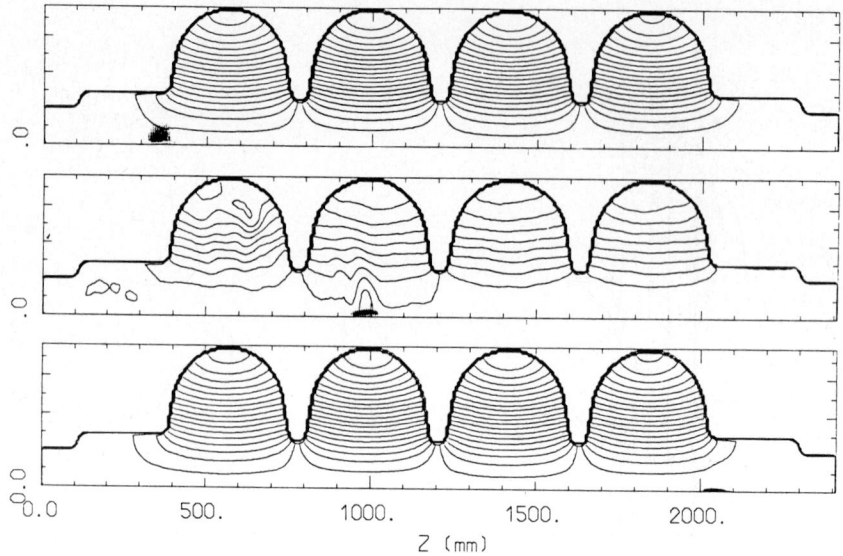

Fig.8 – Acceleration of a 1 MeV, 100 nC, electron-bunch and wake fields excitation in a LEP SC cavity, initially stored with a resonating $\pi\text{-TM}_{010}$ mode. $r \cdot H_\phi$ = constant lines are shown.

REFERENCES

1) - S. Ramo, J. Whinnery and T.V.Duzer, Fields and Waves in Communications Electronics (1965), p.262.
2) - Brodwin, M.E., et al, IEEE, MTT-23 (1975), p.623.
3) - M. Berz and H. Wollnik, NIM, A267 (1988), p.25.
4) - T. Weiland, CERN/ISR-TH/80-07, 1980.

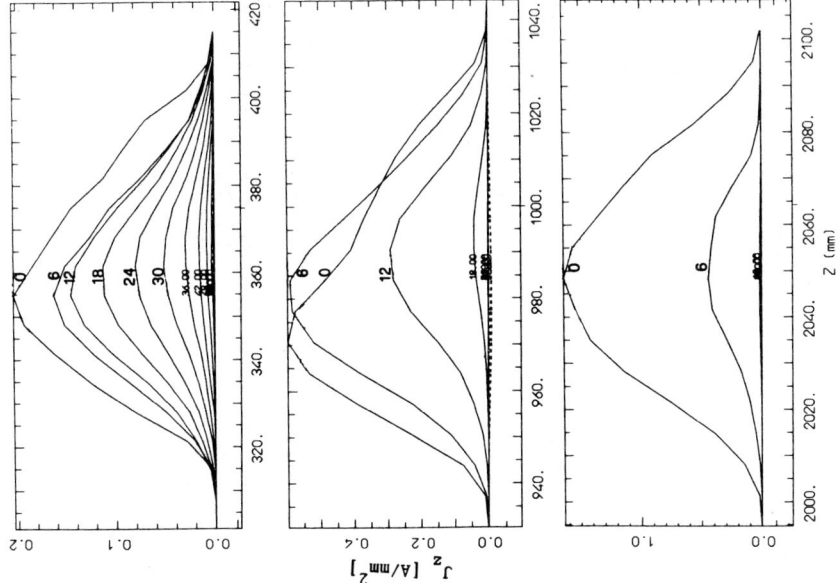

Fig.10 - Current density J_z as a function of z (at some indicated radii) for the case of Fig.8.

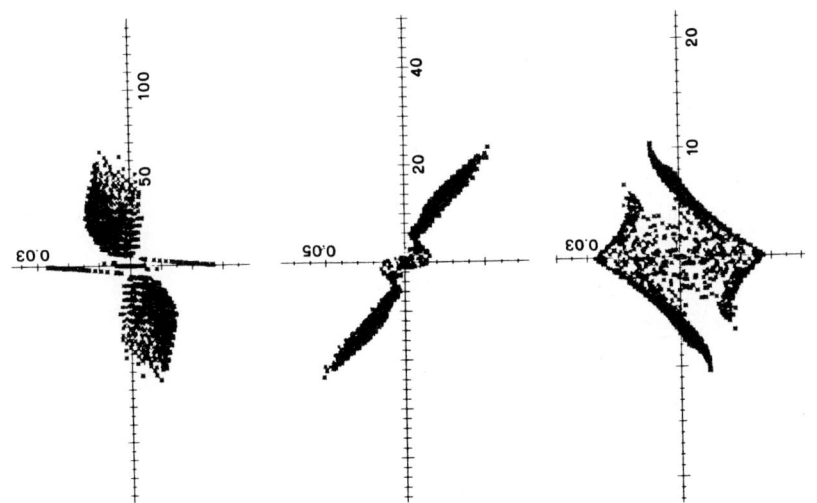

Fig.9 - Transvers phase space of the bunch in the three positions of Fig.8. The bunching effect is apparent.

AUTHOR INDEX

Amaldi, U., 171
Barletta, W.A., 127, 211
Bobin, J.L., 197
Bonifacio, R., 35, 221, 227, 243, 259
Boscolo, I., 221
Casagrande, F., 35, 221, 227
Castelli, F., 243
Cerchioni, G., 35, 221
Corsini, R., 221
De Salvo Souza, L., 35, 221, 243
Esposito, F., 275
Fadini, D., 221
Fedele, R., 275
Ferrario, M., 221, 227
Ferrucci, L., 283, 293

Kim, K.-J., 25
Maroli, C., 221, 259
Miano, G., 275
Muda, L., 293
Pagani, C., 283, 293
Pellegrini, C., 1
Peretti, A., 293
Pierini, P., 35, 221, 227
Piovella, N., 221, 227, 259
Scharlemann, E.T., 95
Serafini, L., 283, 293
Sessler, A.M., 211
Vaccaro, V.G., 275
Van Der Meer, S., 185

SUBJECT INDEX

Accelerator
    Accelerating structure, 283
    High gradient (linac), 127
    Inverse FEL accelerator, 197
    Laser acceleration of particles, 1
    Linear accelerators, 1, 127, 283
    Plasma acceleration of particles, 275
    Plasma accelerator, 197
    RF electron gun, 1
    Two beam accelerator, 185
Beam
    Brightness, 1
    Dynamics, 95, 293
    Emittance, 1, 127, 211
    Self focusing, 211
    Strahlung, 171, 185
    Transport, 1
Beat Wave, 275
Disruption, 171, 185
Free Electron Laser (FEL), 1, 35, 95, 197, 221, 227, 243, 259
    Amplifier, 35
    (high) Gain, 35, 95, 221, 227, 259
    (linear) Theory of, 35, 259
    Superradiance in FEL, 35, 221, 227, 259
Klystrons (relativistic), 127, 185
Linear colliders, 171, 185
    Classical regime of, 171
    Quantum regime of, 171

Luminosity, 171, 185
Optical bistability, 243
Optical guiding, 95
Photocathodes, 1
Plasma
    Plasma channel, 211
    Plasma focusing, 211
    Plasma frequency, 197, 275
    Plasma waves, 197, 275
Radio Frequency (RF)
    RF cavity, 293
    RF electron gun, 1
    RF pulse propagation, 283
    RF pulses, 283
    RF wave guide, 293
Slippage, 35, 227, 259
Synchrotron Radiation (cooperative), 35, 221
Wake fields, 293
Wiggler
    Beam dynamics in, 95
    Plasma wiggler, 211, 275
    Tapered wiggler, 95, 227
    Undulator magnet, 197
    Wiggled plasma channel, 211
    Wiggler focusing, 95
    Wiggler magnet, 35, 95, 211, 221
X-ray laser, 211